This new and exciting interactive resour
quality computer simulations covering essential topics in solid state
physics.

The simulations

Cover x-ray diffraction, phonons, electron states and dynamics,
semiconductors, magnetism, and dislocations. Users can vary different
characteristics and immediately see the results in animations and
graphical displays.

The companion book

Is essential for effective use of the simulations. It guides the user
through hundreds of exercises and examples, illustrates fundamental
physical principles, and contains notes on the relevant physics. The
hardcover edition includes the simulations on CD-ROM (Unix,
Windows, Powermac formats) and a licence for use on a local area
network on a single geographical site. The paperback (without
CD-ROM) is intended for students who have access to the simulations
on a local area network.

This package provides an interactive resource for those studying solid
state physics at advanced undergraduate or graduate level. It will also
be of interest and value to researchers in physics, materials science,
electrical engineering, chemistry and chemical engineering.

Simulations for Solid State Physics

An interactive resource for students and teachers

Simulations for Solid State Physics

An interactive resource for students and teachers

Robert H. Silsbee and Jörg Dräger

Cornell University

CAMBRIDGE
UNIVERSITY PRESS

PUBLISHED BY THE PRESS SYNDICATE OF THE UNIVERSITY OF CAMBRIDGE
The Pitt Building, Trumpington Street, Cambridge CB2 1RP United Kingdom

CAMBRIDGE UNIVERSITY PRESS
The Edinburgh Building, Cambridge CB2 2RU, United Kingdom
40 West 20th Street, New York, NY 10011–4211, USA
10 Stamford Road, Oakleigh, Melbourne 3166, Australia

First published 1997

Printed in the United Kingdom at the University Press, Cambridge

A catalogue record for this book is available from the British Library

Library of Congress Cataloguing in Publication data

Silsbee, Robert H., 1929–
Simulations for solid state physics: an interactive resource for students and
teachers/Robert H. Silsbee and Jörg Dräger.
p. cm.
Includes bibliographical references and index.
ISBN 0521 59094 9 (hardback). — ISBN 0 521 59911 3 (pbk.)
1. solid state physics — Computer simulation. I. Dräger, Jörg.
II. Title
QC176.S487 1997
530.4´1´078553dc21 96–48930 CIP

ISBN 0 521 59094 9 hardback (with CD-ROM)
ISBN 0 521 59911 3 paperback

Contents

Contents

Preface

How would you like to learn solid state physics with a working model to play with? You could watch individual electron trajectories in copper as you turn on electric and magnetic fields. You could see the atomic moments in iron suddenly align with one another as you gradually lower the temperature. You could observe spots come and go in the x-ray diffraction pattern as you change the relative positions of the sodium and chlorine ions in table salt. You could push hard enough on a crystal to make it break and watch how the individual atoms move during the failure. The computer programs, *Solid State Simulations*, offer these possibilities, and more.

The package, consisting of the CD with the *Solid State Simulations* and the accompanying guidebook, *Simulations for Solid State Physics*, is the culmination of a three-year experiment in the use of computer simulations in teaching solid state physics. The project was seeded by an afternoon spent watching the spins in a simulation of the Ising model. On the screen, in visual form, were examples of a surprising variety of physical phenomena: critical fluctuations, critical slowing down, correlation lengths, nucleation, coarsening, and with a little plotting of data, the Curie–Weiss law and the relation of susceptibility to thermal fluctuations. This had to be a useful tool for teaching; and it has been, along with thirteen additional simulations which have been developed over the last three years.

The simulations and guides illustrate topics typically encountered during a one-semester introductory solid state physics course, and are to be used in conjunction with a traditional textbook. Two simulations focus on crystal structure and x-ray diffraction of both perfect and imperfect crystals; two on lattice dynamics and heat capacity; two on dynamics of free electrons; one on band theory; two on the dynamics and transport of band electrons; two on semiconductors; two on magnetism; and one on dislocations. A look at the table of contents will give a more detailed picture of the material included.

For two years, the simulations have provided core material for a Cornell course in solid state physics for advanced undergraduates and graduate students in physical sciences and engineering. Why has the student response been so enthusiastic? The simulations provide a new channel for delivery of information to the brain. They are visual, animated and, most importantly, interactive: the system parameters can be modified by the student. Hypotheses may be tested ("I think this should decrease with temperature. Let's see...") and puzzles may be posed ("Why is that diffraction spot missing?"). Active engagement of the student, an essential ingredient of effective learning, is unavoidable. On returning to the standard texts for theoretical understanding, the student has a visual picture in mind of a system behavior which needs to be understood; or, as a student works with the simulations after reading the text, we hear "Oh, *now* I see what the book was talking about". Also, emphasis has been placed on uniformity and friendliness of the interface: the simulations are easy to use.

A guidebook facilitates the exploration of new territory. This book, *Simulations for Solid State Physics*, provides that guidance for the *Solid State Simulations*. It develops the important conceptual framework for each topic, and leads the student to the regions of the large parameter space which best illustrate the critical ideas. The student participates directly in the conceptual development via the extensive exercises, qualitative and quantitative, which are interwoven with the text.

The Solid State Simulation project continues to develop. A Web page (http://www.ruph.cornell.edu/sss/sss.html), devoted to the support of this package, includes a bulletin board for comments on the material, postings of bug fixes, and additional suggestions for exercises. In time, we hope that additional simulations will become available. The sales royalties will be used to support the Web page, improve the existing simulations, and develop new ones.

We offer our sincere thanks to a number of sources for financial, technical, and psychological support. The development of the simulations was supported by the Alfred P. Sloan Foundation and we particularly thank Frank Mayadas for his continued interest in developing new modes of teaching. Cornell University has provided a new undergraduate computing facility for this and other projects; the facility has been an ideal environment for testing the package. Russ Thompson has written several of the programs, provided elegant algorithms for others, and is additionally responsible for the user-friendly interface which you will be using. Barry Robinson has managed the considerable task of porting the programs from Unix to the Windows and Macintosh platforms and has, on numerous occasions, revived computers stupidly crashed by the senior author. Jim Sethna's Ising program stimulated the initiation of the project, the original ideas for some of the other simulations were his, and he has been an important source of ideas and enthusiasm as the project developed. However busy with his own work, Bruce Roberts always found the time to help, both sharing his invaluable expertise and contributing routines. Kevin Hodgson is a master at turning penciled scribbles into elegant drawings. Becky Jantz cheerfully converted this text from Word into La-TeX. Chris Henley and Don Holcomb have provided many suggestions for the improvement of both text and programs. Dan Vernon programmed many of the conversions of the programs to Tcl/Tk, and Scott Packard helped with much of the final detail work. Rick Durand has carefully screened the manuscript for us and flagged more errors than we would like to admit. Help from the facilities and staff of the Multi-User Computing Facility of the Cornell Materials Science Center has been essential in the preparation of both the simulations and this guidebook. Lastly, we would like to thank all those who have used earlier versions of the simulations and guides, and provided valuable suggestions for their improvement: the students of Physics 454 and Physics 635 at Cornell, Dave Cahill at the University of Illinois, and Yongli Gao at the University of Rochester.

We hope you enjoy using the simulations as much as we have enjoyed writing them; and that they teach you as much physics as we have learned in puzzling over how to make them work.

January, 1997   Robert Silsbee

Ithaca, New York   Jörg Dräger

System requirements

The *Solid State Simulations* (SSS) programs are designed to run under Microsoft Windows 3.1.or later, Apple Power Macintosh running system 7.5 or later, and Unix workstations with X windows.

Windows 3.1 systems: Microsoft Windows 3.1, Windows for Workgroups 3.1, Windows 95, or Windows NT running at least on a 486DX2-66 processor with 8 MB of RAM (16 MB for NT). A Pentium based computer with 12 MB (20 MB for NT) or better is recommended. Windows and Windows for Workgroups 3.1 users need to have Microsoft's win32s version 1.3 (or later) installed. Version 1.3 of win32s is provided on the SSS CD.

Power Macintosh systems: The SSS require a Power Macintosh with system 7.5 or later, with 12 MB of RAM. We recommend running system 7.5.5 with 16 MB.

Unix workstations with X-windows: Executable versions are provided for IBM's RS/6000 line of workstations running AIX 4.1 or 3.2, Sun Microsystems' Sparc line running SunOS 4.1, and systems based on x86 compatible processors running Linux 1.2.13 or later. For other operating systems it should be possible to compile the simulations as all the necessary source code is provided on the CD. An X-display and X11 release 5 or later are needed.

For all platforms a CD-ROM drive and a 1024×768 pixel (or better), 256 color display are required. For more detailed system requirements, and possible exceptions, see Appendix B. For installation instructions see Appendix A.

1

How to use this book

Contents

The heart of the *Solid State Simulation* (SSS) package is the compact disk (CD) containing a set of 14 computer simulations covering a wide variety of topics. Essential to its effective use is this guidebook, which leads you to interesting places in the simulations, sketches some of the relevant physics, exercises your brain muscles, and involves you with some puzzling questions.

1.1   First look at the simulations

1.1.1   Getting started

To become acquainted with the simulations, play with them: click on the buttons and see what happens. For each simulation there is a main panel, typically with a display area on either side, and a column of sliders and menu or toggle buttons in the center. Figure 1.1 shows, as an example, the layout of the program "ising". The displays are sometimes graphs, sometimes arrays of atoms, or electrons, or spins, depending on the simulation. The sliders and buttons define the situation of interest, both the properties of the simulated system (e.g., atom positions and sizes) and the values of external parameters (e.g., electric field or temperature). Animated displays have a RUN/STOP button to turn on/off the animation. Graphical displays may update with every parameter change or by command from a click on a CALCULATE button. Often, additional graphics windows may be opened to give a different perspective on the system behavior.

The suggested random pushing of buttons gives a "feel" for the simulation, but where do you go next, or what if there's a problem? Every simulation has a hypertext (html) browser, accessible by clicking on the

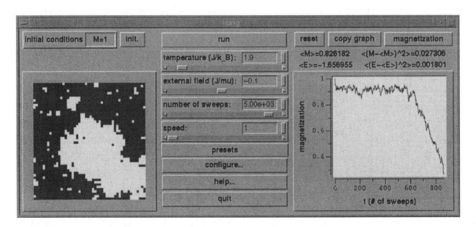

Figure 1.1: The layout of a typical program – "ising".

HELP button, providing resources to allow more effective use of the simulations. From the main help menu you can move to a short description of the simulation and its capabilities as well as a table of contents of the corresponding chapter of the guidebook, which gives a more detailed picture of the physics concepts addressed by the simulation. On a different page is general information about the SSS project. The main help page also gives access to the several features described below. Instructions, both in the annotations and in *Simulations for Solid State Physics*, are written for systems with a three-button mouse. For Mac users, the mouse button serves as the left hand mouse button, the mouse button with the CONTROL key pressed is the right hand mouse button, and (for Mac and PC) the mouse button with the SHIFT key down is the middle mouse button.

1.1.2 Presets

A special feature of the programs is a set of annotated PRESETS for each. CLICKING on a PRESET from within the HELP system sets the system parameters to values chosen to illustrate some particular physics concept. An annotation window contains brief descriptive comments concerning the chosen PRESET, often with a question to challenge the viewer. A quick "tour" through the PRESETS for a program is the easiest way to get a sense of the material discussed. It should be the next step after the random button pushing. The PRESET can also be activated directly from the PRESET menu on the main panel, but the annotation window is neither activated nor updated. The PRESETS are also used extensively in the guides to return the system to conditions appropriate to many of the exercises. Also, if you get stuck on an exercise, check the annotation: you might find a useful hint.

The PRESET system has been set up so that students or instructors can add PRESETS to the preset files or modify the existing presets (see Appendix C). Instructors may wish to set up special sets of PRESETS better adapted to their taste than are ours. The PRESETS are particularly useful when using the simulations as lecture illustrations: a new set of parameters is available with a single click of the mouse. Students may also save their own PRESETS if they find interesting sets of parameters that they wish to recover at a later time.

The animation speeds chosen for the presets were based on performance with our development system. The animations will run faster on some systems, slower on others. Expect to change the speeds or particle numbers in the animations in order to get reasonable performance. You may wish to edit the presets, as described in Appendix C, to adapt the programs to your own needs.

1.1.3   Online help

We have constructed a convenient and intuitive interface, and maintained uniformity from simulation to simulation so that few instructions are needed. There are, however, SPECIAL FEATURES of each of the simulations which are easy to miss with random button clicking. After you first work with a new program to get a sense of what it does, try out the SPECIAL FEATURES given in the HELP files: you'll soon find them useful.

Occasionally a button doesn't work as you expect. The PARAMETER DEFINITIONS should help clarify the problem.

The GENERAL HELP files, accessible from the bottom of the HELP menu, give detailed descriptions of features common to most of the programs, the most important being the various manipulations possible with the graphs. As part of your general introduction to the SSS simulations, be sure to review these files and test the features described. The GENERAL HELP also includes a useful list of values of fundamental constants.

1.1.4   Additional exercises

Also available within the HELP system are ADDITIONAL EXERCISES. These are questions (with answers) based on graphs or displays copied from the programs. (For example, "You see plotted the carrier concentration versus temperature for two semiconductors, A and B. Which has the higher donor binding energy?") These exercises provide individual students with a useful way to check their understanding, groups of students with material to focus their discussion, and instructors, by deleting the ANSWERS files, with material for quiz questions. The limited set included on the CD is primarily to illustrate the capabilities of the system: instructions for adding your own exercises or quiz questions are given in Appendix C. To be useful, the students must be asked to justify the answer: only the justification will reveal whether the material is understood or not.

1.2   Using this guidebook

1.2.1   Fourteen guides

Simulations for Solid State Physics, a set of guides for the SSS programs, is essential to effective use of the simulations. Not using it would be like visiting Paris without a guidebook. You will see something of the city but it will hold little significance. A guide will provide you with background material which gives those places a meaning within a cultural and historical context. The simulation programs give you some taste of the relevant physics; but the guides provide a context and a "story line" within which the programs will have more meaning. For a complete and detailed account you must go even further and work also with one of the conventional texts [1–8], just as a full appreciation of the city of Paris would require study of resources beyond the guidebook.

Most of our programs have a number of system parameters with values to be chosen by the user. The parameter space can be large. It is easy to get lost in meaningless parts of parameter space just as in Paris there are many uninteresting parts of the city. The guides, with the help of the PRESETS, take you to interesting regions of the immense parameter space.

Please don't try to do all of the exercises in detail! If you find none to be trivial then surely you will find some to be ignored until you have developed more sophistication. Be willing to skim the trivial or to delete the advanced as best suits your own ability. Paris is a big city: don't try to see it all at once. However, the material is written as a continuing story line, with much of the story in the exercises. Hence, even though you may have been assigned only a few to complete, be sure as you read to treat all of the exercises as an integral part of the text. Otherwise you miss part of the story. As you walk to the Bon Marché to do some shopping, don't forget to look in the store windows on the way.

Mix up your reading and your computer work. Reading the guidebook will be more meaningful once you have wandered around the city and have the "lay of the land"; and the impact on your mind of a building or neighborhood will be greater if you've first read briefly of its history. In using the SSS package you see more in a simulation if you understand something of the theory; and the reading of the theory, both in this guidebook and in a formal textbook, is easier if motivated by a feature of the simulation that seemed a mystery.

Above all explore. In Paris, go into some markets that are not described in the guide. In the simulations, use the sliders to change the parameters and see what happens. It is often best to predict how the scene will change when you double this or halve that, and see if you are right. You will learn the most by being wrong and figuring out why you were wrong. Take some numerical data by systematic variation of a parameter and design a plot that will yield a straight line; why does it almost work but not quite? Make a quantitative prediction of a variation and see whether it is confirmed.

Often the exercises ask you to take data as some parameter is systematically varied and to plot the results. "How much data?" "How many points?" "The data in the simulation are so noisy I get a different number each time I measure something." Here's your chance to learn what experimental physics is really all about. There are two possible answers to "how much data?" When we have used the material, our answer has been "take enough data that you can convince us that the predicted behavior is confirmed (or contradicted)". It is usually not the details that are of interest but trends. Is it a linear or quadratic dependence? A second approach is to treat the simulations as an experiment in which you *do* want to extract some hard numbers. In this case more data are required and a proper error analysis should be undertaken. Remember, though, that the second approach requires first a "rough run" to get a sense of what the limitations and problems will be in getting useful data from the experiment. Above all, think of a simulation as a physical system you want to understand. Poke it, twist it, to see how it works. Don't expect it to explain itself to you.

1.2.2 Algorithms

The simulations are designed to "look like" physical systems and to be used without the need to understand the algorithms that make the physics work. After some exploration with the simulations, however, many students have shown an interest in what's going on inside the simulations, and instructors may find the algorithms a useful focus for study. An appendix in each chapter gives a brief description of the algorithms and numerical methods used for that program, but without unnecessary details of the programming language.

For a more detailed look, the source code for the programs is available on the CD. Appendix A gives information about the general programming structure of the SSS package. Adventurous students and instructors are encouraged to improve upon the current versions of the programs. The choice of the programming language C and the graphics language Tcl/Tk should make that a manageable task for those who have had some previous programming experience. Please let us know what you have done! E-mail us at `sss@ruph.cornell.edu` or post to the bulletin board at our Web site, `http://www.ruph.cornell.edu/sss/sss.html`.

1.2.3 Notation

Throughout the guidebook, exercises are marked in **bold–face** and their text shown in *slanted typeface*. The easiest exercises are unmarked, the intermediate ones have a single *, and the most difficult a **. Some of the more advanced sections of the guides are also marked with * or **. Exercises purely **C**onceptual in nature are denoted by a **C**, while those requiring some sort of **M**athematical (numerical) analysis are denoted by an **M**.

References to button or menu choices in the simulations are shown in SMALL CAPITAL LETTERS. `Type writer style` is used for referring to computer programs.

The exercises make frequent and continuing reference to the PRESETS.

Suppose you work with PRESET 4, read along in the guide text for a   # 4
while, and then are asked to "reload the PRESET". A glance back at the
margin to find the most recent preset change will tell you which PRESET
you should be using.

Be on the lookout for these slippery road signs in the margin. They
are to alert you to a suggestion or warning in the text about some useful
feature of a program, or a trap that you could easily fall into.

1.3 Who the package is designed for

We envision a variety of ways to use the simulations and accompanying
guides, and have tried to include enough material in the guides to allow
their use at many levels and with different degrees of depth. First, we
suggest that the guides always be used in conjunction with a standard
solid state textbook [1–8]. The guides provide the threads of the theoretical arguments, but are not intended to replace the thorough treatment
of a text. Second, the material should be used selectively. Some of the
advanced material is best ignored in some contexts. In other contexts, the
introductory material should be lightly skimmed and attention focused on
the more difficult issues. Some alternatives for use of the material are the
following:

Lecture course   The material can serve as the core material of a formal lecture course. It has been used successfully for two terms in this
way at Cornell. Roughly half of the assigned homework was drawn from
the exercises in the guides, supplemented by problems on topics not well
suited to simulation. Formal two-hour laboratory sessions each week gave
the opportunity for valuable one-on-one interaction between the instructor and the students, with the simulations serving as the catalyst for a
productive exchange of ideas.

Auto–tutorial course   The same course has been given in a self-study
mode at Cornell. Again the simulations formed the core of the course
material, but with supplementation by material not directly connected to
the simulations.

Supplement   In a formal course in solid state physics at either the
senior or graduate level the material provides a valuable supplement to
the normal fare of lectures, text, and problems. The supplementation
can exploit the simulations as qualitative illustrative material or as the
basis of quantitative problem solving. The exercises in the guides can
be used as a minor part of the homework assignments. The simulations
can also add a new dimension to lectures. We emphasize, however, that
interactive use of the programs by the student is far more valuable than
passive observation of a lecture demonstration.

Independent study   Working through the SSS simulations plus guides in individual study, in combination with a conventional text, provides a well-motivated student with a solid basis for pursuing advanced work in solid state physics.

"Laboratory" experiments   The simulations can define the full content of a "laboratory course". There is plenty of opportunity to work on the techniques of data taking and analysis. The materials offer the instructor a wide range of emphasis from experimental analysis to physics concepts.

Special topics assignment   At Cornell the simulations have provided the starting point for special topic assignments in which the student is asked to probe in greater depth some idea that is suggested by one of the simulations. Often, analysis of data derived from the simulation can be an important component of the report on such an investigation. Possible topics are suggested in the first appendix, *Deeper exploration*, of each chapter.

Overview   The PRESET system (see page 3) allows fruitful "browsing" by people at any level. It serves: neophytes, as an introduction to new and sometimes strange ideas; working professionals as a reminder of things they once knew, or as an introduction to a new area of interest; and advanced students as a self-check on their level of understanding.

Finally, the simulations can be pursued by students working either individually or together in groups of two or three; and it may be carried out either at times of the students' choice or in organized laboratory sessions. If a computer facility is available for teaching, we recommend scheduling formal laboratory sessions. The student–student and the student–instructor interactions which developed in such sessions have been particularly exciting aspects of our own use of the simulations.

2

"bravais" – Crystal structure and x-ray diffraction

Contents

2.1   Introduction

Much of experimental physics is devoted to learning about features of the world around us which are inaccessible to direct observation and manipulation. How are we to learn about the behavior of systems of the atomic scale, eight orders of magnitude smaller than things familiar in our experience? Or astronomical phenomena involving length scales 16 orders of magnitude larger? How is it possible, with instruments of a size that we can manipulate with our hands, to determine the molecular arrangement of thousands of atoms in a biological molecule with typical interatomic distance of angstroms? If the molecule can be persuaded to crystallize into a periodic array, a regular repeating stacking of the molecule, then the x-ray diffraction experiment can do the job. For large molecules, the analysis is sophisticated; but it is based on simple physical principles which are illustrated by "bravais".

In "bravais" we work not with complex biological molecules, but with simple two-dimensional crystals containing only one or two atoms per molecule. The first half of "bravais" introduces a new language, a language which is useful not only in talking of diffraction experiments, but is also relevant to many topics in the theory of solids. A critical idea in developing the language is the idea of representing a crystal as a stacking together of an array of *unit cells*, the content of each cell, called the *basis*, being precisely the same.

With the language established, the second half of the chapter gives many examples of x-ray diffraction patterns which illustrate the various physical ideas which underlie the interpretation of x-ray diffraction data. The ideas of unit cell and basis are reflected in two aspects of the diffraction experiment. The size and shape of the unit cell define possible directions, if any, for diffracted beams; the cell content or basis determines the intensity of any diffracted beams.

2.2   Crystal lattice versus crystal structure

A two-dimensional crystal structure may be defined in terms of a unit cell, a parallelogram with sides of lengths a_1 and a_2 and an interior angle ϕ; and a basis giving the positions of atoms within the cell. The full crystal is generated by tiling the plane with multiple copies of this unit cell. The cell corners in this tiling define the crystal lattice or *Bravais lattice* associated with the crystal structure. Thus the definition of a crystal structure is a two-step process. First, we define a Bravais lattice in terms of the parameters of the unit cell. This lattice has the same translational symmetry as the crystal structure. Then, defining the positions of the atoms within the cell, the *basis*, completes the description.

PRESET 1 shows an example, the *honeycomb structure*. In the picture   # 1
on the left you see the arrangement of the atoms, each with three nearest
neighbors. We associate with the crystal structure a lattice of points
(indicated here by the ×s) which you see by clicking on the LATTICE
button above the display of the atoms. This array is the Bravais lattice of
the honeycomb structure. The parallelogram formed by connecting four
neighboring lattice points is a unit cell of the lattice. The position of the
two atoms in each of the cells, the basis, is specified by their coordinates
with respect to a corner of the cell, in units of the cell edges: ATOM A
with coordinates $[A_1 \ A_2]$ is at the position $\mathbf{R}_A = A_1 \mathbf{a}_1 + A_2 \mathbf{a}_2$ where $\mathbf{a}_1$
and $\mathbf{a}_2$ are the vectors defining the edges of the unit cell. The structure
is defined by the combination of a Bravais lattice and a basis.

Move either atom around by dragging the handle of one of its coordinate
sliders, e.g., A1 or A2. The crystal structure is obviously being changed
but the Bravais lattice remains fixed.

Exercise 2.1 (C) *Remove the second atom by clicking on the* ATOM B
*toggle. The result is obviously a different structure from the first. Is the
associated Bravais lattice the same or different?*

A smallest unit cell which can be used to define a crystal structure is
called a *primitive unit cell*. Its shape is not unique, but its area is.

Exercise 2.2 (M**) *Find a new pair of cell edges a_1', a_2' and an angle ϕ',
and two pairs of atom coordinates with which "bravais" will generate this
same honeycomb structure, but rotated by 90° from the picture in* PRESET
*1. Note that in "bravais" since you're stuck with $\mathbf{a}_1$ being horizontal, your
unit cell will have a different shape from that of the* PRESET.

PRESET 2 gives a structure with a rectangular unit cell. It is called   # 2
a *centered rectangular* structure, with one atom at the corner of the cell
and the other atom at the center of the cell $[\frac{1}{2} \ \frac{1}{2}]$. As with the honeycomb,
"bravais" has chosen to define this structure in terms of a cell with two
atoms.

Exercise 2.3 (M) *Show that the unit cell is* not *a primitive cell. What
are a possible pair of cell edges and the angle between them which give a
primitive cell for the structure of* PRESET 2?

The original specification in terms of the larger unit cell is referred
to as *non-primitive indexing*. In the example of PRESET 2 it has the
advantage over the *primitive indexing* that the coordinate vectors $\mathbf{a}_1$ and
$\mathbf{a}_2$ define an orthogonal coordinate system. The most common examples
of the use of non-primitive indexing are the face centered cubic (fcc) and
body centered cubic (bcc) structures: the advantages of the orthogonal
coordinate system outweigh the simplicity of the primitive description.

Exercise 2.4 (C*) *Restore the honeycomb structure by selecting* PRE-
1   SET 1 *again. Can you find a smaller unit cell, with only a single atom
in the cell, which can be used to define this structure? Explain why you
cannot do this. The honeycomb structure, despite its apparent simplicity,
requires a two-atom basis.*

Be sure that the difference between the two previous examples is clear.
A primitive basis must contain one atom for each inequivalent atom in the
structure. In Exercise 2.3 the environments of the two atoms in the non-
primitive indexing were equivalent; for the honeycomb, though all atoms
are trigonally coordinated, the orientations of the triangle of neighbors are
different. The three-dimensional diamond structure is similar to the honeycomb: all atoms are tetrahedrally coordinated carbons, but there are
two distinct orientations of the tetrahedra, hence two atoms per primitive
cell, as for the honeycomb.

3   **Exercise 2.5 (M)** *For* PRESET 3, *give "bravais" a more convenient but
non-primitive choice of lattice plus basis that gives the same structure and
orientation. What is the most obvious primitive cell to use (not available
to "bravais" without rotation of the crystal)?*

2.3   Lattice planes

In the preceding section we defined the crystal structure in terms of a unit
cell and its contents or basis. Much of our intuitive thinking about the
x-ray diffraction experiment represents the crystalline array as a stacking
of planes of atoms. The Bragg condition for x-ray diffraction, discussed
in Section 2.5, is expressed in terms of planes of atoms and the spacing
between the planes. In this section and the next we develop the language
for defining the planes making up the crystal lattice and their description
in terms of the parameters of the corresponding Bravais lattice.
We need a way to specify, for a given crystal structure, the various sets
4   of lattice planes used to represent the associated lattice. In PRESET 4 you
see two atomic planes drawn in the real-space picture. (For now, ignore
the red and green dots on the right.) These are two of a set of equally
spaced and parallel planes which, in this particular example, contain all
of the atoms of the crystal. Sets of planes are conventionally defined in
terms of *Miller indices*. To find the Miller indices of a set of planes:

1. find the plane which goes through the origin of the Bravais lattice
   (a black instead of red ×);
2. find the intercepts, on the coordinate axes, of the nearest plane of
   the set ($\frac{1}{3}$ and $\frac{1}{2}$ in this example);

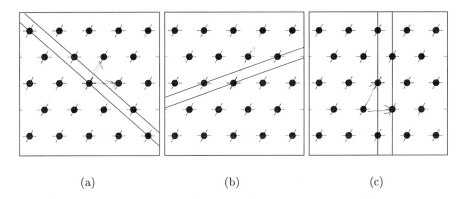

(a) (b) (c)

Figure 2.1: A triangular lattice with three different pairs of lattice planes for identification. (Exercise 2.7)

3. the inverses of these intercepts (3 2) are the Miller indices of this set of planes.

"bravais" gives the Miller indices of the selected planes below the window on the right. Clicking on the SHOW PLANES box toggles on and off the display of the planes and typing in the MILLER INDICES defines the selected planes.

Exercise 2.6 (C) *Check out a number of other $(h\,k)$ values until you're comfortable with indexing the planes. Enter $(0\,1)$ for $(h\,k)$ and be sure you understand how this one works.*

Exercise 2.7 (M) *What are the Miller indices of the three sets of planes in Figure 2.1? Assume $\mathbf{a_1}$ to be directed horizontally to the right and ϕ to be $60°$.*

Exercise 2.8 (M) *Return to PRESET 1 and click on SHOW PLANES, and on the real-space LATTICE, to show two of the set of (1 1) planes. Note that "bravais" draws the planes through the lattice points, not the atoms. There happen to be no atoms on the planes drawn here. Change the basis to: ATOM A at $[0\,0]$ and ATOM B at $[\frac{1}{3}\,\frac{1}{3}]$. Now half of the atoms lie on the planes. Enter $(1 - 1)$ (usually written $(1\,\bar{1})$) for $(h\,k)$. What has happened to the positions of the atoms relative to the planes? Perhaps it is safer not to associate the planes defined by the Miller indices too closely with planes of atoms.* # 1

Although introduced in terms of planes of atoms, the Miller indices should be thought of as defining sets of *lattice planes* rather than *atomic planes*. This avoids ambiguities which arise in structures with more than a single atom per unit cell, as in the honeycomb example. "bravais" has adopted this approach and draws the planes through the lattice points, wherever the atoms might be.

If the Miller indices contain a common divisor n, only every nth plane of the corresponding set of planes will contain lattice points. The higher order Bragg reflections may be thought of as first order reflections from these more closely spaced lattice planes. The diffraction is a consequence of the Fourier components of the electron density with periodicities equal to the interplanar spacing of the higher order planes.

Note that, in writing Miller indices, we adopt the standard notation in which the minus sign for a negative value appears above, not in front of, the index. "The set of $(1\,\bar{1}\,1)$ planes" is read as "the set of one, one-bar, one planes".

2.4   Reciprocal lattice

Many of the properties of crystals and many of the theoretical techniques used to describe crystals derive from the periodicity of crystalline structures. This suggests the use of Fourier analysis as an analytical tool. In the analysis of periodic time varying fields (e.g., acoustic signal analysis, radio signal analysis) we often do much of our work in the frequency domain rather than in the time domain. In analogy with the time $\Leftrightarrow$ frequency duality, for crystal problems there is a corresponding *real space* $\Leftrightarrow$ *reciprocal space* or *wave-vector space* duality. Many concepts are best understood in terms of functions of *wave vector* just as an amplifier response is typically characterized in terms of its *frequency* dependence. We prefer to describe a wave with wavelength λ as a plane wave with wave vector $\mathbf{k}$ of magnitude $|\mathbf{k}| = 2\pi/\lambda$ and direction perpendicular to the wave fronts. The space of the wave vectors is reciprocal space, the analog of the frequency domain for the time problem.

Recall that a periodic signal in time with period T has Fourier components only at angular frequencies $\omega = 2\pi n/T$. For crystals, any property (e.g., the electronic charge density) with periodicity defined by the periodicity of the associated Bravais lattice will have Fourier components only at a discrete set of wave vectors. This set of wave vectors defines the lattice reciprocal to the Bravais lattice of the crystal. In PRESET 5 "bravais" shows in the right hand display the lattice reciprocal to that on the left. The *reciprocal lattice* is defined by a unit cell with coordinate vectors, $\mathbf{b}_1$ and $\mathbf{b}_2$. The $\mathbf{b}$s (with units of inverse length) are defined such that $\mathbf{b}_i \cdot \mathbf{a}_j = 2\pi\delta_{ij}$. Note the similarity of this set of relations to the simpler relation for the time domain problem $\omega T = 2\pi$. The added complexity is because we are concerned with spatial variations in two or three dimensions instead of just one: hence we need a vector quantity $\mathbf{k}$

5

for the analog of the scalar frequency variable ω.

Exercise 2.9 (C) PRESET 5 *shows a rectangular real-space lattice and its reciprocal, with no atoms. Vary the cell parameters a_1 and a_2 and verify that the qualitative behavior of the reciprocal lattice is appropriate to maintain the relation $\mathbf{b}_i \cdot \mathbf{a}_j = 2\pi\delta_{ij}$.*

Now vary the angle ϕ. First verify that the orthogonality condition is maintained. Then look carefully at the magnitudes of b_1 and b_2. Explain why they both increase as ϕ departs from 90°, though the magnitudes of a_1 and a_2 remain fixed.

Fortunately, we usually work with orthogonal coordinate vectors in which case $b_i = 2\pi/a_i$; but don't count on it!

Exercise 2.10 (M*) *Set $a_1 = 3$ Å, $a_2 = 5$ Å, and $\phi = 60°$. Calculate the values of b_1 and b_2 and the angle between them. Check "bravais" as follows. You can measure the distance between two points in either space by dragging the cursor between the two points with the left mouse button depressed. Readouts give the length and orientation of the line. Has "bravais" got the reciprocal lattice right?*

Exercise 2.11 (C) *Return to PRESET 1 and click on LATTICE above the right hand frame to display the lattice reciprocal to that of the honeycomb structure. Remove one of the atoms in the unit cell by clicking on ATOM B. Explain why nothing changes in the reciprocal lattice.*   # 1

So far the reciprocal lattice seems a mathematical construct with no physical content. Let's see how the reciprocal lattice can give a concise representation of the conditions for x-ray diffraction. The connection with x-ray diffraction can be seen by relating the geometry of the reciprocal lattice to the sets of planes discussed earlier. In two dimensions we denote a vector of the reciprocal lattice by the symbol $\mathbf{G}_{hk}$, the pair of indices $(h\,k)$ denoting the particular reciprocal lattice point of interest. Expressed in terms of the vectors $\mathbf{b}_1$ and $\mathbf{b}_2$ we have $\mathbf{G}_{hk} = h\mathbf{b}_1 + k\mathbf{b}_2$. The critical relationships between sets of lattice planes and $\mathbf{G}$s are:

1. the normal to the planes with Miller indices $(h\,k)$ is parallel to the reciprocal lattice vector $\mathbf{G}_{hk}$;
2. the perpendicular distance d_{hk} between adjacent lattice planes of the set $(h\,k)$ is equal to $2\pi/|\mathbf{G}_{hk}|$.

These relations require some exploration. With PRESET 1 and both the reciprocal LATTICE and SHOW PLANES toggles activated, DOUBLE CLICK on one of the reciprocal lattice points. The corresponding $(h\,k)$ values are displayed below and a pair of the $(h\,k)$ planes appear in real space. Again, if the $(h\,k)$ have a common divisor n, only every nth plane will contain

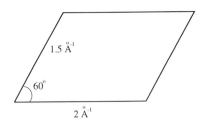

Figure 2.2: A unit cell of the reciprocal lattice. (Exercise 2.14)

lattice points. The extra planes correspond to higher Fourier components of the electron density, or to higher order Bragg reflections.

Exercise 2.12 (C) DOUBLE CLICK *on a number of reciprocal lattice points to check qualitatively the two statements above. Look for the two critical features:*

1. *the vector from the origin of the reciprocal lattice to the* $(h\,k)$ *reciprocal lattice point is perpendicular to the lattice planes shown at the left;*

2. *the higher the* $(h\,k)$ *values the more closely spaced are the planes.*

Exercise 2.13 (M*) *With* $\mathbf{G}_{31}$ *in reciprocal space and the* $(3\,1)$ *planes in real space displayed, use the mouse dragging technique to measure:*

1. *the magnitude and orientation of* $\mathbf{G}_{31}$*;*

2. *the perpendicular distance between the adjacent* $(3\,1)$ *planes;*

3. *the orientation of the normal to the* $(3\,1)$ *planes.*

Verify the relations between the $\mathbf{G}$*s and the lattice planes.*

Exercise 2.14 (M) *Figure 2.2 defines the cell of a reciprocal lattice. What are the parameters defining the cell of the associated real-space lattice?*

2.5   Diffraction conditions

So far we've indicated how to define a crystal structure in terms of the associated lattice and a basis, and discussed relations between the sets of lattice planes and the reciprocal lattice. Given a crystal of unknown structure, how do we determine the microscopic arrangement of the atoms? The x-ray analysis falls naturally into two parts, the determination of the lattice and the determination of the basis, which we discuss in that order.

Figure 2.3 illustrates a beam of x-rays incident from the left on a set of crystalline planes with interplanar spacing d. Constructive interference

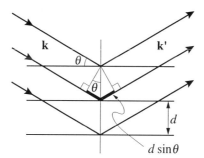

Figure 2.3: Construction for the Bragg condition. Heavier lines indicate the extra path length for reflection from second plane.

of the waves reflected from two adjacent planes requires the difference in path length to be equal to an integral number of x-ray wavelengths. This is the Bragg condition for x-ray scattering,

$$2d \sin \theta = n\lambda. \tag{2.1}$$

θ is the angle of the incident x-ray beam with respect to the diffracting planes and λ the x-ray wavelength. In the two dimensions used by "bravais", the analogs of the atomic planes are atomic lines; we will call them planes to connect more directly to the three-dimensional terminology.

Exercise 2.15 (M) *Obtain the Bragg condition (2.1) by calculating the difference in path length for the two rays illustrated in Figure 2.3 and requiring it to be an integral number of wavelengths, the condition for constructive interference.*

We prefer to express the condition for diffraction in the language of reciprocal space rather than real space. Instead of speaking of x-rays of wavelength λ propagating in some specified direction, we speak of x-rays with wave vector $\mathbf{k}$, where $|\mathbf{k}| = 2\pi/\lambda$ and the direction of $\mathbf{k}$ is the direction of propagation. The Bragg condition (2.1) may be concisely expressed as the combination of the requirement of elastic scattering, $|\mathbf{k}| = |\mathbf{k}'|$, already implicit in the Bragg argument, and the requirement that the change in wave vector in the scattering be equal to a reciprocal lattice vector. This condition,

$$\Delta \mathbf{k} \equiv \mathbf{k}' - \mathbf{k} = \mathbf{G}_{hk}, \tag{2.2}$$

is called the Laue condition. It is only for this limited set of $\Delta \mathbf{k}$s that the waves scattered from the different unit cells in the crystal will interfere with one another constructively. Incidentally, don't fall into the notational trap of confusing the Miller index k with the magnitude k of the wave vector. The distinction should be obvious from context!

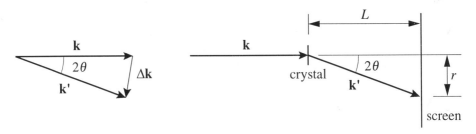

Figure 2.4: Geometry of the diffraction experiment.

Exercise 2.16 (M*) *Show that the elastic condition combined with the Laue condition imply the Bragg law (2.1) for x-ray diffraction. In the Bragg condition, there is the integer n denoting higher order diffractions. Explain how these are naturally included by the Laue formulation which gives the diffraction conditions in terms of the reciprocal lattice. (Hint: to get the higher order, $n > 1$, Bragg reflections, think about the interpretation of $\mathbf{G}$s with hkls with common divisors, e.g., (4 8 2).)*

 # 6   In PRESET 6 "bravais" simulates a two-dimensional diffraction experiment. On the right of the monitor screen is the diffraction pattern for the honeycomb structure on the left. Figure 2.4 illustrates the geometry of the experiment. We are looking at the crystal edge on. The incident x-ray beam, with wave vector $\mathbf{k}$, comes from the left and is normal to the two-dimensional array of atoms. The beam is scattered through the double Bragg angle 2θ, leaves the crystal with new wave vector $\mathbf{k}'$, and hits the screen. For small θ the diffraction peak positions on the screen are simply related to the scattering vectors $\Delta\mathbf{k}$ by the relation $r/L \approx \Delta\mathbf{k}/|\mathbf{k}|$. The screen, for small diffraction angles, is a map of the reciprocal space of the crystal. The right hand display of "bravais" gives the picture that would be seen on the observation screen but with distances appropriately scaled to $\Delta\mathbf{k}$. The size of the spots is chosen for convenience of display: the physics of the width of diffraction spots is not addressed by "bravais".

Exercise 2.17 (C) *Switch on the display of the reciprocal LATTICE as a reminder that the positions of the diffraction peaks give a scaled map of the reciprocal lattice. Move the atoms around with respect to one another, or remove and replace one of the atoms in the basis. The intensities change but the diffraction spots remain on the reciprocal lattice points. The positions of the diffraction peaks are determined by the Bravais lattice of the crystal, not by the basis.*

The first message, then, is that the positions of the diffraction peaks are determined by the parameters of the Bravais lattice associated with the crystal structure. Typically the first step in a structure analysis is the determination of the Bravais lattice. There are established techniques [10, 11] for translating a two-dimensional recording, e.g., on photographic film, of a diffraction experiment to a three-dimensional space lattice. In the next two sections we will discuss the physics determining the intensities of the diffraction peaks.

Exercise 2.18 (M*) *Suppose the diffraction pattern of* PRESET 6 *had been obtained in a real experiment, with the x-ray wavelength equal to 0.1 Å. Use the geometry defined in Figure 2.4 to determine the crystal to screen distance L that would give the pattern at the actual size scale, in centimeters, seen on your screen.*   honeycomb

Exercise 2.19 (M**) *Use the Ewald construction [1, 6, 11] to analyze a geometry different from that of Figure 2.4, one in which both the incident x-ray beam and the detector lie in the plane of the crystal. In this case there is no simple mapping of the reciprocal lattice onto the observation plane in the experiment. Show that, for a monochromatic beam, it would be very unlikely for there to be any allowed diffraction.*

Exercise 2.20 (M**) *Use the Ewald construction [1, 6, 11] to remove the assumption of small Bragg angle in the experiment illustrated in Figure 2.4. Show that the diffraction peaks give a non-linear scaled map of the two-dimensional reciprocal lattice. What is the relation between $\Delta\mathbf{k}$ and the spot position on the screen? (Hint: Let the z-axis be in the direction of $\mathbf{k}$. The Laue condition restricts the x- and y-components of the scattering vector but not its z-component. In the three-dimensional reciprocal space for this experiment, what is the locus of the values of $\Delta\mathbf{k}$ allowed by the Laue condition? How is this representation of the Laue condition then combined with the elastic condition?)*

2.6   Atomic form factor

The Bravais lattice defines, via the corresponding reciprocal lattice, the *scattering vectors*, $\Delta\mathbf{k} \equiv \mathbf{k}' - \mathbf{k}$, for which we might expect to observe diffraction peaks. The relative intensities of these peaks are determined by the contents of the unit cell, the basis. For a crystal with a single atom per unit cell the intensities are governed by the scattering by that single atom. Choose PRESET 7 to recover a monatomic rectangular crystal structure. Notice that the intensities of the x-ray peaks fall off with increasing magnitude of $\Delta\mathbf{k}$. This is a consequence of the destructive interference of components of the wave scattered from different parts of the atom when $|\Delta\mathbf{k}|^{-1}$ becomes comparable with the size of the atom.

7

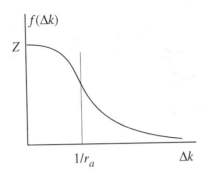

Figure 2.5: Atomic form factor.

This destructive interference is characterized by the "atomic form factor" $f(\Delta\mathbf{k})$ with the essential properties illustrated in Figure 2.5:

1. in the limit of small $\Delta\mathbf{k}$ (small angle scattering) the scattered amplitude is proportional to the total number of electrons in the atom, the atomic number Z;

2. the scattered amplitude falls off for values of $|\Delta\mathbf{k}|$ comparable to the inverse of the atomic radius r_a.

In the simulation "bravais" uses the Gaussian,

$$f(\Delta\mathbf{k}) = Ze^{(-\frac{1}{2}|\Delta\mathbf{k}|^2 r_a^2)}, \tag{2.3}$$

for the atomic form factor, where r_a is the atomic RADIUS. Recall that the scattered intensity measured in the x-ray experiment is proportional to the square of the amplitude, $I(\Delta\mathbf{k}) \propto |f(\Delta\mathbf{k})|^2$. Thus it is the atomic form factor which governs the decrease in scattered intensity with increasing $|\Delta\mathbf{k}|$ in PRESET 7. The choice of a Gaussian electron density is one of convenience; it does not give a good representation of an atomic electron density.

Exercise 2.21 (C) *Change the* SIZE *of atom A by a factor of 2 (both larger and smaller) and verify that the envelope of the intensities of the diffraction peaks varies as it should.*

If you want to make a more quantitative check, note that in the real-space display in "bravais", the radius of the circles representing the atoms is three times the RADIUS given by the slider.

Exercise 2.22 (C) *What, roughly, are the ratio r_A/r_B of the sizes of the atoms and the ratio a_A/a_B of the lattice constants of the crystals which gave the two diffraction patterns illustrated in Figure 2.6?*

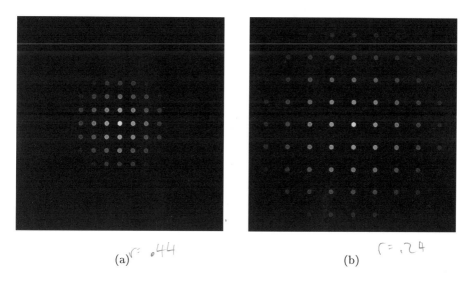

(a) $r = .44$ (b) $r = .24$

Figure 2.6: Diffraction patterns for two crystals with different lattice constants and different atom sizes. (Exercise 2.22)

2.7   Structure factor

If there is a single atom per unit cell of the crystal, the diffracted intensities give us information about the size of the atom and, as we shall see with "laue", about the electron distribution in the atom, as well. More interesting is the case in which there are a number of atoms in the unit cell: hundreds of thousands in crystals of important biological molecules but, fortunately for us, at most two in "bravais". Now we must include the effect of interference among the waves scattered from each of the atoms in the unit cell. The bookkeeping is expressed by the *structure factor*. For our case of two atoms per unit cell in a two-dimensional crystal it takes the form,

$$I(\Delta \mathbf{k}) = I(h, k)$$
$$\propto |S(\Delta \mathbf{k})|^2 = \left| f_A(\Delta \mathbf{k}) e^{i\Delta \mathbf{k} \cdot \mathbf{R}_A} + f_B(\Delta \mathbf{k}) e^{i\Delta \mathbf{k} \cdot \mathbf{R}_B} \right|^2 \quad (2.4)$$
$$\propto |S(h, k)|^2 = \left| f_A e^{2\pi i (hA_1 + kA_2)} + f_B e^{2\pi i (hB_1 + kB_2)} \right|^2. \quad (2.5)$$

The intensity for scattering with a change in the wave vector $\Delta \mathbf{k}$ is proportional to the square of the scattered amplitude, which in turn is the sum of the contributions from the two atoms A and B. They depend upon the individual atomic form factors and on exponential phase factors which define the relative phases of the scattered waves from the two atoms at positions $\mathbf{R}_A$ and $\mathbf{R}_B$. Since we are interested in these intensities only at the particular values of $\Delta \mathbf{k} = \mathbf{G}_{hk} = h\mathbf{b}_1 + k\mathbf{b}_2$ it is convenient to express the results in terms of the h and k instead of $\Delta \mathbf{k}$, giving the form (2.5) of the structure factor. To get Eq. (2.5) we have written $\mathbf{R}_A = A_1\mathbf{a}_1 + A_2\mathbf{a}_2$, etc., and used the relation $\mathbf{a}_i \cdot \mathbf{b}_j = 2\pi\delta_{ij}$.

Exercise 2.23 (M*) *Show how to obtain the form (2.5) of the structure factor from Eq. (2.4).*

2.7.1   Extinctions

8   PRESET 8 demonstrates the effect of the interference between the waves scattered from the two atoms in the cell. It is like PRESET 7 but with a second atom, identical to the first, at the center of the rectangular cell. Click on the ATOM B button to remove and add ATOM B while watching the diffraction pattern. What if you delete ATOM A instead of ATOM B? Include the reciprocal LATTICE display. It, of course, does not change when the second atom is added or deleted. The loss of one half of the spots is the consequence of the destructive interference of the waves scattered from the two distinct atoms in the cell.

With both atoms present, click on the SHOW PLANES button and then double click on the (1 0) reciprocal lattice point (a point of zero diffracted intensity). In the crystal structure display, note that there are as many atoms midway between the lattice planes as there are on them. Recall Figure 2.3 which we used to obtain the Bragg condition for diffraction. A quick evaluation should convince you that the waves scattered from the atoms between the planes are 180° out of phase with those scattered from the atoms in the planes. We call it an "extinction" when, for a reciprocal lattice point, there is complete destructive interference among the atoms in the unit cell so that no diffraction peak is seen.

The structure factor gives a systematic way to calculate the relative intensities of the different diffraction peaks.

Exercise 2.24 (M*) *Evaluate the structure factor for this case of identical atoms at positions $[0\,0]$ and $[\frac{1}{2}\,\frac{1}{2}]$. From the structure factor develop a rule that will tell for what $(h\,k)$ pairs the structure factor will be zero. Does that rule match the pattern seen in "bravais"?*

Exercise 2.25 (C) *Check out several low index planes and compare the numbers of atoms in the lattice planes with the numbers midway between the planes. How do these comparisons relate to the presence or absence of peaks at the corresponding reciprocal lattice points?*

2.7.2 Atomic inequivalence

In PRESET 8 the two atoms were identical: same ATOMIC NUMBER and same RADIUS. What happens if they are different in nature? In "bravais" you can change the ATOMIC NUMBER Z. As you do, the color of the atom changes to indicate that the atoms are becoming inequivalent. "bravais" can also destroy the equivalence of the atoms by changing their RADII independently of one another. Be aware in these exercises that "bravais" normalizes the diffracted intensities to keep the central spot at a fixed brightness. When you change the value of a parameter, you may see unexpected changes in some intensities which reflect the renormalization of the whole pattern with the parameter change.

Exercise 2.26 (C) *Vary the* ATOMIC NUMBER *of one of the atoms and watch the behavior of the diffraction pattern. What happens to the extinctions we noted in the original version of* PRESET 8*?*

Exercise 2.27 (M*) *If the ratio of* ATOMIC NUMBERS *of the two atoms is 3/1, predict the ratio of the intensities of the strong and weak reflections?*

Exercise 2.28 (C*) *Be sure the* ATOMIC NUMBERS *have been reset to be the same and now change the* RADIUS *of one of the atoms. Again, how does the intensity of the extinctions vary as the* RADIUS *of one of the atoms is varied. Explain the qualitative details of that variation. (Hint: check carefully the behavior of the atomic form factor (2.3) to see how it depends upon both radius and atomic number.)*

Exercise 2.29 (C**) *Figure 2.7 shows the pattern for two atoms at the positions of* PRESET 8*. The atoms have different* ATOMIC NUMBER *and different* RADIUS*. Find a convincing argument that the atom with the higher atomic number has the larger radius.*

Exercise 2.30$_A$(C*) *The contrast of intensities of the different diffraction peaks for KCl is very different from that for NaCl, though the structures, apart from lattice constant, are the same. In fact it is easy to misinterpret the KCl diffraction pattern as that of a simple cubic structure. Explain why. (Hint: check out the positions of these elements in a periodic table.)*

Exercise 2.31 (M**) *Adjust the* ATOMIC NUMBER *and the* RADIUS *of one of the atoms in "bravais" until there is a ring of strong extinctions for some magnitude of Δk, but only partial extinctions for smaller and larger Δk. If the atomic form factors are represented as in Eq. (2.3), find a condition on the radii r_A and r_B and the atomic numbers Z_A and Z_B that will give an extinction at $\Delta k \approx 4 \text{ Å}^{-1}$ but not elsewhere.*

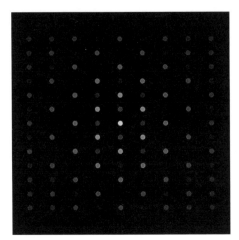

Figure 2.7: Diffraction pattern for a diatomic structure with atoms of different RADIUS and ATOMIC NUMBER. (Exercise 2.29)

9   **Exercise 2.32 (C**)** PRESET 9 *returns our favorite structure, the honeycomb. The interference between the two atoms in the cell gives a systematic variation in intensities of the spots. Try to vary the RADII and ATOMIC NUMBERS in such a way as to reduce the intensities of the weaker spots to zero. Explain clearly why you cannot. (Hint: you should find Eq. (2.5) useful.)*

2.7.3 Positional information

In the examples based on PRESET 8 the second atom is specially placed in the cell so that for certain reflections (those for which $h + k = $ odd) the scattering from the two atoms is 180° out of phase. What happens if the atoms are identical but they are moved with respect to one another so that the interference is no longer strictly constructive or destructive?

8   Reopen PRESET 8 and change the position of ATOM B to $[\frac{1}{5} \; \frac{1}{5}]$. Now the intensities of the peaks have a slowly varying sinusoidal modulation or *envelope*. This envelope, proportional to the square of the structure factor (2.5), is the diffraction intensity of the contents of the unit cell. Being the interference pattern of the waves scattered from the two atoms, the envelope should remind you of a two-slit diffraction pattern.

Exercise 2.33 (C) *Use the slider to move one of the atoms around and be sure you can explain the variation in diffraction pattern which you observe. How would you expect the period of the modulation or envelope to depend upon the separation of the atoms? Is that what happens? As the atoms move further apart the impression of the smooth modulation* *is lost and at certain positions the smooth modulation becomes a simple "yes or no": a "no" is one of the extinctions noted earlier.*

Exercise 2.34 (M*) *Use the press and drag technique to measure the perpendicular distance from one valley of the envelope function to the next and the orientation of the lines of extinction (the valleys). Remembering that this envelope is determined by the interference between the two atoms, predict the relation between the separation of the atoms (magnitude and direction) and the properties of the diffraction pattern you've just measured. Was your prediction correct?*

Exercise 2.35 (M*) *Predict the relative intensities of the peaks for one atom at the origin and the other at $[\frac{1}{3} \frac{1}{2}]$. Does "bravais" agree?*

Note when you calculate the structure factors that the phase factors in Eq. (2.5) depend only on the indices defining the diffracting planes and the atom positions in the cell (normalized to the cell dimensions). The atom–atom interference condition for a peak defined by a particular $(h\,k)$ is *in*dependent of the parameters defining the size and shape of the unit cell except for a dependence due to the Δk dependence of the atomic form factors. Confirm this by varying a_1, a_2, and ϕ. The positions of the spots move with the reciprocal lattice, but the intensities are determined only by the $(h\,k)$ indices except for some smooth variation related to the atomic form factors!

2.7.4 Structure determination

We have now all of the ingredients we need to determine the structure of an unknown sample. The first step is the analysis of the *positions* of the peaks in a diffraction pattern. These positions, with a suitable translation based on the parameters of the experiment, geometry and x-ray wavelength, give us the reciprocal lattice which in turn tells us the Bravais lattice of the unknown material: hence the parameters defining the unit cell.

The second step is to measure the *intensities* of a large number of the peaks. Their relative values are given by the square of the structure factor and we might hope to determine the atom positions in some direct fashion from the intensities. That turns out to be harder than one might think. The usual solution is not elegant. Crudely speaking, we make a guess for the structure, calculate the structure factor based on that guess, and compare the predicted set of intensities with the observed ones. The assumed atomic positions are then changed slightly, the intensities recalculated and compared. The outcome of this comparison suggests further changes and the process is iterated until no further improvement

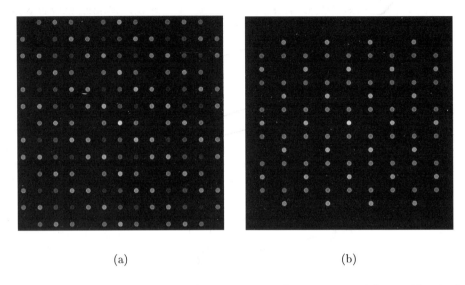

(a)   (b)

Figure 2.8: Diffraction patterns from two crystal structures, with two identical atoms per unit cell. (Exercises 2.36 and 2.37)

is made. If the agreement is good, we hope to have the correct structure. If it is poor we make a new guess and repeat the process. With only two atoms per cell, this is a viable approach and, with a little practice, would give answers fairly quickly. For a biological molecule with up to a million atoms it is a formidable task, even for a computer [13]. Success depends on the intuition of the biochemist who provides the computer with the critical first guess as to a possible structure. Here are a couple of two-atom examples with which to practice.

Exercise 2.36 (M*) *If one of the two equivalent atoms giving the diffraction pattern in Figure 2.8(a) is at the origin of a square cell, where is the other? With these positions in "bravais", see if you get this pattern.*

Exercise 2.37 (M**) *If one of two equivalent atoms giving the diffraction pattern in Figure 2.8(b) is at the origin of a rectangular cell, where is the other? Try these positions in "bravais" and see if you get this pattern.*

2.8   Summary

Two features of the x-ray diffraction experiment, the wave-like propagation of the x-rays and the periodicity of the crystalline lattice, drive us to a language based on the idea of wave-vector space or reciprocal space. Associated with a crystal structure is a reciprocal lattice, defined by the crystal real-space lattice but independent of the way atoms are arranged within the unit cell defining the lattice. This reciprocal lattice defines, through the Laue condition (2.2), a set of scattering vectors $\Delta\mathbf{k} = |\mathbf{k}' - \mathbf{k}|$ for which there is constructive interference of the waves scattered from the various unit cells in the crystal.

It is the arrangement of the atoms within the unit cell which determines the intensities of the various diffracted beams allowed by the Laue condition. The structure factor gives a formal expression for the effects of the interference among the waves scattered by the several atoms in the cell. Its square, evaluated for a particular allowed scattering vector, gives the intensity of the corresponding diffracted beam. Within the structure factor is information about the relative positions of the atoms, and through their atomic form factors, their atomic numbers and atomic radii. Though there is no direct way to deduce a crystal structure from the diffraction pattern, techniques have been developed which allow the deciphering of the structure of exceedingly complex structures [13], the most famous example being the double helix of the DNA molecule.

2.A   Deeper exploration

Diffraction game   This is a project for two. One sets up a structure using "bravais", and the other has to deduce that structure from the diffraction pattern; you then trade roles. After a few cycles, work together to formulate, if you can, a *systematic* scheme for going from the diffraction pattern to the structure. Remember that structure implies information about both the atom location and the atomic electron distribution.

2.B   "bravais" – the program

"bravais" illustrates the relationship between a two-dimensional crystal structure and the corresponding diffraction pattern.

2.B.1   Crystal structure (real space)

A crystal structure is given by an underlying Bravais lattice together with a basis that describes the atomic arrangement within each unit cell. The (two-dimensional) Bravais lattice in "bravais" is specified by two primitive lattice generating vectors $\mathbf{a}_1$ and $\mathbf{a}_2$. In "bravais" the vector $\mathbf{a}_1$ is taken to be parallel to the (horizontal) x-axis and has length a_1 (in Å); the vector $\mathbf{a}_2$ forms an angle ϕ with $\mathbf{a}_1$ and has length a_2 (in Å). For simplicity, the basis in "bravais" is limited to two atoms, A and B. The position $\mathbf{R}$ of each basis atom within the unit cell is specified in terms of the generating vectors of the primitive lattice, $\mathbf{a}_1$ and $\mathbf{a}_2$. For example, A_1 and A_2 give the fractional distance along the basis vectors $\mathbf{a}_1$ and $\mathbf{a}_2$, or

$$\mathbf{R}_A = A_1 \, \mathbf{a}_1 + A_2 \, \mathbf{a}_2. \tag{2.6}$$

Each atom is characterized by an ATOMIC NUMBER Z and a RADIUS r (in Å). "bravais" assigns a color to each atom according to its ATOMIC NUMBER on a sixteen-step color scale from yellow ($Z = 0$) to red ($Z = 100$), and a size given by *three* times its RADIUS r (the factor 3 is chosen for better visibility of the atoms).

2.B.2   Diffraction pattern (reciprocal space)

An x-ray diffraction pattern can be computed as the (two-dimensional) Fourier transform of the (two-dimensional) real-space electron density of the crystal.[1] For the crystal structures in "bravais", however, it is possible to use a simpler (and faster) method to determine the x-ray diffraction pattern, using the representation of the x-ray diffraction pattern by a reciprocal lattice with a Bragg peak associated with each lattice point. The intensity of the Bragg peaks is proportional to the magnitude squared of the structure factor, and both the reciprocal lattice and the structure factor can be calculated analytically, as shown below.

Reciprocal space lattice   A two-dimensional reciprocal lattice is generated by two lattice vectors $\mathbf{b}_1$ and $\mathbf{b}_2$. These two reciprocal-space vectors are related to the generating vectors of the real-space lattice, $\mathbf{a}_1$ and $\mathbf{a}_2$, by $\mathbf{b}_i \cdot \mathbf{a}_j = 2\pi \delta_{ij}$, $i, j = 1, 2$, where δ_{ij} is the Kronecker delta, $\delta_{ij} = 1$ for $i = j$ and 0 otherwise. From our choice of generating vectors for the real-space lattice it follows that $\mathbf{b}_2$ is parallel to the (vertical) y-axis and $\mathbf{b}_1$ forms an angle $\phi - 180°$ with $\mathbf{b}_2$.

Atomic form factor   The electronic density of each atom in "bravais" is assumed to be Gaussian. The atomic form factor $f(\mathbf{G})$ of ATOM A (B) at the reciprocal lattice vector $\mathbf{G}$ is taken to be proportional to the Fourier transform of that electron density, and thus given in terms of the atomic RADIUS $r_a = r_A$ ($r_a = r_B$) and ATOMIC NUMBER $Z = Z_A$ ($Z = Z_B$) by Eq. (2.3). "bravais" allows the ATOMS A and B to be "toggled" on or off. If an ATOM is not selected, its form factor is zero.

Structure factor   The intensity of a Bragg peak associated with a reciprocal lattice vector $\mathbf{G}$ is proportional to $|S(\mathbf{G})|^2$, where $S(\mathbf{G})$ is the

---

[1]This is essentially how the (one-dimensional) diffraction pattern is computed in "laue".

(complex) structure factor. The structure factor at a reciprocal lattice vector $\mathbf{G} = h\mathbf{b}_1 + k\mathbf{b}_2$ is given by Eq. (2.5), with the atomic form factors $f_A(\mathbf{G})$ and $f_B(\mathbf{G})$ defined in Eq. (2.3). A_i and B_i determine the positions of atoms A and B within the unit cell as given in Eq. (2.6). Thus the magnitude of S is given by

$$|S| = \sqrt{f_A^2 + f_B^2 + 2f_A f_B \cos\left[h(A_1 - B_1) + k(A_2 - B_2)\right]}. \qquad (2.7)$$

(Note that the magnitude squared of S, and thus the intensity of the Bragg peaks, is proportional to the squared atomic form factor f^2 for a single atom in the unit cell ($f_B = 0$, say) and zero for no atoms in the unit cell ($f_A = f_B = 0$).)

To display the intensities of the Bragg peaks associated with the lattice points, "bravais" assigns each reciprocal lattice vector $\mathbf{G}$ a "color" proportional to the magnitude of the structure factor $|S(\mathbf{G})|$. The color is given on a sixteen-step grey scale from white (maximum value of $|S|$) to black ($|S| = 0$). The normalization is chosen so that the diffraction spot at the origin of reciprocal space $\mathbf{G} = (0,0)$ is always white. Note that the grey scale is taken to be proportional to the magnitude of S rather than its square to obtain a more useful dynamic range for the display.

2.B.3   Lattice planes and Miller indices

"bravais" illustrates the connection of the wave vectors $\mathbf{G}$ of the reciprocal lattice with lattice planes of the real-space lattice. To each reciprocal lattice vector $\mathbf{G}_{hk} = h\mathbf{b}_1 + k\mathbf{b}_2$ with h and k having no common divisor, corresponds a set of lattice planes denoted by the Miller indices $(h\,k)$. A particular $(h\,k)$ may be selected either by DOUBLE-CLICKING near a reciprocal lattice point with the left-hand mouse button (or by CLICKING with the right-hand mouse button) or by typing the indices in the appropriate entry boxes. The corresponding reciprocal lattice point is highlighted in green (the origin of reciprocal space is highlighted in red) and two of the set of lattice planes are drawn in the real space display. The two planes in real-space are perpendicular to the reciprocal space vector $\mathbf{G}_{hk}$ and have separation $2\pi/|\mathbf{G}_{hk}|$. One of the planes is always drawn through the (arbitrary) origin of the real space lattice. When $(h\,k)$ have a common divisor, the same construction is used and some of the lattice planes will contain no lattice points. The periodicity of this set of planes corresponds to higher order Bragg reflections.

3

"laue" – Diffraction in perfect and imperfect crystals

Contents

3.1 Introduction

The motivation for the development of dramatically improved sources of x-rays is made evident by the wealth of information potentially available from a variety of techniques of x-ray diffraction. The complex structure determinations mentioned in "bravais", so important in biology, are only the tip of an iceberg. X-ray techniques can characterize thermal motion in solids, grain size and preferential orientations in polycrystalline material, the degree of order in ordered alloys, and homogeneous and inhomogeneous strains introduced by a wide variety of crystalline imperfections.

Sections 3.2–3.4 of "laue" illustrate many of the same issues that were discussed in "bravais", but in one dimension instead of two, and often in a more quantitative fashion. The Fourier transform algorithm used by "laue" allows calculation of the diffraction from imperfect crystals, and is used in Sections 3.5–3.7 to illustrate spatial modulation of diffraction patterns by lattice waves, thermal scattering, the effect of disorder in alloys, and quasicrystal diffraction. The later sections may be profitably explored without the earlier ones if the concepts developed in "bravais" are clear. On the other hand, "bravais" need not be considered a prerequisite to "laue".

3.2 Bragg condition

The geometry of the one-dimensional experiment is less intuitive than that of the two- and three-dimensional cases. Figure 3.1 illustrates a line of atoms, lattice constant a, with an x-ray beam of wave vector $\mathbf{k}$ (with $|\mathbf{k}| = k = 2\pi/\lambda$) incident from the lower right at an angle θ with respect to the line perpendicular to the string of atoms. We look for the condition for constructive interference for scattering the beam into the direction $\mathbf{k}'$, making the same angle θ with respect to the perpendicular. We might imagine these directions being defined by the position of a source and detector, and ask for the beam intensity as the angle θ is varied. The condition that the difference in path lengths for the rays scattered from atoms A and B (the bold line in Figure 3.1) should be an integral number of wavelengths defines the θ values at which to expect diffracted beams.

Exercise 3.1 (M) *Show that this condition for constructive interference may be stated as $2a \sin\theta = h\lambda$, where h is an integer.*

This is a one-dimensional specialization of the Bragg condition for x-ray diffraction which is conventionally written as

$$2d \sin\theta = n\lambda. \tag{3.1}$$

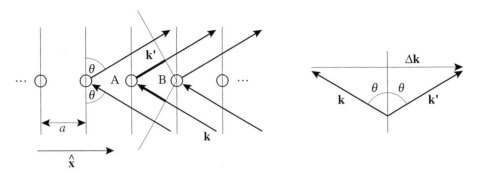

Figure 3.1: Bragg condition for a row of atoms with lattice constant a and equal angles of incidence and diffraction. The length of heavier line gives the extra path length for rays scattered from A relative to those scattered from B.

In one dimension the lattice constant a and the "interplanar spacing" d are the same. We use h as the integer to designate higher order diffractions since it is the one-dimensional analog of the hkl indices used to denote lattice planes in three dimensions.

Formal analysis is more conveniently done in terms of wave vectors than wavelengths. A critical conclusion of the *kinematic theory* of x-ray diffraction is that the amplitude for the diffraction of an x-ray beam with initial wave vector $\mathbf{k}$ to a final state with wave vector $\mathbf{k}'$ is proportional to the Fourier transform of the electron density $\rho(\mathbf{r})$ of the scattering medium [1, 6, 11]. In n dimensions this statement takes the form,

$$A(\mathbf{k}' \leftarrow \mathbf{k}) \equiv A(\Delta\mathbf{k}) \propto \int_{\text{sample}} \rho(\mathbf{r})\, e^{i\Delta\mathbf{k}\cdot\mathbf{r}} d\mathbf{r}, \qquad (3.2)$$

where $\Delta\mathbf{k} \equiv \mathbf{k}'-\mathbf{k}$ is the *scattering vector*. If $\rho(\mathbf{r})$ is periodic, as in an ideal crystal, then the Fourier transform (for an infinite crystal) is non-zero only for a discrete set of values of $\Delta\mathbf{k}$, namely the *reciprocal lattice vectors*. "bravais" assumes an ideal crystal and works with analytic expressions to determine the scattering intensities, which are proportional to the squared magnitudes of the amplitudes. "laue" explicitly calculates the Fourier transform and is thus able to give results for imperfect as well as ideal crystals. The intensities, displayed in the right hand plot, are normalized to give 1 for the height of the central peak for the MONATOMIC CRYSTAL, and $(1 + Z_B/Z_A)^2$ for the DIATOMIC CRYSTAL.

The primitive reciprocal lattice vector for the one-dimensional crystal of "laue" is $\mathbf{b} = 2\pi\hat{\mathbf{x}}/a$ where $\hat{\mathbf{x}}$ is the unit vector along the string and the general reciprocal lattice vector is $\mathbf{G} = h\mathbf{b}$. The *Laue condition* for constructive interference may be expressed as $\Delta\mathbf{k} \equiv \mathbf{k}' - \mathbf{k} = \mathbf{G}$ or, for our one-dimensional scattering experiment, $\Delta k \equiv |\mathbf{k}' - \mathbf{k}| = G = 2\pi h/a$.

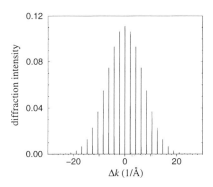

Figure 3.2: Diffraction from a MONATOMIC CRYSTAL. (Exercise 3.5)

Exercise 3.2 (M*) *Demonstrate the equivalence of the Laue condition,* $\Delta k \equiv k' - k = G = 2\pi h/a$, *and the Bragg condition (3.1) for x-ray diffraction for our one-dimensional problem. (You will need to use the elastic condition that x-ray energy and hence the magnitude of the wave vector is unchanged by the scattering, or* $|\mathbf{k}'| = |\mathbf{k}|$.)

In PRESET 1 of "laue" you see the electron density of a portion of a 64-atom crystal on the left and the corresponding diffraction pattern on the right. The abscissa of the diffraction graphs is Δk. This PRESET is for a MONATOMIC CRYSTAL of atoms with interatomic spacing $a = 4$ Å. ZOOM in on one or a few of the atoms to see their SHAPE which is GAUSSIAN in this PRESET. The discreteness of the sampling is evident. A more refined sampling, via the CONFIGURE menu, improves the curves, but lengthens the time to CALCULATE new plots.   # 1

Exercise 3.3 (M) *Verify that the peaks occur at the values of* Δk *appropriate to the parameters in the panel display. ZOOM a small* Δk-*interval to get better resolution.*

Exercise 3.4 (C) *Increase the LATTICE CONSTANT by a factor of 2. What is the effect on the diffraction pattern?*

Exercise 3.5 (M) *From the diffraction pattern in Figure 3.2 determine the lattice constant* a *of the corresponding crystal.*

Exercise 3.6 (M**) *In Figure 3.1 we have arbitrarily restricted the scattering to be in the plane of the paper, and the scattered beam to be at the same angle with respect to the atomic chain as the incident beam. Modify the arguments of the Ewald construction [1, 6, 11] to predict the nature of the diffraction in a real three-dimensional experiment. That is, with a source position defining the incident wave vector* $\mathbf{k}$, *in*

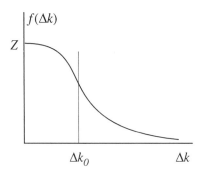

Figure 3.3: Atomic form factor.

what directions does the Laue condition allow a scattered beam. (Hint: for the chain of atoms show that the one-dimensional Laue condition is satisfied for a scattering vector $\Delta\mathbf{k}$ going from the origin to one of a set of planes in reciprocal space. Look first at the cases $\theta = 0°$ and $90°$ – the geometry is simpler to envision.)

3.3 Atomic form factor

3.3.1 Atomic scattering

The argument for the Bragg condition (3.1) above is given in terms of scattering from point atoms. It implies that we should see diffraction peaks for all values of $\Delta k = 2\pi h/a$. What is seen in the diffraction plots is a decreasing intensity with increasing Δk. This is because of the interference among the waves scattered from electron density in different parts of the finite-sized atom. If the x-ray wavelength is large compared with the atomic size, waves scattered from different parts of the atom will interfere constructively. In this case the scattered amplitude, from a single atom, is proportional to the number of electrons and hence the atomic number Z. However, if the wavelength, more properly the inverse of the magnitude of the scattering vector $|\Delta k|^{-1}$, is small compared with the atomic size, there will be substantial destructive interference. The Δk dependence of the scattered amplitude is described by an atomic form factor defined, in one dimension, by

$$f(\Delta k) \equiv Z \frac{\int_{\text{atom}} \rho(x)\, e^{i\Delta kx} dx}{\int_{\text{atom}} \rho(x) dx} = \int_{\text{atom}} \rho(x)\, e^{i\Delta kx} dx, \qquad (3.3)$$

where $\rho(x)$ is the electron density of the atom and Z its atomic number, the total number of electrons. Figure 3.3 shows the typical qualitative dependence of the atomic form factor on Δk.

Exercise 3.7 (M*) *Argue from the definition of $f(\Delta k)$ in Eq. (3.3) that, for small values of Δk, the atomic form factor is equal to the atomic number Z. Suppose σ is a measure of the size of the atom, i.e., $\rho(x)$ becomes small for $|x| > \sigma$. Then what is the rough value of Δk_0 which characterizes the width of the plot, Figure 3.3, of the form factor versus scattering vector?*

"laue" can calculate the diffraction pattern of a single atom as well as of a crystal. Further it gives several choices for the electron density of the atoms. PRESET 2 gives the diffracted intensity for a SINGLE ATOM   # 2 for which the ATOM SHAPE is the GAUSSIAN,

$$\rho(x) = \frac{Z}{\sigma\sqrt{2\pi}}\, e^{-\frac{x^2}{2\sigma^2}}. \tag{3.4}$$

Because this PRESET gives the diffraction from a single atom, the pattern does not show the spiky character resulting from the interference among the many atoms in the chain. Rather it is a diffuse scattering pattern, with scattered intensity for a wide range of directions, or for all values of Δk that are not too large.

Exercise 3.8 (C) *Vary the SIZE of the atom in PRESET 2 to check the qualitative prediction in Exercise 3.7 of the relation between the atom size and the width in Δk of the atomic form factor. (Don't be fooled by the auto-scaling. Check in the GRAPHS section of the HELP files to learn how to use the ZOOM to negate the auto-scaling.)*

Exercise 3.9 (M*) *Calculate the one-dimensional atomic form factor for the electron density in Eq. (3.4). Verify that the width of the diffraction pattern for PRESET 2 is properly given by "laue". (Remember, that the intensity is proportional to the square of the form factor!)*

3.3.2 Envelope of the diffraction pattern

What can we now say about the diffraction pattern of the chain of atoms? We must somehow combine with one another the two following ideas: the Bragg condition, describing the interference between atoms; and the atomic form factor, describing the interference of waves scattered from different parts of the same atom. The formal result for the intensity of the scattered wave in this one-dimensional case is the proportionality

$$I(\Delta k) \propto |f(\Delta k)|^2 \sum_h \delta\left(\Delta k - \frac{2\pi h}{a}\right), \tag{3.5}$$

where $\delta(k)$ is the Dirac delta function. This result can be summarized by the statements:

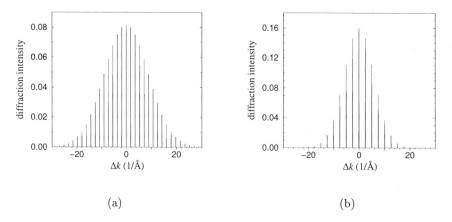

Figure 3.4: Diffraction patterns for two crystals with different LATTICE CONSTANTS and different atomic SIZES. (Exercise 3.11)

1. the intensity $I(\Delta k)$ is non-zero only when Δk is a multiple of $2\pi/a$ (the Laue condition);

2. the intensity of the peak at $\Delta k = 2\pi/a$ is proportional to the square of the atomic form factor evaluated at this Δk.

ϱ **Exercise 3.10 (C)** *Verify the two statements above as follows. Reload* PRESET 2, CALCULATE, *and* COPY DIFFRACTION. *Now select the* MONATOMIC CRYSTAL *from the* MATERIAL *menu,* CALCULATE *again, and then* STEAL DATA *to compare the envelope of the crystal diffraction pattern with the single-atom diffraction. Are the values of Δk at the peaks in the diffraction graph correct? Does the atomic diffraction pattern match the envelope of the peaks? It may be useful to highlight one or the other of the curves by moving the cursor onto the appropriate key at the upper right of the graph window.*

Exercise 3.11 (M) *Figure 3.4 shows two diffraction patterns. What are the* LATTICE CONSTANTS *of the two crystals? What is the ratio of the* SIZES *of the atoms in the two crystals?*

3.3.3   Other atomic shapes*

Exercise 3.12 (C*) *Atom "sizes", as determined from typical lattice constants, are much larger than the few tenths angstroms that "laue" likes to use. Why might the smaller value still be appropriate for a simulation of the x-ray experiment? (Hint: atomic electron densities are not Gaussian.*

Table 3.1: Atomic electron densities ρ for the different ATOM SHAPES in "laue". R is the atom's position, Z its atomic number, and σ or SIZE, the root mean square width of its electron density. The PSEUDO ATOM is represented by an electron density of the core electrons, a Gaussian of width $\sigma_c = \sqrt{2/17}\sigma$, superimposed on an electron density of the valence electrons, a Gaussian of width $\sigma_v = \sqrt{32/17}\sigma$. All four ATOM SHAPES agree in their zeroth, first, and second moments: i.e., $\int_{-\infty}^{\infty} dx\, x^n\, \rho(x)$ is the same for each $\rho(x)$ for $n = 0, 1, 2$.

ATOM SHAPE	Electron density $\rho(x)$ of a single atom
GAUSSIAN	$(Z/\sqrt{2\pi}\sigma)e^{-\frac{(x-R)^2}{2\sigma^2}}$
EXPONENTIAL	$(Z/2\sigma)e^{-\frac{|x-R|}{\sigma}}$
TRIANGULAR	$\begin{cases} (Z/6\sigma^2)\left(\sqrt{6}\,\sigma - |x - R|\right) & \text{if } |x - R| < \sqrt{6}\,\sigma \\ 0 & \text{otherwise} \end{cases}$
PSEUDO ATOM	$(Z/2\sqrt{2\pi}\sigma_c)e^{-\frac{(R-x)^2}{2\sigma_c^2}} + (Z/2\sqrt{2\pi}\sigma_v)e^{-\frac{(R-x)^2}{2\sigma_v^2}}$

Where is most of the electronic charge in an atom compared with what we think of as the size of an atom? Remember the shell structure of the atomic electrons.)

We chose the GAUSSIAN ATOM SHAPE for the electron distribution in our atom not because it is a good representation of the shape of a real atom but because it is easy for you to find its Fourier transform. "laue" includes options for several ATOM SHAPES as listed in Table 3.1: these are chosen so that each has the specified number of electrons Z and the same second moment σ^2, with σ defined by the SIZE slider.

Exercise 3.13 (C*) *Reload* PRESET 2 *and* COPY DIFFRACTION. *Then change the* ATOM SHAPE *from* GAUSSIAN *to* PSEUDO ATOM *and* CALCULATE *again.* STEAL *this second graph to allow a careful comparison in the separate graph window. Explain why the diffraction pattern of the* PSEUDO ATOM *extends to larger* Δk *than that of the* GAUSSIAN *atom.*

Exercise 3.14 (C*) *Have "laue"* CALCULATE *the diffraction pattern using several of these atomic shapes.* COPY DIFFRACTION *for each to get copies for intercomparison. Can you identify differences in these patterns which you can associate with differences in the electron distributions of* the different shaped atoms? (See GRAPHS in the HELP files for ways to highlight or change colors of graphs.)

This example suggests that diffraction techniques, if carried out with sufficient care, can characterize the detailed electron distribution of single atoms (and, with a generalization of the arguments, of molecules) as well as determine the structures of crystals.

Exercise 3.15 (M**) *The* PSEUDO ATOM *has been represented as the sum of two Gaussians, one representing the core electrons and one the valence electrons (see Table 3.1). To understand the strange choice of numbers for σ_c and σ_v, verify the following properties of the* PSEUDO ATOM:

1. *the core and the valence shell contain the same number of electrons;*
2. *the valence shell has four times the width of the core;*
3. *the contribution to the electron density at the origin is dominated, by a factor of 4, by the core electrons.*

Exercise 3.16 (C*) *Set up "laue" to compare the* SINGLE ATOM *electron densities and diffraction patterns for the* GAUSSIAN *atom and the* PSEUDO ATOM. *Be sure they have the same number of electrons Z and* SIZE.

1. *Explain why the scattered intensity at $\Delta k = 0$ is the same for both atoms, though the two peak heights in the electron density plots are different.*
2. *Why is there more scattering at large Δk for the* PSEUDO ATOM?

Exercise 3.17 (M**) *Calculate the atomic form factor (3.3) analytically for each of the four one-dimensional electron distributions given in Table 3.1. Relate the qualitative differences seen in the diffraction graphs (see Exercise 3.14) of these four kinds of* ATOM SHAPES *to the analytic expressions for the form factors.*

3.4   Structure factor

What new ideas come into play if there is more than one atom in the unit cell? As in the case for the monatomic crystal, the Bragg argument (or an equivalent) accounts for the interference with one another of waves scattered from *different* unit cells of the crystal. Next, we need to account for the interference of waves scattered from the different atoms *within* the cell. The formal result,

$$I(\Delta k) \propto |S(\Delta k)|^2 \sum_h \delta \left(\Delta k - \frac{2\pi h}{a} \right), \qquad (3.6)$$

is much like that for a single atom in the cell. It is obtained by replacing the atomic form factor $f(\Delta k)$ in Eq. (3.5) by the *structure factor* $S(\Delta k)$,

$$S(\Delta k) \equiv \sum_j e^{i\Delta k x_j} f_j(\Delta k), \qquad (3.7)$$

where the sum is over the different atoms in the unit cell, with the jth atom, at position x_j, having the atomic form factor $f_j(\Delta k)$. The square of the structure factor gives the envelope of the intensity pattern for the scattering from the contents of a single cell of the crystal.

3.4.1 Identical atoms

Let's consider first the diffraction from an isolated PAIR OF ATOMS.

Exercise 3.18 (M) *Calculate the structure factor (3.7) and intensity, $|S(\Delta k)|^2$, for the scattering by an isolated pair of identical atoms separated by a distance of 0.4 Å (not 0.4 a) in terms of their common atomic form factor. (Note that the* ATOM SEPARATION *listed by "laue" is in units of an arbitrarily chosen lattice constant a, not in Å.) Compare your predicted positions of the zeros of this intensity function with the pattern in the diffraction graph of* PRESET 3. *(Warning: because of the finite k-space grid, the diffraction nulls will be "clean" only if the* ATOM SPACING *is chosen equal to* (LATTICE CONSTANT)/*(even integer), as explained on page 59. The resolution of the two atom diffraction pattern can be enhanced by increasing the* LATTICE CONSTANT a, *decreasing the* ATOM SPACING d/a, *and increasing the* POINTS SAMPLED PER ATOM *in the configure menu, each by the same power of 2, perhaps 8. Remember, however, to return to* PRESET 3 *before proceeding.)*

\# 3

COPY DIFFRACTION for the pair of atoms in PRESET 3 to save for comparison. Now assemble a chain of such atom pairs by choosing DIATOMIC CRYSTAL from the MATERIAL menu.

Exercise 3.19 (C) STEAL DATA *to compare the diffraction pattern of the* DIATOMIC CRYSTAL *with that of the* ATOM PAIR. *Change the position of the second atom to other values of $d/a = 1/2n$, with n an integer, and verify qualitatively that "laue" is correctly calculating the diffraction pattern.*

If you move the second atom to the middle of the cell you will see that half of the diffraction peaks disappear. For odd values of h in $\Delta k = 2\pi h/a$, the scattering from the two atoms interferes fully destructively to give zero intensities. We refer to *extinctions* in the scattering pattern when the structure factor dictates zero intensity even though the Bragg (Laue) condition is satisfied.

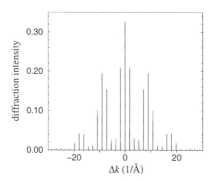

Figure 3.5: Diffraction for a DIATOMIC CRYSTAL with two identical atoms in the unit cell. (Exercise 3.20)

Note that we are interested in the structure factor *only* at the special values of the argument $\Delta k = 2\pi h/a$. It is convenient, then, to express the structure factor in terms of h instead of Δk;

$$S(\Delta k) = f(\Delta k) + f(\Delta k)e^{i\Delta k(a/2)} \Rightarrow$$
$$S(h) = f\left(\frac{2\pi h}{a}\right)\left(1 + e^{i(2\pi h/a)(a/2)}\right) = f\left(\frac{2\pi h}{a}\right)\left(1 + e^{i\pi h}\right) \quad (3.8)$$

The term $(1 + e^{i\pi h})$ in (3.8), from the interference between the atoms, takes on only two possible values: 2 if h is even and 0 if h is odd. The possible relative intensities are then 4 and 0 times $|f|^2$. (Remember that $I \propto |S|^2$.) This gives a succinct way of summarizing the pattern we just saw, with extinctions for odd values of h. What if the second atom is not at such a special position?

Exercise 3.20 (M) *Figure 3.5 illustrates the diffraction pattern of a DI-ATOMIC CRYSTAL in which the two atoms are identical. What is the ATOM SPACING between the atoms relative to the LATTICE CONSTANT? Why is there not complete extinction for any of the peaks?*

Exercise 3.21 (M*) *Calculate the effect of the interatom interference on the diffraction pattern of the diatomic crystal if the second atom is at the positions $\frac{1}{6}$, $\frac{1}{5}$, $\frac{1}{4}$, $\frac{1}{3}$, and $\frac{2}{3}$ of the lattice constant. As suggested in Eq. (3.8), express the results not in terms of Δk but in terms of the hs for which the interference term takes on one or another of its few possible values.*

Note that there are extinctions for some choices of the ATOM SPACING, but not for others. For the second atom at the position $1/m$ in the cell, what condition on the integer m determines whether or not there will be extinctions?

Exercise 3.22 (M**) *Suppose the second atom is at the position b/c, in units of the lattice constant a, with respect to the first, where b and c are positive integers with no common divisor and $b < c$. Neglect the variation with Δk of the atomic form factor.*

1. *Express the number of different intensities in the diffraction pattern in terms of b and c.*
2. *Of the $(c - 1)$ possible choices for b, how many distinct diffraction patterns are possible?*

3.4.2 Different atomic numbers

What happens if the two atoms are not identical? Now the structure factor becomes a little more complicated. Let's look at a simple case. Still using PRESET 3 with the MATERIAL set to DIATOMIC CRYSTAL, place the second atom in the center of the cell, midway between two atoms of the first kind. Without varying the atom sizes, try some different values of the atomic number of the second atom relative to the first.

Exercise 3.23 (C) *What is the diffraction pattern for a large $(\gg 1)$ ratio of ATOMIC NUMBERS ZB/ZA? Or for a small ratio $(\ll 1)$? Verify the conditions for the extinction that the second atom must be in the correct position and have the same atomic number as the first.*

Exercise 3.24 (M) *What is the predicted ratio of the intensities of successive peaks for the case $Z_B/Z_A = 3/1$? (Ignore effects of the Δk dependence of the atomic form factors.) Has "laue" got it right? (Note that you can get quantitative values for the peak intensities by placing the cursor on the diffraction peak and pressing the left mouse button.) How would the pattern differ if the atomic number ratio were set equal to $1/3$ instead of $3/1$?*

Exercise 3.25 (M*) *Figure 3.6 shows the diffraction pattern for a DIATOMIC CRYSTAL. What is the ratio of the ATOMIC NUMBERS of the two atoms? What is the SPACING between the atoms relative to the LATTICE CONSTANT?*

Exercise 3.26 (C**) *The physics of neutron diffraction [25] is closely related to that of x-ray diffraction. There are important differences, however. One is that the amplitude for neutron scattering from an atom depends upon the relative orientation of the neutron spin and any unpaired electron spin on the atom. Imagine the simplest possible case in which the scattering amplitude has the opposite sign for scattering from an atom with total spin up and one with total spin down. "laue" can simulate a one-dimensional antiferromagnet, with an alternating . . .-up-down-up-down-. . . arrangement of atomic moments, by choosing $Z_B/Z_A = -1$. If*

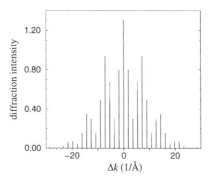

Figure 3.6: Diffraction for a DIATOMIC CRYSTAL with atoms of different ATOMIC NUMBERS. (Exercise 3.25)

the atoms are uniformly spaced, what is your prediction for the diffraction pattern? How would it differ from the pattern if all spins were aligned, the ferromagnetic state? What does "laue" give for the patterns?

3.4.3   Different sizes*

There's one more variable to play with, the atom SIZES. Now the dependence on Δk of the form factors $f_j(\Delta k)$ differs for the different atoms, and the strength of the interference between the atoms will depend on the value of Δk. In PRESET 4 the ATOMIC NUMBERS are the same but the atomic SIZES are different. As a consequence, near the center of the diffraction pattern there is one spacing of the diffraction peaks, and in the wings there is a different spacing.

4

Exercise 3.27 (C*) *Explain clearly what is going on! (Hint: convince yourself that the atomic form factors for the two types of atoms will be the same in the limit of small Δk but not for large Δk. For which will the atomic form factor fall off more rapidly with Δk, the wider or the narrower atom? Be sure you can justify your answer.)*

Exercise 3.28 (M*) *For the diffraction pattern in Figure 3.7, make the best estimate you can for the ATOM SPACING, the relative ATOMIC NUMBERS, and the relative SIZE of the two atoms in the unit cell.*

These examples may seem trivial, and perhaps they are, but the ideas are the same as those used in the analysis of crystals of large organic molecules [13]. Imagine trying to work with a crystal with 10^5 or 10^6 atoms per unit cell. Although information about the structure of the molecule is contained in the relative intensities of the many diffraction

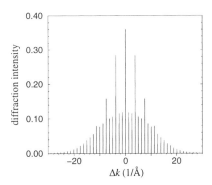

Figure 3.7: Diffraction for a DIATOMIC CRYSTAL in which the atoms have both different SIZE and different ATOMIC NUMBER. (Exercise 3.28)

peaks, there is no straightforward procedure for going from the intensities to the structure [13]. A fast computer is essential to calculate the structure factor for a hypothetical model, and to compare the prediction with diffraction data. Discrepancies between prediction and data are used to suggest changes to improve the model.

3.5   Order–disorder*

So far we have talked about perfect crystals. What happens if the crystals don't have the atoms arranged in the precise fashion that we've been talking about? Let's restrict our discussion to two sources of imperfection, compositional disorder and positional displacements. We'll look first at compositional disorder.

Suppose we have two types of atoms, A and B, and that at low temperatures they arrange themselves regularly in the array: ... ABABAB.... At higher temperatures this neat arrangement may become disturbed as atoms begin to change places with one another in response to thermal agitation. For example, sometimes two or three like-atoms may be next to each other. At very high temperatures the A and B atoms may be arranged over the lattice sites in a completely random fashion.

"laue" models such a system in the following way. The sites of a uniform chain are labeled alternately A sites and B sites. An ORDER PARAMETER p is introduced which can be given a value between -1 and 1. The program assigns A atoms to A sites with probability $\frac{1}{2}(1+p)$ and to B sites with probability $\frac{1}{2}(1-p)$. Similarly, B atoms are assigned to B sites with probability $\frac{1}{2}(1+p)$ and to A sites with probability $\frac{1}{2}(1-p)$. If p is 0 this system is fully disordered: any given site is equally likely to be occupied by an A atom or a B atom. If p is ± 1, it is fully ordered.

What will be the diffraction pattern? We introduce the idea of *average atoms* of two types, A$'$ and B$'$. The atomic form factors of the average atoms are weighted averages of the form factors of the real atoms, with weight factors based on the probabilities of occupation. A hypothetical crystal is constructed from these average atoms, placing the A$'$ and B$'$ atoms on alternate sites. Its structure factor is

$$S = \frac{1}{2}\left\{[(1+p)f_A + (1-p)f_B] + e^{i\pi h}\left[(1-p)f_A + (1+p)f_B\right]\right\}. \quad (3.9)$$

This average structure factor will properly describe the intensities of the diffraction peaks [10, 11, 12]. However, in addition to the scattering described by this structure factor there will be *diffuse scattering* distributed more or less continuously over the range of Δk.

The diffuse scattering may be thought of as scattering from the difference between the real structure and the average structure. The scattering amplitude from any site of the *difference structure* is equally likely to be positive or negative, destroying any systematic coherence in the interference of waves scattered from different sites. Scattering from the difference structure is equally likely at any value of Δk.

5   **Exercise 3.29 (C)** PRESET 5 *puts the second atom at the center of the cell with the same* SIZE *as the first but with a ratio of* ATOMIC NUMBERS *of 3/1. The diffraction pattern, corresponding to the ordered state ($p = 1$), should be no surprise. Now, before you* CALCULATE *again, predict what you will see for $p = 0$, the disordered state. Only then are you allowed to* CALCULATE *and check out your prediction. Compare as well the graphs of the electron densities for the two cases.*

For the ordered case ($p = 1$) there are diffraction peaks which do not appear for the disordered crystal ($p = 0$). These are called *superlattice peaks*. What is the superlattice? The average crystal in the fully disordered state has a lattice constant which is $a/2$ since the two sites in the original cell are, on average, equivalent. In the ordered state, the two sites are inequivalent and the lattice constant of the structure is the full a. It is this lattice with the larger lattice constant which is referred to as the superlattice.

Exercise 3.30 (M*) *Derive an expression for the intensities of the superlattice peaks as a function of the* ORDER PARAMETER p.

Exercise 3.31 (C) *If you* ZOOM *in* PRESET 5 *to obtain much enhanced sensitivity on the intensity scale, can you see the extra diffuse scattering between the diffraction peaks? Is it there if $p = 0$? If $p = 1$? Or if $p = 0.5$?*

Exercise 3.32 (M**) *Estimate theoretically the intensity of the diffuse scattering and compare with "laue" for different values of the* ORDER PARAMETER *p. (Hint: use the idea of the difference structure. What is the magnitude of the scattering amplitude from a difference atom. If the difference atoms are scattering incoherently, what will be the resultant intensity? Warren [11] or Guinier [12] can help.)*

The scattered intensity, integrated over all Δk, is independent of the details of the arrangement of the atoms: a sort of conservation of intensity theorem. A check will show that the intensities of those lines appearing in the disordered structure do not change substantially as the ordering develops. (There may be some random changes because "laue" will sometimes use more A than B atoms and sometimes the other way around. This effect becomes fractionally small as the number of atoms becomes large.) In contrast, the superlattice peaks have p-dependent intensities that complement the diffuse intensity keeping the total integrated intensity independent of p.

Because the intensities of the superlattice lines give a direct measure of both the ordered structure and the degree of ordering, the x-ray technique provides a powerful method of studying ordering transitions. The idea is applicable to the study of both alloys, which form ordered arrangements of the constituents, and ferrimagnetic and antiferromagnetic systems, in which the ordering of the atomic magnetic moments can be monitored using neutron diffraction.

Exercise 3.33 (M**) *Explain how neutron diffraction [25] could be used to determine the degree of alignment in a partially ordered antiferromagnet of the sort described in Exercise 3.26. Use "laue" to construct the diffraction patterns for several different degrees of antiparallel alignment.*

3.6 Atomic displacements and thermal disorder

In real solids there are many causes for the displacement of atoms from their ideal positions: inhomogeneous strain fields, dislocations, periodic distortions (e.g., the charge density wave), thermal motion, etc. Diffraction techniques provide a means to characterize these sources of disorder. We will focus on the effects of thermal motion.

3.6.1 Periodic distortion*

As a preliminary to the thermal problem, consider the case of a static periodic distortion of the atom positions. For example, placing the nth atom in the chain at the position $x_n = na + c\cos(qna)$ defines a periodic distortion with WAVE VECTOR q (expressed in units of π/a, *not* $2\pi/a$ or Å^{-1}) and with WAVE AMPLITUDE c, the maximum displacement of an atom from its perfect lattice site. Such a periodic modulation is obtained in "laue" by choosing PHONON from the MODULATION menu. The electron

 # 6　distribution shown on the left in PRESET 6 illustrates this case with a WAVE VECTOR $q = 0.375\ \pi/a$ and an AMPLITUDE $c = 0.4\ a$. Here, c is chosen large to make the distortions clearly visible in the electron density plot, though too large to give an interpretable diffraction pattern. Verify that the interatomic spacing varies periodically with a wavelength equal to $5.3\ a \approx 21$ Å as appropriate to the choice of $q = 0.375\ \pi/a$. (The variation in heights in the density plot is a consequence of the discrete sampling.)

Section 3.B sketches a derivation of the diffraction pattern for a crystal with such a periodic disturbance. The derivation gives the leading terms in a perturbation expansion in which the small parameter is the product of the distortion amplitude and the scattering vector, $c\Delta k$. The important lowest order results of this derivation are:

1. the diffraction peak at $\Delta k = 2\pi h/a$ develops two *sidebands* or *satellites* separated from the peak by $\pm q$, the wave vector of the distortion;

2. the intensity of the satellites, relative to the associated main peak, is $(c\Delta k/2)^2$.

Exercise 3.34 (M) *To get a more interpretable diffraction pattern, reduce the distortion* AMPLITUDE *to $c = 0.04\ a$, i.e., $c = 0.16$ Å, and note that each of the low order Bragg peaks has developed a pair of satellite peaks. (Don't forget to* ZOOM.*) Compare the intensities of the satellites relative to the main line as the order of the Bragg peak (h) increases. Change the* WAVE VECTOR *of the distortion and verify that the positions of the satellite peaks are correct. For clean results, the discrete sampling requires the choice $q =$ (even integer)$(\pi/a)/$(*NUMBER OF CELLS*).*

Be careful! Here, as elsewhere in these simulations, many parameters will be entered into the program as a multiple of some other parameter. For example, to set the amplitude of the periodic distortion equal to 0.12 Å enter the value 0.03, *not* 0.12. This is because c is measured in units of a, not Å. Whenever you calculate something which seems inconsistent with the program, check to see how the relevant parameters are entered. Also, as you are taking data from the simulations, be sure to check the units given by the sliders.

Exercise 3.35 (M) *Verify quantitatively that the intensities of the first order peaks, the innermost pair, relative to their associated main peaks, vary as predicted at the end of Section 3.B. You can check this both*

by varying the AMPLITUDE *of the distortion and by sampling peaks of different order (different h values).*

Additional predictions, if the theory in Section 3.B is carried further, include:

1. higher order satellites split by multiples m of q from the parent line;
2. intensities of the higher order peaks relative to their associated main peak varying with c and Δk as $\propto (c\Delta k)^{2m}$ in the limit of small $c\Delta k$;
3. a decrease in intensity of the Bragg peaks by a factor $e^{-(c\Delta k)^2/2}$.

Exercise 3.36 (M**) *Make quantitative checks of these predictions.*

In accord with the first two of these predictions we see additional satellites appearing in the diffraction pattern, their intensity increasing with increasing Bragg order and amplitude of the modulation. The third prediction derives from the requirement that the diffracted intensity, integrated over the full range of Δk, must be independent of the positions of the atoms. A distortion which introduces new satellites must "steal" the intensity for those satellites from a nearby Bragg peak. We have already seen this theorem at work in the exchange of diffraction intensity between the superlattice lines and the diffuse scattering in the example of the order–disorder transition.

The presence of static charge density wave distortions in some materials is demonstrated by the existence of satellite peaks of this nature. You can read in reference [16] about some striking x-ray work based on monitoring the intensity and position of satellite peaks generated by charge density waves. Similarly, satellite peaks in systems with spin density wave states may be observed in neutron diffraction as well as certain antiferromagnetic structures which have associated periodicities which are not commensurate with the crystal structure.

3.6.2   Thermal motion*

We have seen the consequences of a periodic modulation of the lattice for the diffraction pattern. This provides a starting point for the discussion of the effect of the thermal motion of the atoms. In the harmonic model of the lattice dynamics of a crystal the normal modes of motion of the atoms are traveling waves, in which the atoms move by small amounts with respect to their equilibrium positions. The random thermal motion of the atoms is described as a linear combination of these traveling-wave normal modes. Each wave-like distortion is of very small amplitude, proportional at high temperature to the square root of the temperature, but there are many of them. In the diffraction pattern, each contributes very weak satellites at positions relative to the Bragg peaks which depend on its wave vector in the manner we've just explored. From the arguments of Section 3.B, summarized on page 46, the intensity of each satellite relative to its associated Bragg peak is proportional to the square of the amplitude of the lattice wave. By equipartition, at high temperature this squared amplitude is proportional to the temperature. The sum of all of these unresolved satellites is called *thermal diffuse scattering*.

The lattice waves or phonons are, of course, not static as we have assumed. However, for most x-ray experiments, it is a good approximation to think of the distorted lattice as stationary on the time scale of the scattering event. This is *not* the case for neutrons where an extension of these ideas forms the basis for determining the frequencies of the lattice waves by inelastic neutron scattering.

"laue" simulates the thermal distortions by displacing the atoms from their lattice sites by random amounts, distributed as a Gaussian with rms width SIGMA THERMAL. This is in the spirit of the *Einstein model* for the atom motion in a solid. PRESET 7 selects the RANDOM choice from the MODULATION menu, and shows the electronic density distribution associated with thermal motion with SIGMA THERMAL = 0.3 *a*. The random displacements are evident in the crystal graph. To see an interpretable diffraction pattern, reduce the amplitude of the thermal fluctuations to SIGMA THERMAL = 0.03 *a* and COPY DIFFRACTION for comparison with the diffraction graph for MODULATION set to NONE.

7

Exercise 3.37 (C*) *How has the randomization produced by the "thermal" or* RANDOM *displacements changed the character of the diffraction peaks? Be sure to* ZOOM *in on the peaks in the two graphs. Are the peaks in the "thermally disordered" crystal any broader than in the perfect crystal? What has happened to the intensities of the peaks? (Remember when you zoom in that there is no significance to the width of the narrow triangular peaks. The Fourier transform is computed at a discrete set of Δk values. A triangle corresponds to an intensity pattern which is finite at one point and zero at the two neighboring points, and yields no information about intermediate values of Δk.)*

The intensity loss from the Bragg peaks can be described quantitatively by the *Debye–Waller factor*,

$$e^{-2W} = e^{-\sigma^2_{\text{thermal}}\Delta k^2}, \qquad (3.10)$$

where σ_{thermal} is the rms displacement of the atoms from their perfect-lattice position. It gives the magnitude of the diffraction intensity remaining in the Bragg peaks after transfer of intensity to the thermal diffuse scattering by the thermal motion of the atoms.

Exercise 3.38 (M**) *On page 47 we gave the result, for a single lattice* *wave of peak amplitude c or rms amplitude $c/\sqrt{2}$, that the intensity of the diffraction peak is reduced to $e^{-c^2 \Delta k^2/2}$ times its ideal intensity. Using this result, deduce the expression for the Debye–Waller factor (3.10). Think of the thermal motion as the result of many such lattice waves whose displacement functions are orthogonal to one another.*

Exercise 3.39 (M*) *Treat "laue" as an experimental system for which you wish to test the validity of the Debye–Waller factor. Take appropriate data sets and devise plots to verify the functional dependences of the intensities on both Δk and $\sigma_{thermal}$?*

Exercise 3.40 (M**) *Show that in the Einstein model of the lattice motion the Debye–Waller factor (3.10) in the high and zero temperature limits becomes*

$$e^{-2W} = \begin{cases} e^{-kT\Delta k^2/M\omega_E^2} & \text{for } T \to \infty \\ e^{-\hbar\Delta k^2/2M\omega_E} & \text{for } T \to 0 \end{cases}, \qquad (3.11)$$

where ω_E is the Einstein frequency characterizing the atomic motion and M is the atomic mass. (Don't forget the quantum mechanical, zero point motion of the atoms.)

3.6.3 Long range crystalline order**

We call attention to a serious question concerning the validity of the Einstein model we have just used to represent the thermal motion. It treats the motion of each atom independently of the others: each atom is bound to its lattice site by a harmonic potential. To illustrate the difficulty we use "laue" to generate an alternative model for the effects of thermal motion. In the LIQUID model used in PRESET 8 the atoms are arranged as follows. One atom is put down. Then the next is placed at a distance $a \pm$ (a small random displacement) from the first. A third is at $a \pm$ (a small random displacement) with respect to the second, and so on until the whole sample is built. "laue" constructs the LIQUID model in this fashion with the "small displacements" distributed as a Gaussian with standard deviation SIGMA LIQUID.

\# 8

Exercise 3.41 (C**) COPY *the crystal graph for this* LIQUID *case using* SIGMA LIQUID $= 0.4\, a$. *Then return to* PRESET 7, *set* SIGMA THERMAL *to* $0.28\, a$ *(this gives mean square fluctuations in nearest neighbor distance which are about the same as in the liquid model). Compare the electron densities for the two models. If you were given the two crystal graphs could you tell which was from the* LIQUID *model and which from the* CRYSTAL *model with* RANDOM *displacements?*

\# 7

Now reduce the σ for each of the models by a factor of 10 and compare the diffraction patterns. (You will need to COPY DIFFRACTION *for at least one of the graphs.)* ZOOM *in to see what has happened to the Bragg peaks. In the* CRYSTAL *model, the peak is diminished in height but remains a narrow spike in a sea of thermal diffuse scattering. In the* LIQUID *model, the peak has disintegrated into a mish-mush of peaks extending over some width in* Δk. *How can it be that apparently indistinguishable electron distributions give qualitatively different diffraction patterns?*

What is the critical difference between the two models? The difference in the diffraction patterns suggests where to look. What is required to give a narrow diffraction peak? There must be long range coherence. Recall that for a diffraction grating, the width of the spectral lines is inversely proportional to the length of the grating. A similar result holds for Bragg peaks: the width of a peak is inversely proportional to the size of the coherently diffracting region. We can consider the whole crystal to be scattering coherently only if knowledge of the position of the atoms at one end of the crystal gives us knowledge of the positions of the atoms at the other end, to within roughly a wavelength of the x-rays.

For the crystal model that coherence is clear. All atoms are within about σ_{thermal} of the precisely defined lattice sites. This is *not* the case for the LIQUID model. As we move down the line, the uncertainties in the positions of successive atoms relative to the first accumulate as in a random walk.

Exercise 3.42 (M**) *Show that at the nth atom in the* LIQUID *model, the accumulated deviation* $\delta(n)$ *from the lattice position is given by the relation* $\delta(n) \approx \sigma_{liquid}\sqrt{n}$.

The uncertainty in the relative positions of atoms n sites apart is of order a when $n \approx (a/\sigma)^2$. For two atoms separated by more than $(a/\sigma)^2$ cells, knowing the position of one gives basically no useful information about the location of the other. The picture is clearly quite different from that for the thermally perturbed crystal model where *all* atoms are near their corresponding lattice points.

These two examples leave us with an interesting question: which is the more appropriate model to describe real crystals? The answer is subtle. In a real one-dimensional system, the liquid model is appropriate, *not* the crystal model. The Bragg peaks are broadened by the thermal (and even the zero point) motion. In three dimensions, on the other hand, the long range coherence is maintained, despite the presence of thermal motion, and the crystal model is more appropriate: the Bragg peaks are not broadened. These results provide a warning about extending one-dimensional results to three-dimensional systems. Often features are qualitatively, not just quantitatively, different as the dimensionality is changed.

Exercise 3.43 (M**) *Use the Debye model for thermal motion to esti-mate* $\delta^2(|R|) \equiv \langle [u_x(R) - u_x(0)]^2 \rangle$ *at high temperature in one, two, and three dimensions, both at high T and at $T = 0$. Here, $u(R)$ is the displace-ment from equilibrium of the atom at lattice site R. Introduce phonon coordinates and express δ^2 as an integral over the normal mode frequen-cies. Focus your attention on the low frequency behavior of the integrand. (Do not concern yourself with issues of mode polarization: keep the es-timate as simple as possible.) Your result should confirm the statement above that the long range coherence is maintained in three dimensions ($\delta^2(R)$ is independent of R for large R) but that in one dimension $\delta^2(R)$ is proportional to R as suggested by the random walk argument.*

3.7 Quasicrystals*

For your amusement we have included, in PRESET 9, one more arrange-   # 9
ment of atoms, called a *quasicrystal* [14]. The atoms are put down in a deterministic, but rather strange fashion. The spacing between adjacent atoms is one of two distances, a and b. In the crystal graph you see an irregular alternation of the two spacings, which happen to be in the ar-bitrary ratio $b/a = 1.27$. The sequence of large and small spacings may appear to be random but it is not. You can generate the pattern in the following way.

Exercise 3.44 (M*) *Write a sequence of rows of as and bs according to the following rules. Start the sequence by writing the first line as b. Write a new line of as and bs below the one you have just written, putting under each b, the pair of letters ab (in that order) and beneath each a, just a single b. Follow the rule until you have generated enough as and bs to convince yourself that there's no pattern that's easy to see. "laue" generates a long enough sequence to define positions for the appropriate number of atoms in its algorithm. This arrangement of sites is called a Fibonacci chain.*

This array is called quasiperiodic by the mathematicians. If you look at the electron density it doesn't look all that different from the examples of the LIQUID and the perturbed CRYSTAL models for the thermal motion. However, take a look at its diffraction pattern. It has sharp diffraction peaks, just like those of the periodic crystal, but the peaks seem to have sidebands and the sidebands have more sidebands and, if you ZOOM in, you find that there are more sidebands on the sidebands as you look with more sensitivity.

Exercise 3.45 (C*) *Starting at 1.05, gradually increase the* SPACING RATIO *to see systematic variations in the pattern. Also the patterns at ratios of small integers* $(2, \frac{3}{2}, \frac{4}{3}, \frac{5}{4})$ *appear initially to be simpler and to have their own systematics. However, as you* ZOOM *in, again you see the hierarchy of satellites on satellites that is characteristic of the diffraction patterns from quasicrystals.*

You may also change the MODULATION of the QUASICRYSTAL from DISPLACEMENT to TYPE. Here the atoms are equally spaced, but there are two types of atoms assigned to the regular lattice sites according to a rule like that given above.

Some years ago these quasicrystal diffraction patterns would have been a mathematical curiosity. Now, however, three-dimensional analogs of this example have been discovered and extensively studied [14]. The evidence, from diffraction studies, of five- and ten-fold symmetry axes aroused considerable interest in the community of crystallographers. Although there was considerable skepticism of the first reports of these materials, because of basic theorems of crystallography which prohibit such point symmetries for periodic structures, the results are now firmly established. More recently, examples of one-dimensional quasiperiodicity have been artificially made in semiconductor structures [15].

3.8   Summary

The intensities of diffraction peaks of perfectly ordered crystals yield valuable structural information. These intensities depend upon the relative positions of the atoms in the unit cell, as well as upon the atomic numbers of the atoms and the electron distributions within each of the atoms. For neutron diffraction other parameters influence the scattering amplitude, such as the orientation of any unpaired electron spin and the nuclear isotope. Ingenious techniques provide routes to recover structural information from the diffraction patterns.

In addition, diffraction techniques are able to give much information about the deviation from perfection of crystal structures. Periodic distortions of the atomic array give satellites on the Bragg peaks. Representing the thermal motion of the crystal as the superposition of lattice waves or phonons we obtain a natural picture for understanding the presence of the thermal diffuse scattering. The growth of thermal diffuse scattering with increasing temperature requires a decrease in the Bragg intensities: the satellites steal intensity from the main peaks. The decrease in intensity with increasing temperature is described by the Debye–Waller factor (3.10). Another application is the study of the temperature dependence of ordering transitions, including magnetic ordering and compositional ordering in alloys.

3.A Deeper exploration

Debye–Waller factor Warren [11] and Guinier [12] describe the physics of the Debye–Waller factor in detail. Work with one or both of them to devise an experiment which uses "laue" to illustrate some aspect of this physics.

Conserved intensity? In discussing the disorder scattering and the thermal diffuse scattering we have made made assertions about the "conservation of intensity". Explore experimentally with "laue" to see whether you can confirm or deny this assertion. Can you construct a theoretical argument to justify the statement, or to show that it needs qualification?

Quasicrystals Use "laue" to explore the x-ray diffraction by quasicrystals more carefully than the brief taste given in Section 3.7. References [14, 15] can serve as a starting point.

3.B Periodically modulated crystal

The picture we have developed for x-ray diffraction is so firmly based on the idea of crystal periodicity it seems hard to know how to approach the problem of the crystal containing a periodic modulation. The method suggested here is a perturbation approach in which we suppose the amplitude of the modulation to be small, small in a sense we will define shortly.

Considering point atoms for illustration, we write the electron density for the crystal with a periodic modulation as

$$\rho(x) = \sum_n \delta[x - na + c\cos(qna)], \tag{3.12}$$

which says there is a point atom, a delta function electron distribution, at the positions

$$x_n = na - c\cos(qna). \tag{3.13}$$

The x_n are positions near the normal lattice sites na but displaced from them by an amount which varies periodically as $\cos(qna)$. The amplitude for scattering with a change in wave vector Δk is proportional to the Fourier transform of the electron density,

$$A(\Delta k) \quad \propto \quad \int dx\, e^{-i\Delta kx} \sum_n \delta[x - na + c\cos(qna)] \qquad (3.14)$$

$$\propto \quad \sum_n e^{-i\Delta k[na - c\cos(qna)]} \qquad (3.15)$$

$$\propto \quad \sum_n e^{-i\Delta kna} \left[1 + i\Delta kc\cos(qna) + \cdots\right]. \qquad (3.16)$$

The exponential in Eq. (3.15) was factored into the product of two exponentials and the second of these was expanded in a power series valid under the assumption $\Delta kc \ll 1$. This assumption requires the amplitude of the displacements to be small compared with the inverse of the change in wave vector of the x-ray (more crudely, small compared with the x-ray wavelength). Next, by writing the cosine as half the sum of a positive and a negative exponential, you show that Eq. (3.16) may be rewritten

$$A(\Delta k) \propto \sum_n \left[e^{-i\Delta kna} + \frac{ic\Delta k}{2}\left(e^{-i(\Delta k+q)na} + e^{-i(\Delta k-q)na} \right) + \cdots \right].$$

$$(3.17)$$

The first term in Eq. (3.17) is familiar: it's the usual lattice sum which is large only if $\Delta k = 2\pi h/a = G$, a reciprocal lattice vector (h = integer). Thus we recover the usual Bragg reflections which, in this approximation, are unaltered by the presence of the periodic displacement of the atoms from their normal sites. Similarly, the second and third terms in Eq. (3.17), when summed, are large only when $\Delta k = (2\pi h/a) - q = G - q$ or when $\Delta k = (2\pi h/a) + q = G + q$ respectively. For wavelengths long compared with the lattice constant, the qs will be small on the scale of the reciprocal lattice constant $2\pi/a$.

The bottom-line predictions, then, are of additional diffraction peaks:

1. which have intensity $c^2\Delta k^2/4$ relative to that of the nearby Bragg peak; and
2. which are spaced a distance q on either side of the Bragg peak.

3.C "laue" – the program

"laue" computes the electron density and diffraction pattern of a one-dimensional array of atoms.

3.C.1 "laue" algorithm

The one-dimensional crystal is represented by its real-space electron density which is generated from the superposition of atomic electron densities.

The corresponding diffraction pattern is then computed as the square of the Fourier transform of that electron density.

Electron density "laue" simulates a one-dimensional crystal as an array of atoms which can be periodically ordered, or for which the perfect periodicity can be destroyed in a RANDOM, periodic (PHONON), or QUASIPERIODIC fashion. Each atom in that array is characterized by its electron density ρ_i. The electron density ρ of the whole "crystal" is simply a superposition of the electron densities ρ_i of the N individual atoms,[1]

$$\rho(x) = \sum_{i=0}^{N-1} \rho_i(x). \qquad (3.18)$$

The individual contributions $\rho_i(x)$ to that electron density can be GAUSSIAN, EXPONENTIAL, TRIANGULAR, or PSEUDO ATOM functions, centered at each atomic position. The detailed forms of these ATOM SHAPES are given in Table 3.1 on page 37, while Section 3.C.2 describes how the atomic position R, and the width σ and amplitude Z of the electron density depend on the chosen parameters MATERIAL and MODULATION.

"laue" samples the electron density (3.18) only at a discrete set of equally spaced points in order to compute the diffraction pattern using a Fast Fourier Transform (FFT) method: given the NUMBER OF UNIT CELLS N, the NUMBER OF POINTS SAMPLED PER ATOM N_s, and the length L of the crystal ($\approx N \times$ LATTICE CONSTANT), "laue" generates a grid of length L with $N \times N_s$ ($2N \times N_s$ for the DIATOMIC CRYSTAL or PAIR OF ATOMS) equally spaced points and carries out the summation (3.18) only at these grid-points j. The contribution to the electron density of each single atom is added into the sum (3.18) only up to some CUTOFF[2] parameter N_c: only those single atom electron densities ρ_i whose centers lie within the grid point range $(j - N_c)$ to $(j + N_c)$ contribute to the electron density $\rho(j)$ at a grid-point j. In the case that $j - N_c < 0$ or $j + N_c > NN_s$, "laue" uses periodic boundary conditions, i.e., it identifies in the sum (3.18) $\rho_i(0)$ with $\rho_i(NN_s)$ etc. Because the FFT routine in "laue" (see below) requires the number of grid-points to be a power of 2, both the NUMBER OF UNIT CELLS and the number of POINTS SAMPLED PER ATOM are restricted to be a power of 2.

Diffraction pattern To compute the diffraction pattern (intensity) of the electron density (3.18), "laue" uses an FFT routine. The intensities are normalized to give a height for the central peak of 1 for the MONATOMIC

---

[1]For crystals with two atoms per unit cell (DIATOMIC CRYSTAL and PAIR OF ATOMS), the crystal consists of N unit cells with $2N$ atoms.

[2]The CUTOFF is specified relative to N_s, $N_c =$ CUTOFF $\times N_s$.

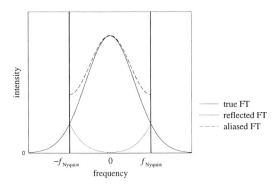

Figure 3.8: The effect of aliasing when Fourier transforming discretely sampled data. The Nyquist frequency is indicated by the two vertical lines. The part of the Fourier transform (FT) that lies outside the frequency interval specified by the Nyquist frequency is folded back into that interval.

CRYSTAL and $(1 + Z_B/Z_A)^2$ for the DIATOMIC CRYSTAL. A thorough explanation of FFT methods, see Burden et al. [65], is beyond the scope of this section, and we only point out one important consequence (related to aliasing) of discretely sampling data which one is likely to encounter when working with "laue".

The number of POINTS SAMPLED PER ATOM N_s determines the rate at which "laue" samples the electron density ρ: a large N_s results in a small grid spacing Δ and thus in a high sampling rate $1/\Delta$. The maximum spatial frequency, called the *Nyquist frequency*, which is resolved by a Fourier transform of such discretely sampled data is $f_{\text{Nyquist}} = 1/2\Delta$. Any part of the power spectrum outside the frequency range $-1/2\Delta$, ..., $1/2\Delta$ is folded back into that range. This (unwanted) effect, called *aliasing*, is illustrated in Figure 3.8. The diffraction pattern in "laue" shows the *aliased* rather than the *true* power spectrum, and only for a sufficiently large N_s and sufficiently broad atoms (large σ) do the two agree.

We can test the effect of aliasing directly with "laue", as shown in Figure 3.9. Choose an UNMODULATED MONATOMIC CRYSTAL, let the LATTICE CONSTANT be equal to 4 Å, the SIZE OF ATOM A be equal to 0.025 Å, and the ATOM SHAPE be GAUSSIAN. The settings in the CONFIGURE menu should be NUMBER OF UNIT CELLS: 64; POINTS SAMPLED PER ATOM: 64; and CUTOFF: 0.5. The diffraction pattern corresponding to this configuration seems to level off to a constant intensity value of about 0.052 at a Δk of about 50 Å^{-1}. To show that this is only an aliasing effect, you can check by increasing the number of POINTS SAMPLED PER ATOM to 128, and thus doubling the Nyquist frequency. Now

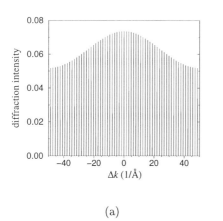
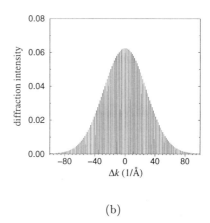

(a)   (b)

Figure 3.9: The effect of aliasing in "laue". The NUMBER OF SAMPLED POINTS is 64 in (a) and 128 in (b).

the diffraction pattern looks more or less like a Gaussian, with the peak intensity falling off to near zero at $\Delta k = 100$ Å^{-1}. Note that the peaks at around $\Delta k = 50$ Å^{-1} now have an intensity of about 0.013 which is only a quarter of what we found when we sampled 64 points per atom: aliasing in the Fourier transform caused the amplitudes of the two peaks near $\Delta k = 49$ Å^{-1} and $\Delta k = 50$ Å^{-1} to be added, thus resulting in an intensity four times too large. Consequently, if the intensity does not fall off to zero within the computed (displayed) region you should either choose a larger number of POINTS SAMPLED PER ATOM or increase the SIZE of the atoms.

3.C.2 Parameter choices in "laue"

The ATOM SHAPE, which determines the electron density of each atom, together with the chosen MATERIAL and MODULATION, allows computation of the electron density of the whole crystal according to Eq. (3.18). If only one type of atom is present in the crystal structure (e.g., for the MONATOMIC CRYSTAL) then the electron density function for each atom is taken to have area $Z = 1$ and width σ (specified by the SIZE OF ATOM A). If a second type of atom is present (e.g., for the DIATOMIC CRYSTAL) the width of its electron density is given by the SIZE OF ATOM B, the atomic number Z_A of atom A is taken as 1, and the ratio of the atomic numbers is ATOMIC NUMBER ZB/ZA. The intensities of the diffraction pattern are normalized to give a value for the forward scattering intensity, $I(\Delta k = 0) = |1 + Z_B/Z_A|^2$.

Pair of atoms   For a PAIR OF ATOMS the electron density of atom A is centered at $R = 0$ Å, and the electron density of atom B is centered at ATOM SEPARATION (in units of the LATTICE CONSTANT a).

Monatomic crystals   The atoms in a MONATOMIC CRYSTAL with MODULATION NONE are periodically ordered with the interatomic spacing given by the LATTICE CONSTANT a. If the MODULATION is a PHONON each atom is shifted from its original lattice position by the amount $\Delta x = c \cos(qx)$, where the WAVE VECTOR q is specified in units of π/a and the WAVE AMPLITUDE c in units of a. If the MODULATION is RANDOM each atom is shifted from its original lattice position by a random amount Δx, where the distribution of the Δx is Gaussian with a standard deviation SIGMA THERMAL (given by "laue" in units of a).

Diatomic crystals   For a DIATOMIC CRYSTAL the notion of an *atom* is replaced by the notion of a *unit cell* which contains two atoms A and B whose electron densities can differ in their ATOMIC NUMBER and SIZE. The unit cells in a DIATOMIC CRYSTAL are periodically ordered with the LATTICE CONSTANT a giving the size of the cell. If DISORDER is NONE the A atoms are located at the origin of the cells and the B atoms a distance ATOM SEPARATION (in units of a) away from the A atoms. If DISORDER is chosen to be ORDER/DISORDER then the probability for an A atom to be located at the origin of any given cell is $\frac{1}{2}(1 + p)$, where p is the ORDER PARAMETER. Correspondingly the probability for an A atom to be located a distance ATOM SEPARATION away from the origin of that cell is $\frac{1}{2}(1 - p)$. The probabilities for a B atom to be located at the origin of the cell, and at a distance ATOM SEPARATION away from the origin, are $\frac{1}{2}(1 - p)$ and $\frac{1}{2}(1 + p)$ respectively. Note that the order parameter is such that $p = 0$ corresponds to a totally random arrangement of atom types A and B in the crystal while $p = \pm 1$ corresponds to the perfectly ordered diatomic crystal.

Liquids   The atoms in a LIQUID are separated *on average* by the LATTICE CONSTANT a. The interatomic distance fluctuates randomly around this average value by an amount Δx. The distribution of the Δx is Gaussian with a variance SIGMA LIQUID (given in units of a). To build up a liquid, "laue" puts the first atom at position $R = 0$ Å, and each successive atom i a distance $a \pm \Delta x$ away from atom $(i - 1)$. Note that in contrast to the thermally distorted CRYSTAL the long range order is destroyed in a LIQUID.

Quasicrystals The atoms in a TYPE MODULATED QUASICRYSTAL are ordered periodically with a spacing given by the LATTICE CONSTANT a. Their electron density is determined in a quasiperiodic fashion according to the Fibonacci chain. Such a chain consists of a sequence of 0s and 1s and is created recursively through an algorithm known as an *inflation algorithm*. In each recursive step all 0s in the sequence are replaced by 1s and all 1s by 01s. The initial sequence contains only a 1. In that way "laue" generates a Fibonacci chain of length N, where N is the number of atoms in the quasicrystal. The electron density of the ith atom in a TYPE MODULATED QUASICRYSTAL is characterized by the SIZE OF ATOM A if the corresponding position in the Fibonacci chain contains a 0. Otherwise the electron density is characterized by the SIZE OF ATOM B. The ratio of integrated electron densities is given by the ratio Z_B/Z_A of the respective ATOMIC NUMBERS.

The atoms in a DISPLACEMENT MODULATED QUASICRYSTAL are ordered quasiperiodically. The spacing between the ith and the $(i + 1)$th atom is given by the LATTICE CONSTANT a if the ith position in the Fibonacci chain is 0. Otherwise the separation is given by a times the SPACING RATIO.

3.C.3 Bugs, problems, and solutions

The reciprocal space graph can only display 32768 xy points. Thus the product of the NUMBER OF UNIT CELLS and the number of POINTS SAMPLED PER ATOM is restricted to be at most 32768. (For reasons we do not fully understand, for Windows 95 or Windows 3.1x this number is reduced to 8192.) Remember also that NUMBER OF UNIT CELLS and POINTS SAMPLED PER ATOM must each be an integral power of 2.

The effect of the discrete sampling of the electron density is noticeable in the real-space display for materials with non-periodically spaced atoms such as a LIQUID. Because the electronic density of an atom can be centered as far as half a grid spacing away from the nearest grid-point, the electronic density peaks can appear to have different heights although all atoms have the same atomic number. This effect becomes more pronounced the smaller the number of POINTS SAMPLED PER ATOM and the narrower the peaks.

Because of the finite resolution in reciprocal space, expected zeros in the diffraction patterns for the two-atom diffraction may not appear. For example, the discrete Fourier transform of a crystal with one unit cell of length a has values only at $\Delta k = 2\pi h/a$ for integers h. Two atoms separated by an ATOM SEPARATION d have zeros in their diffraction patterns at $\Delta k = (2m + 1)\pi/d$ for integers m. Only if the ATOM SEPARATION is a suitable fraction of the LATTICE CONSTANT, namely $d = a/2n$ with

n an integer, will all zeros in the diffraction pattern be cleanly shown. The resolution of the two atom diffraction pattern can be enhanced by increasing the LATTICE CONSTANT a, decreasing the ATOM SPACING d/a, and increasing the POINTS SAMPLED PER ATOM (in the configure menu), each by the same power of 2, perhaps 8. (You may be required to reduce the NUMBER OF CELLS in order to make these changes.)

The finite resolution in reciprocal space also causes the diffraction peaks for a crystal to have finite width. Note that the discrete Fourier transform has values only at $\Delta k = 2\pi h/Na$, where N is the number of cells in the crystal and h some integer. Because "laue" displays the intensity by connecting the intensities at a discrete set of Δk with straight lines, the apparent line width is of the order of π/Na. This apparent line width happens to be the same as the diffraction broadening due to the finite size of the crystal (with the line width in "laue" approaching zero for an infinite crystal, $N = \infty$).

4

"born" – Lattice dynamics in one dimension

Contents

4.1   Introduction

In most of the discussion of x-ray diffraction we pictured the crystalline solid as a set of atoms at fixed positions in some sort of regular or semi-regular arrangement. Such a picture is inadequate for the discussion of nearly any property of the material. The room temperature electrical resistance of metals is the result of thermal disorder: motors and magnets with aluminum windings can be made more efficient by cooling them to liquid nitrogen temperature. The burn on your finger when you touch a hot stove is the result of the large amplitude of motion of the atoms in the stove top.

The atoms in a solid are in constant thermally induced motion and we require models with which to discuss this motion. Many of the critical concepts can be conveniently illustrated in one dimension and in "born" we study in detail the dynamics of a linear chain of atoms. Impurity and anharmonic effects are also discussed briefly. "born" considers only the classical motion of the atoms in the chain with no reference to the quantization.

Although the description of the motion of the order of 10^{23} atoms in strong interaction with one another seems a daunting task, a quite satisfactory treatment of the problem for a crystalline material is remarkably straightforward. Two ideas are central.

First, we assume that the atomic arrangement that has minimum energy is the periodic array of the sort discussed in "bravais". We then expand the potential energy of the system in powers of the displacements of the atoms from these equilibrium positions. The starting point for most discussions is the *harmonic approximation* in which terms in this expansion of order higher than quadratic are ignored. For most solids at room temperature this is a good approximation which can be further refined by perturbation methods. Within the harmonic approximation, a theorem of classical mechanics tells us that an exact solution to the equations of motion is possible in principle. The trick is to find special linear combinations, called *normal coordinates*, of the atomic displacement coordinates. Expressed in terms of the normal coordinates, the potential and kinetic energies of the system contain no cross terms, only terms quadratic in each. The equations of motion then reduce to a set of independent harmonic oscillator equations which are easily solved. These solutions are called the *normal modes* of the system.

Second, the perfect periodicity of the ideal crystal allows us (with the added assumption of periodic boundary conditions) to find the required normal coordinate transformation and the normal mode frequencies of the system, despite the large size and apparent complexity of the system. The most general motion of the crystal can then be written as a linear superposition of these normal modes. The normal modes correspond to traveling plane wave solutions and we explore some of the properties of these solutions here for the one-dimensional chain.

4.2 Plane waves – infinite medium

First we look at the properties of an infinite chain of atoms. Later we will impose boundary conditions at the ends of a finite piece of the chain to find the normal modes. The atoms, of mass M, are coupled by springs of force constant K. The display in "born" indicates a transverse wave on the chain: this is easier than a longitudinal wave both to display and to visualize. However, it may be easier to think of the dynamics in terms of a longitudinal wave. In this chapter we will remain ambivalent: all remarks are appropriate to waves of either polarization. You are welcome to think of the simulation as representing either a transverse wave or a schematic plot of a longitudinal wave with the displacements plotted vertically instead of horizontally. If you prefer to think in terms of transverse motion on the chain, the force constant in the equations below is *not* the spring constant of the coupling springs, but the ratio of the externally applied tension in the chain to the lattice constant, $K = (\text{tension})/(\text{lattice constant})$.

Exercise 4.1 (M*) *Verify explicitly the statement above. (Hint: fix two nearest neighbors of an atom and ask for the restoring force for both longitudinal and then transverse displacements of the atom from its equilibrium position. To get a first order restoring force in the transverse case, you must assume the springs are in tension.)*

You will note that "born" works in dimensionless units for all of the parameters. That is, the PACKET WIDTH is expressed in units of the lattice constant a, the WAVE VECTOR in units of π/a, and the time in units of the step time Δt of the integration algorithm. None of the relevant constants K, M, a, or Δt are specified. If you prefer to work in units with dimensions, the frequencies and distances become appropriate to waves in crystals with the choices $a = 3$ Å and $\Delta t = 10^{-14}$ s.

4.2.1 Traveling waves

PRESET 1 shows a possible solution to Newton's equations for a short   # 1
segment of an infinite chain. (The static display gives the initial atomic displacements in the "born" simulation. Click on the RUN button to start the motion.) It is a traveling wave solution,

$$u_n(t) = A\cos(qna - \omega t), \tag{4.1}$$

with u_n the displacement from equilibrium of the nth atom in the chain. The wave is characterized by an AMPLITUDE A, a WAVE VECTOR $q = 2\pi/\lambda$, and an angular frequency ω. The frequency and WAVE VECTOR may *not* be independently chosen: if one is specified, the other is determined by the dispersion relation for the chain,

$$\omega(q) = 2\sqrt{K/M}\left|\sin\left(\tfrac{1}{2}qa\right)\right|. \tag{4.2}$$

In contrast with the ideal string which will propagate a wave of any frequency, the dispersion relation (4.2) implies that, whatever the choice of wave vector q or wavelength, the angular frequency can never be greater than $2\sqrt{K/M}$.

Exercise 4.2 (M) RUN *to have the graph display the time dependence of the displacement of the center atom of the chain. From this graph determine the angular frequency ω of the wave. (Take the unit of time in "born" to be 10^{-14} s.) Vary the* WAVE VECTOR *q and make a plot of the dispersion relation $\omega(q)$ vs. q. Let q vary over the range $0–\pi/a$. (Note that this is entered as a number in the range 0–1 in "born" since q there is expressed in units of π/a.)*

In a three-dimensional medium, the solutions for very long wavelengths, $q \ll \pi/a$, correspond to the propagation of sound waves. How are the parameters characterizing the continuum model, elastic moduli and density, related to the atomic scale model we use in "born"?

Exercise 4.3 (M*) *Typical sound velocities in solids are $10^5 – 10^6$ cm/s, and atomic separations in solids are $2 - 3$ Å. Use this information, along with the one-dimensional model, to estimate a typical maximum frequency ω_{max} for vibrational frequencies in solids. Express the result not only in s^{-1}, but also in other energy related units, such as eV and kelvin, using $\hbar\omega \Leftrightarrow h\nu \Leftrightarrow eV \Leftrightarrow kT$. (Hint: how do you determine the velocity of sound from an $\omega(q)$ plot?)*

Exercise 4.4 (M) *Taking the atomic mass to be 50 amu (atomic mass units) and the time step to be $\Delta t = 10^{-14}$ s, determine from data from "born" the spring constant of the interatomic forces. Express that spring constant in units of $(eV/Å^2)$. Why do you not need information about either the amplitude of the wave or the lattice constant?*

A guess at the value for the spring constant K would start with the idea that, for relative displacements of the order of the atomic radius (~ 1 Å), the cost in energy should be of the order of the binding energy of the valence electrons ($\sim$few eV). This argument would give for K a value of a few $eV/Å^2$. Is this anywhere close to your estimate for the "born" model?

Choose TRIPLE from the DISPLAY menu and RUN again, briefly. This graph gives the displacement versus time for three successive atoms in the chain. They move in sinusoidal motion with equal amplitudes but different phases.

Exercise 4.5 (C) *One of the colored traces is for the central atom in the chain. The other two are for the two atoms to its right. Which color trace goes with which atom?*

We seem to have ignored q values outside the range $-\pi/a < q < \pi/a$. What happens if you type in a WAVE VECTOR $q > 2\,(\pi/a)$?

Exercise 4.6 (C) *Determine from "born" the value of ω for some choice of $q > 2\,(\pi/a)$? What are the two q values in the range $-\pi/a < q < \pi/a$ for which $\omega(q)$ has this same value? Verify that one of these q values differs from the chosen q by the subtraction of $2n\pi/a$ where n is an integer.*

Exercise 4.7 (M) *Show, with the help of the dispersion relation (4.2), that $\omega(q) = \omega(q + 2\pi/a)$.*

How can the propagation frequency for two waves of different wavelengths possibly be the same? "born" suggests the answer in its display of the initial condition. The display shows both the initial displacements of the atoms, described by the discrete function $u_n = A\sin(qna)$, and the continuous function $u(x) = A\sin(qx)$.

Exercise 4.8 (C) *Set the WAVE VECTOR to 0.3 (π/a); you see a clear sinusoidal displacement function for the atoms which follows the continuous sine function very well. Now type into the slider window a new value, 2.3 (π/a). Watch the screen as you hit the ENTER key. Why does the continuous curve change while the atomic displacements remain fixed? (The demonstration may be clearer if you use the CONFIGURE menu to change the NUMBER OF ATOMS to 20.)*

The *same* set of atomic displacements may be represented by any one of the set of wave vectors, $q_n = q + 2\pi n/a$ where n is an integer. You should recognize $2\pi/a$ as the primitive vector of the reciprocal lattice for this one-dimensional crystal. This redundancy implies that the dispersion relation for these wave solutions is fully specified if it is specified over the range of q from $-\pi/a$ to $+\pi/a$. The redundancy in q, often expressed as a periodicity in *reciprocal* space of certain physical quantities, is a consequence of the periodicity in *real* space of the physical system. The range from $-\pi/a$ to $+\pi/a$ of q is called the *first Brillouin zone* for this problem. You will hear a great deal about Brillouin zones in exploring many topics of solid state physics.

Exercise 4.9 (M) *Verify by calculation the statement of the preceding paragraph, that all wave vectors of the set $q_n = q + 2\pi n/a$ give the identical set of atomic displacements.*

The traveling waves in "born" clearly appear to be moving from left to right. The velocity of the wave is given by the familiar result, $v_\phi = \omega/q$. We call this the *phase velocity*. A subscript ϕ has been added since we will soon see another possible velocity of interest.

Exercise 4.10 (M) *You can measure the phase velocity of the traveling wave. Start with a* WAVE VECTOR *of 0.3 (π/a) or less. Find a peak or valley of the wave for the initial configuration.* RUN *the simulation until the peak has traveled some distance in the displayed grid and then stop. The* TIME *counter gives the elapsed time and the velocity can be determined. Measure the phase velocity for a couple of values of q in the first Brillouin zone and compare with the standard result $v_\phi = \omega/q$.*

We must question the physical significance of this phase velocity when we recall that many distinct q values may be used to describe the same lattice wave. For a periodic medium such as a crystal, in contrast with a homogeneous medium, the phase velocity is clearly *not* a uniquely defined concept.

4.2.2   Energy and momentum*

Associated with the traveling wave is motion of the atoms and stretching of the springs. Hence we can associate an energy with the presence of the wave (measured relative to the state when all of the atoms are at rest).

Exercise 4.11 (M*) *For a given amplitude A of the wave, calculate the kinetic, the potential, and the total energy per unit length of the chain. Take the lattice constant to be 3 Å. To get useful expressions, be sure to take either time or space average values of the energies you compute.*

Combine these expressions with the dispersion relation $\omega(q)$ to show the equality of the potential and kinetic energy densities, which the virial theorem shows to be a characteristic of the free normal mode motion of any harmonic system.

Since there is energy density associated with the traveling wave, and it is clearly moving left to right, we may expect it to carry energy along the chain. The ratio of energy flow (energy/time) to energy density (energy/length) will give a velocity with which the energy is carried.

Exercise 4.12 (M)** *Calculate the time average of the work done by a spring on the atom to its right. This will be the same at any point along* the traveling wave and will be the rate of energy transfer down the chain. (You might equally well ask for the work required by an energy source driving the first atom in a semi-infinite chain to maintain a traveling wave on the chain of amplitude A.) Calculate the velocity of energy propagation

$$v_{\text{energy}} = (\text{energy flow})/(\text{energy density}). \qquad (4.3)$$

Note that the velocity of energy propagation (4.3) is *not* the phase velocity. In fact, your expression for the rate of energy transfer down the chain should yield zero if $q = \pi/a$. How can it be that the traveling wave $A\sin(qna - \omega t)$, with $q = \pi/a$, does not carry energy?

Exercise 4.13 (C) *Use* PRESET 1 *and change the* WAVE VECTOR *in "born" to* $q = \pi/a$ *and watch the motion. Does it surprise you now that the rate of energy flow is zero?*

The preceding few exercises considered the propagation of energy down the chain. Is there, in a similar way, propagation of momentum down the chain?

Exercise 4.14 (M*) *The momentum delivered to an object is the time integral of the force exerted on it. Show that the time average of the force exerted by a spring on the mass on its right is zero. Note that this result obtains irrespective of whether the waves are transverse or longitudinal.*

There is energy transfer down the chain, but *not* momentum transfer. In solid state physics you will frequently see reference to energy and pseudo-momentum, or crystal-momentum, selection rules. This exercise is a reminder that, however much those rules are suggestive of momentum, there is *no* mechanical momentum, in the usual sense, associated with these waves.

4.2.3 Standing waves

Select PRESET 2. The initial displacements you see are nearly identical # 2
to those of PRESET 1. Click on RUN or press the SPACE BAR and you see not a *traveling wave* pattern but a *standing wave*.

Exercise 4.15 (C) *What is different about the initial conditions that gives rise to these two very different types of motion?*

Exercise 4.16 (M) *What are the initial conditions required to give the standing wave behavior? The traveling wave behavior?*

Exercise 4.17 (C) RUN *and use the graph to compare the relative amplitudes and phases of the motion of the successive atoms in the chain. How does this differ from the behavior for the traveling wave.*

Exercise 4.18 (M*) *For the standing wave, calculate the work done on the right half of the chain by the left half, just as in Exercise 4.12 for the traveling wave. Show that, in the case of the standing wave, the rate of energy flow along the chain is zero for any value of q.*

As for the traveling wave there is energy stored in the standing wave, though the pictures are somewhat different. For the traveling wave we see for any cell of (atom + spring) that the time averages of the kinetic and potential energies are equal, and there is a flow of the total energy along the chain. For the standing wave the regions with large atomic displacements have a large time average kinetic energy. The regions of small atomic displacements have highly stretched springs and hence a large time average potential energy. The energy is transferred back and forth between these two kinds of regions but does not propagate along the chain.

The behavior of waves on a chain of atoms is very similar to the physics of waves on the ideal stretched string. Indeed, in solids the motion of the atoms at low frequencies is well described by ideas of continuum elasticity theory. The velocities of the wave motions of the atoms for long wavelengths are the velocities of sound propagation in the crystals. At high frequencies, when the wavelength becomes comparable to the lattice constant, we see variation in phase velocity with frequency and the inability of the system to support propagating waves above a maximum frequency.

4.3   Normal modes of finite chain

The atoms in crystals are in constant motion about their average position as a consequence of thermal excitation and zero point motion. "born"

 # 3   simulates this motion crudely in PRESET 3. The initial positions of the atoms are chosen as RANDOM displacements from the lattice sites and released by a click of the RUN button. To understand the thermal properties of solids we require a mathematical description of the motion you see. Watching the motion on the screen it seems a formidable task: in fact it's straightforward. The essential idea is that the complex thermal motion may be represented as a linear combination of the harmonic, and unfrightening, waves that "born" was generating for us in the first PRESETS.

It is convenient to work in terms of a finite, but large, crystal. In the harmonic approximation, implicit in all but Section 4.7 of this chapter, the motion of the system can be described in terms of a linear combination of normal modes. The normal modes for the chain are easy to find by combining the solutions for the infinite medium already discussed with appropriate boundary conditions.

Reactivate PRESET 2 and watch the end atoms. If we think of the chain on the screen as the full crystal, not as a segment of an infinite crystal, then the motion is appropriate to the FIXED choice for the BOUNDARY conditions. # 2

Exercise 4.19 (M) *Derive a relation for the values of wave vector q (in units of π/a) for which standing wave solutions will satisfy the boundary conditions for a chain of length N, where the N atoms include the two fixed atoms at either end. Drag the* WAVE VECTOR *slider and watch the readings that "born" is willing to accept. Do they correspond to the values you've just calculated? Each of these q values yields a* STANDING WAVE *motion on the chain which is a normal mode of motion of the chain with* FIXED BOUNDARY CONDITIONS.

Exercise 4.20 (M) *Recall that q values of magnitude greater than π/a give no new solutions of the equations of motion for the discrete chain and hence are of no interest. How many q values, with $-\pi/a < q \leq \pi/a$, are there which give distinct standing wave solutions which satisfy the boundary conditions?*

If we consider only the longitudinal motion of the atoms, or motion restricted to a single transverse direction, there should be $(N-2)$ normal modes. The "-2" is because of the two fixed end atoms. If this is not what you obtained, review your arguments.

Thus we have found the $(N-2)$ normal modes of motion for the finite chain, for a single polarization, and we could use these to describe fully the most general motion of the chain in this polarization. Usually we adopt different boundary condition, namely the *periodic boundary conditions* (PBC). We imagine the chain wrapped in a ring so that the Nth atom is connected by a spring to the first atom. This gives a finite one-dimensional crystal with no boundaries. One common expression of the PBC is to place the requirement that $u_{n+N} = u_n$, where u_n is the displacement function of the nth atom in an infinite chain. They may also be imposed by modification of the equations of motion for the end atoms of the chain as described on page 83. This boundary condition may be satisfied by either the traveling or standing wave solutions for the infinite medium with suitable choice of q. For most problems the traveling wave solutions are the more convenient.

Exercise 4.21 (M) *What are the q values in the range $-\pi/a$ to π/a for which traveling wave solutions satisfy the PBC. Check that these are the* WAVE VECTORS *allowed by "born" in* PRESET 1. *How many allowed q values are there? Does this number match the expected number of degrees* # 1

*of freedom which, with no fixed atoms, becomes just N? Remember,
though, to consider the role of the center of mass motion.*

By choosing appropriate solutions of the equations of motion for the
infinite chain we are able to satisfy a special set of boundary conditions,
the PBC, for a finite crystal. This selected set of solutions is the set of
normal mode motions for the finite crystal, and can serve as the basis for
the description of arbitrarily complex free motion of the system.

4.4   Wave packets

We often prefer to consider the thermal excitation of solids, not in terms
of plane waves which extend throughout the crystal, but rather in terms
of wave packets which are localized linear combinations of plane waves
of nearly the same frequency. Later you will meet the idea of phonons,
quantized states of the vibrational waves, and these are frequently treated
in a particle language. Understanding the properties of classically propagating wave packets will be helpful in making that transition.

\# 4   PRESET 4 constructs a TRAVELING PACKET for which the initial displacements are given by

$$u_n = A \cos[q(n - n_0)a] \, e^{\frac{-(n-n_0)^2 a^2}{2\sigma_0^2}} , \tag{4.4}$$

where σ_0 is the PACKET WIDTH, n_0 the number of the atom on which it
is centered (PEAK POSITION), and A its AMPLITUDE.

Exercise 4.22 (C) PRESET 4 *sets the* BOUNDARY *conditions to* PERI-
ODIC. RUN *and you will see that the meaning of PBC is much more
evident for the packet than for the infinite wave if you let the packet
propagate to the end of the chain. Switch the* BOUNDARY *conditions to*
FIXED *to see the contrasting behavior.*

Exercise 4.23 (C) *Change the* BOUNDARY *conditions to* FIXED *and explain why, during reflection from the end of the chain, the displacement
amplitude a few atoms from the end of the chain is roughly twice the
maximum displacement at the center of the original packet.*

4.4.1   Traveling versus standing packets

Reselect PRESET 4 but with the INITIAL condition changed to STANDING
PACKET. Here the initial displacements of the atoms are the same as for
the TRAVELING PACKET. RUN with this initial condition and see what
happens.

Exercise 4.24 (C) *The initial displacements for the* STANDING PACKET *and the* TRAVELING PACKET *are the same, yet the subsequent motion is dramatically different. Explain qualitatively what must be different about the two initial conditions,* STANDING PACKET *and* TRAVELING PACKET.

Exercise 4.25 (C*) *Be sure the* BOUNDARY *conditions are* PERIODIC. *Let the program* RUN *until the two packets "wrap around the ends" and meet again. What feature of the equations of motion, in the harmonic approximation, insures that the packets can pass through each other without interaction?*

Seeing the packets pass through one another may give a little more confidence in the ideas that the different normal mode solutions are independent of one another, and that the use of the superposition principle is valid. It also shows the impossibility of trying to make a standing packet, which really remains still, by superposing an appropriate set of standing waves. Such a packet simply peels apart into a pair of traveling packets.

Exercise 4.26 (M**) *What initial velocities must be given to each of the atoms to generate the proper initial conditions for the* TRAVELING PACKET. *(Hint: don't make hasty guesses, they're probably wrong. One approach is to use, as a starting point, the Fourier representation of the packet given in the next subsection, Eq. (4.7).)*

4.4.2 Group velocity

Exercise 4.27 (C) *Reselect* PRESET 4, RUN, *and follow one of individual peaks of the cosine function (not the center of the packet or peak of the envelope function) as the packet propagates to the right. With a little attention you can pick up the same peak as the packet wraps around and comes back in at the left hand end. You will see that the peak gradually moves ahead within the envelope and ultimately is lost out the front end of the packet.*

The velocity of the point you were following is the *phase velocity* of the wave, while the velocity with which the overall packet moves is the *group velocity*. To measure the group velocity, RUN the program long enough that the packet passes the center atom a second time (increase the SPEED to a high value to save time). The time between successive passes of the packet allows calculation of the group velocity. For the lower group velocities you may want simply to measure the time required to move between some pair of lines on the displayed grid (ten lattice constants between vertical lines).

Exercise 4.28 (M) *Choose a number of q values, determine the phase and group velocities and plot the variation of each with q. You will want to vary the* PACKET WIDTH *as well as the* WAVE VECTOR. *Don't spend time working at q values below about 0.2 (π/a). Take only enough points to establish clearly the behavior of both.*

The phase velocities should be the same as determined from the infinite traveling waves. The group velocities should be equal to the slope of the dispersion curve. A puzzling result is that the group velocity goes to zero at $q = \pm\pi/a$.

Exercise 4.29 (M) *Use the data you have taken for $\omega(q)$ and for $v_g(q)$ to verify qualitatively the expression*

$$v_g = d\omega(q)/dq \qquad (4.5)$$

for the group velocity. Note that it confirms the observation that at $q = \pi/a$ the group velocity goes to zero.

Exercise 4.30 (C) *Reselect* PRESET *4 and change the* WAVE VECTOR *q to π/a and* RUN. *What happens?*

Ultimately the packet falls apart but it doesn't seem to go anywhere! The group velocity indeed seems to be zero at the boundary of the Brillouin zone, $q = \pi/a$. However, we've seen a manifestation of this already. If we talk about energy propagation, we see that there is energy associated with the packet and localized within it. For the packet, the velocity with which energy is carried must be the group velocity and, as Exercise 4.12 shows, the velocity for energy propagation goes to zero at $q = \pm\pi/a$.

Any motion on the chain can be written as a linear combination of the normal mode solutions. The packet you were watching in PRESET 4 can be written in the form

$$u_n(t) = \text{Re}\left\{ \frac{\sqrt{2\pi}\sigma_0 A}{Na} \sum_{j=-N/2}^{N/2} e^{i[\frac{2\pi j}{Na}(n-n_0)a - \omega(\frac{2\pi j}{Na})t]} e^{-\sigma_0^2(\frac{2\pi j}{Na}-q)^2/2} \right\}.$$

$$(4.6)$$

The real part of the first exponential in any single term of the summand is a plane wave solution, with wave vector $q' = 2\pi j/Na$, for the chain. The Gaussian gives the amplitudes of the different plane wave contributions. The packet is made up of a number of plane waves with wave vectors in the vicinity of q and is centered on the site $n = n_0$ in the chain.

If we assume that σ_0 is large enough that the Gaussian term is zero at $q' = 2\pi j/Na = \pm\pi/a$, we can extend the limits on the sum to infinity, and if σ_0 is small enough that many terms contribute to the sum, we may replace the sum by the integral,

$$u_n(t) = \mathrm{Re} \left\{ \frac{A\sigma_0}{\sqrt{2\pi}} \int_{-\infty}^{\infty} dq' \, e^{i[q'(n-n_0)a - \omega(q')t]} \, e^{-\sigma_0^2(q'-q)^2/2} \right\}. \qquad (4.7)$$

Exercise 4.31 (M*) *Find explicitly quantitative bounds on σ_0 which allow the conversion of Eq. (4.6) to Eq. (4.7). Evaluate the integral in (4.7) for $u_n(t)$ at $t = 0$ and show that it is the same as the expression (4.4) which was used to define the initial displacement pattern for a packet.*

We can develop an approximate expression for the time evolution of the packet by replacing the $\omega(q')$ in the plane wave exponent in (4.7) by the first three terms in its Taylor series expansion,

$$\omega(q') = \omega(q) + (q' - q)\frac{d\omega}{dq} + \frac{1}{2}(q' - q)^2\frac{d^2\omega}{dq^2} + \cdots. \qquad (4.8)$$

Exercise 4.32 (M**) *Keeping only the first two of these three terms and without evaluating the integral over q', show formally that the group velocity of the packet is given by $v_g = d\omega(q)/dq$.*

If in Exercise 4.12 you worked out an expression for the velocity of energy flow for the infinite traveling wave, compare that result with this new expression for the group velocity.

4.4.3 Dispersion*

Watch the evolution of PRESET 5 and let the packet RUN around through the PERIODIC BOUNDARY CONDITIONS until it has passed the central atom three or four times. You see that the envelope shape and width do not remain constant. The packet broadens with time, referred to as *dispersion*, and the envelope becomes asymmetric.   # 5

Exercise 4.33 (C*) COPY GRAPH *for later comparison. Now halve the* WIDTH *of the packet and* RUN *for the same time interval. At the longer times, which packet is narrower? Is it the packet that started narrower or broader?*

To make a narrower packet requires a wider range of frequencies, the uncertainty principle, if you like. But with a wider range of frequencies, the effects of dispersion become more important. Further manipulation with Eqs. (4.7) and (4.8) shows that the width of the packet evolves in time as

$$\sigma(t) = \sqrt{\sigma_0^2 + \left(\frac{d^2\omega}{dq^2}\right)^2 \frac{t^2}{\sigma_0^2}}. \qquad (4.9)$$

Note that the term in t^2 depends on the curvature of the dispersion relation and is inversely proportional to the square of the initial width of the packet.

Exercise 4.34 (M**) *Keeping all three terms in the expansion (4.8) of $\omega(q)$, use Eq. (4.7) to verify relation (4.9) for $\sigma(t)$.*

Exercise 4.35 (M*) *Using experiments from "born", show that expression (4.9) properly describes the time dependence of the packet width.*

Packets on optical fibers also propagate with degradation due to dispersion. High data rates require short packets; but short packets at the transmitter imply long packets at a distant receiver. What is the best compromise?

Exercise 4.36 (C*) *Try several initial PACKET WIDTHS in "born" to show that, indeed, there is some intermediate initial width σ_0 which gives a narrower packet after the first round trip than both a very long (non-dispersive) packet and a very short (highly dispersive packet).*

Exercise 4.37 (M*) *Put yourself in the position of the engineer who must decide on the best packet width to choose for a signal link of length L. For a given slope and curvature of the dispersion relation $\omega(q)$, what packet width gives the highest data rate? How does that rate depend upon L?*

Exercise 4.38 (M*) *Look at the graph of the central atom displacement as the packet goes by after the first round trip. It shows an asymmetry with a sharper leading edge and more extended trailing edge. The formal expression for the dispersed pulse gives a shape, at a fixed time, which is a symmetric Gaussian. Explain the source of the asymmetry.*

Exercise 4.39 (M**) *A careful look at the leading and trailing edges shows a different period of the oscillation. It is tempting, but not correct, to ascribe this to the faster phase velocity of the low frequency components relative to the high frequency ones. What is the source of the variation of the period of the oscillation as the packet moves by the central atom? (Hint: this requires more manipulation of Eqs. (4.7) and (4.8).)*

Exercise 4.40 (M**) *While you do something else, let the program RUN under PRESET 5 until it accumulates the order of 30 000 time steps. (Warning: your computer might run out of memory trying to store 30 000 sets of data in the graph. Running at high SPEED alleviates the problem by committing less of the data to memory.) After an initial phase in which the packet gradually disintegrates into mush, interesting patterns emerge until finally a series of packets recur with nearly the same width as the*

original packet. Can you track down the source of this recurrence? Show that the ratio of the recurrence time T_R to the period of the oscillation T_P is given by $T_R/T_P = (N/\pi)^2$ where N is the number of atoms in the chain. (Hint: remember that this is a finite system and hence has a discrete normal mode spectrum. If your answer is off by a factor of 2 you probably have the essential idea!)

4.5 Pulse propagation*

PRESET 6 illustrates another possible initial condition. "born" can simu- # 6
late pulse propagation as well as packet propagation. All that is necessary
is to choose the WAVE VECTOR q of the center of the packet to be zero.
Let the program RUN long enough for the graph to accumulate six or so
trips of the traveling pulse past the center atom. (Use a fast SPEED to
save time.) The shape of the pulse is nicely preserved. Now COPY GRAPH
to save the data for comparison. Repeat the procedure for pulse WIDTHS
of 4, 2, and 1 lattice constants and COPY GRAPH each, to generate a set
of graphs. For the shorter pulses the pulse shape is clearly not preserved:
there are decaying oscillations generated on the trailing edge of the pulse.
This is sometimes referred to as ringing because of the similarity with the
ringing of a struck bell.

Exercise 4.41 (M**) *For several of the shorter pulses, measure the ringing frequency in the extreme trailing edge of the pulses. How does it depend on the pulse width? Construct a qualitative argument that lets you estimate that frequency knowing the pulse width. (Hint: what Fourier components are required to represent the initial pulse shape?)*

Exercise 4.42 (C*) *Explain why we see oscillations trailing behind the pulse but the leading edge of the pulse remains clean. This is in contrast with the packet where there was tailing on both ends of the packet. (Re-member that the graph is watching a single atom versus time. It is not a snapshot of the atomic displacements versus position: hence later times are to the right.)*

4.6 Impurities

4.6.1 Scattering

In the model of the perfect harmonic crystal wave packets travel infinitely
far without scattering (though narrow packets do disperse with time).
Real crystals contain impurities, and these impurities, particularly at low
temperature, limit the distance a wave packet (a phonon) can propagate

7   without scattering. In PRESET 7 "born" has replaced the central atom in the chain with an atom of IMPURITY MASS smaller than the mass of the rest of the atoms in the chain.

Exercise 4.43 (C) RUN *to send a wave packet towards the impurity atom.* STOP *when the transmitted and reflected packets are well separated and note the relative amplitude of the transmitted and the scattered packet. Repeat for several different* IMPURITY MASSES. *Draw some qualitative conclusions about the dependence of the scattering amplitude on the ratio of the impurity mass to the host mass.*

Exercise 4.44 (M*) *Try a number of different values for the* IMPURITY MASS *in order to define quantitatively the limiting scattering amplitude for mass ratios which are both very large and very small compared with the mass of the host atoms. (You can get* IMPURITY MASSES $> M_{host}$ *by dragging the slider beyond the right hand end of the track, or by typing in a value.)*

Exercise 4.45 (M**) *If you are confident of your algebra or comfortable with programs such as Mathematica or Maple, develop a theoretical relation for the dependence of scattering amplitude on mass ratio and compare it with the "born" experiment.*

4.6.2   Localized modes*

The presence of light mass impurities, in addition to scattering the lattice waves of the host crystal, can give rise to *localized modes*. These are normal modes of motion of the crystal which decay exponentially with distance from the impurity instead of extending throughout the crystal.
8   Let's see an example. PRESET 8 is again for a crystal with a blue impurity, but with its MASS, for the moment, set equal to the host mass. It illustrates the extreme effects of dispersion if we try to prepare a very localized packet with WAVE VECTOR $q = \pi/a$. The packet is so narrow that it requires contributions from the whole spectrum of modes to construct it. The different phase velocities for these various waves result in very rapid dispersion of the energy throughout the chain as you see when you RUN "born".

Now change the IMPURITY MASS to 0.5 M_{host} and RUN again. The only change was to reduce the mass of the impurity atom to $\frac{1}{2}$ of that of the host atoms. There is now no propagation but also no significant dispersion: the energy remains highly localized at the impurity.

Exercise 4.46 (M) *Measure the frequency of this localized mode. How does it compare with the frequency of the highest modes of the perfect* crystal $\omega_{\max} = 2\sqrt{K/M}$? *Recalling the dispersion relation (4.2), you can measure* $\omega_{\max}$ *with a* RUN *of* PRESET 1 *with* $q = \pi/a$. #1

We seem to have said that, for the monatomic chain, there can be no motion of the chain at frequencies higher than some cut-off frequency, $\omega_{\max} = 2\sqrt{K/M}$. There are no physically allowed solutions at higher frequency for the free motion of the homogeneous perfect chain, but if a semi-infinite chain is driven at a higher frequency, it must respond. How?

Exercise 4.47 (M*) *For* $\omega > \omega_{\max}$, *perfectly valid mathematical solutions to the equations of motion for the infinite linear monatomic chain exist. See if you can find them. (Hint: try* $q = i\kappa + \pi/a$.) *Why are these not physically allowed solutions for the free motion of an infinite chain?*

Exercise 4.48 (C*) *Suppose the end of a semi-infinite chain is driven with a sinusoidal force varying with frequency* ω *for the two cases: first, for* $\omega > \omega_{\max}$ *and then, for* $\omega < \omega_{\max}$. *Qualitatively what will be the response of the chain for each of these two cases?*

Exercise 4.49 (M**) *Show, by calculating the time average work done by a spring on a neighboring atom, that the solution for* $\omega > \omega_{\max}$ *carries no energy down the chain.*

Exercises 4.47 and 4.49 show that there are solutions to the equations of motion corresponding to the motion of a half-chain driven at a frequency $\omega > \omega_{\max}$ characterized by:

1. successive atoms vibrating out of phase with each other;
2. an exponentially decaying envelope;
3. no propagation of energy along the chain.

These exponentially decaying waves are similar to the *evanescent waves* of optics and microwave engineering.

The solution to Exercise 4.47 suggests the nature of the localized mode. Think of the light mass as vibrating against its neighbors at a high frequency. The motion of the light mass then drives the two half-chains in this evanescent wave with the spatial decay. Since, at frequency $\omega > \omega_{\max}$, there can be no energy carried away from the localized region, the motion persists undamped indefinitely.

Exercise 4.50 (C*) *Choosing the* IMPURITY MASS $= 0.5\ M_{\text{host}}$ *in* PRESET 8 *to verify experimentally the first two properties of the high frequency solutions using the* QUINTUPLE *choice from the* GRAPH *menu. This allows you to follow the motion of five of the atoms in the chain. The first atom is the impurity atom: the rest are the four adjacent atoms to the right of the impurity.* #8

Exercise 4.51 (M*) *Vary the* IMPURITY MASS *in steps of factors of 2, changing the* WIDTH *as you go, and for each mass, measure the frequency of the impurity mode.* (IMPURITY MASS/WIDTH) *pairs that work fairly well are:* $(0.025/0.3)$, $(0.05/0.3)$, $(0.1, 0.4)$, $(0.2, 0.5)$, $(0.4/0.6)$. *Sketch an appropriate log–log plot to determine the power law dependence of the frequency of the local mode upon the impurity mass.*

Interpret the dependence you see without getting into long involved calculations. Look first at the case $M_{\text{impurity}} \ll M_{\text{host}}$. *Then comment on what happens for larger* M_{impurity}.

Exercise 4.52 (M**) *Calculate the frequency of the localized mode and its dependence on the impurity mass. (Hint: use the properties of the evanescent wave solutions, together with Newton's second law for the impurity mass.)*

Exercise 4.53 (M**) *Try setting the* IMPURITY MASS *to* $3M_{\text{host}}$. *Is there a heavy mass localized mode? Can you think of a reason why the light mass gives a localized mode but the heavy mass does not? If you could reduce the spring constants on either side of an impurity mass, then you would see an in band resonance or pseudo-localized mode [26].*

Impurity modes in real crystals often give rise to sharp absorption features in infra-red spectra. The spectroscopy of these modes is a useful tool in characterization of impurity species. The modes are also of importance in technical optics: they have potential as the active element for infra-red lasers but can give deleterious energy absorption as unwanted impurities in propagating media.

4.7   Anharmonicity**

Our analysis has been based on the assumption of harmonic forces. What if the forces are anharmonic? The most important consequence is to give a mechanism for the scattering of lattice waves from each other, or *phonon–phonon scattering*, an issue *not* addressed in these simulations.

\# 9   An amusing theoretical consequence, however, is illustrated by PRESET 9. The PRESET initially gives the response of the perfectly harmonic crystal to a too narrow packet, just as a reminder of the harmonic response and the implications of the dispersion in the $\omega(q)$ relation.

"born" includes the anharmonic forces by using the interaction potential between neighboring atoms:

$$V(u_n, u_{n+1}) = \tfrac{1}{2}K(u_n - u_{n+1})^2 + \tfrac{1}{4}K'(u_n - u_{n+1})^4. \qquad (4.10)$$

Exercise 4.54 (M*) *For a value of* $K'/K = 4$, *what is the force due to the quartic term in the potential relative to that due to the harmonic*

term if the relative displacement of the two atoms is $|u_n - u_{n+1}| = 0.3$?
What if the relative displacement is 1.0?

In the following examples, the displacement AMPLITUDE A of the atom at the center of the disturbance is large compared with that of its neighbors. Thus Exercise 4.54 tells us, for the central atom, that $K'A^2$ relative to K is a measure of the magnitude of the quartic force term relative to the harmonic.

Exercise 4.55 (C) RUN PRESET 9 *with the* QUARTIC CONSTANT K' *changed to about 4 and watch the fun. To be sure we haven't left a light mass at the center of the chain, move the center of the packet somewhere else and try again. How does the frequency of the anharmonic mode compare with ω_{max} of the harmonic crystal? How do the amplitudes and the relative phases of the motion at successive atoms fall off as you move away from the point of large excitation?*

Both here and with the light mass impurity mode you do see a slight disturbance propagating away from the local mode. The initial conditions for "born" don't quite match the normal mode displacement pattern of the local mode, and a small amount of the initial energy is dissipated into the usual extended modes. As for the example of the light-mass impurity mode, for each value of $K'A^2$ there is an optimum width in the initial conditions which minimizes this lost energy.

The anharmonic mode and the localized impurity mode both have frequencies greater than ω_{max} and a displacement pattern that falls off rapidly with position. The similarity suggests the following explanation of the anharmonic mode. If one atom, any atom in the chain, is set into oscillatory motion of large amplitude ($K'A^2 \geq K$), then the frequency of vibration of that one atom against its neighbors is higher than it would be for the harmonic crystal, because of the quartic term in the potential. Recall that the pure harmonic potential is special in having a frequency which is independent of amplitude. The nearby atoms, however, have a smaller amplitude of motion and are more nearly harmonic. But we know how the harmonic chain responds to an excitation of frequency $\omega > \omega_{max}$. The response is the evanescent wave which carries away no energy. Thus this excitation gets stuck wherever it started and never decays! It remains an open question whether modes related to this are to be found in real physical systems [28].

Exercise 4.56 (M*) *Watch the behavior of the system as you make trials with different* AMPLITUDE A *and* K'. *If either is made too small, the packet disperses as for the harmonic case. See whether you can establish an approximate, but quantitative, criterion on the parameter $K'A^2/K$ to* give a fairly well-defined anharmonic mode. Verify that it is this combination and not AK'^2/K or some other combination of the two which is relevant.

Exercise 4.57 (M**) *Determine from "born" how the frequency ω_A of the anharmonic mode varies with* AMPLITUDE *A and with K'. Use the virial theorem to predict the limiting behavior (when $K'A^2 \gg K$) of this dependence and test it against the simulation results.*

4.8   Summary

"born" reveals to us many features of the atomic motion in crystals. Most importantly we have seen how, in principle, to find the normal modes of a periodic array of a very large number of atoms. The essentials which allow a solution are the translational symmetry of the structure, the assumed harmonicity of the forces, and the use of periodic boundary conditions. At long wavelengths the solutions of the equations of motion resemble the standard results for wave propagation in continuous media, with standing and traveling wave solutions, wave packets, pulses, and so on. As the wavelength approaches the lattice constant, the discreteness of the mass distribution become important. The frequency is no longer proportional to the wave vector, the group and phase velocities become different, and there is a maximum frequency for which energy can be propagated along the chain. Finally "born" has given hints of more complex behavior in imperfect crystals. Impurities can give rise to scattering of the waves and to normal modes of motion which are localized near the impurity.

4.A   Deeper exploration

Thermal motion   The RANDOM choice of the INITIAL conditions is intended to be suggestive of the thermal motion of the atoms. Consider the implications of the *equipartition theorem* for the motion of the atoms on the chain if one of the atoms is of different mass from the others. What theoretical statements can you make about the thermal velocity and the thermal amplitude of the motion of the light atom relative to that of the others? Follow up with either quantitative or qualitative observations from "born".

Impurity modes   Selected sections from a general review of optical studies of the vibrational properties of disordered solids [26] will assist in addressing any of the following issues.

Carry out a quantitative study, both theoretical and experimental, of the mass dependence of the frequency of the light mass impurity mode.

The initial conditions in "born" for the impurity modes are not quite properly chosen. This is evident from the disturbance which initially propagates away from the impurity, especially when the mass ratio is not large. Issues to consider are: the appropriate set of initial conditions to insure that *no* energy is lost to the continuum modes; the best approximation to these that can be made with the Gaussian envelope of the initial disturbance used by "born"; and the fractional loss of energy to the host modes if the Gaussian width is far from its optimum value.

The more ambitious, who are willing to modify the source code, could introduce a more complicated impurity in which the force constant to the nearest neighbors, as well as the impurity mass, could be adjusted. With a heavy mass and soft springs coupling to the neighbors, an *in band resonance mode* [26] occurs. This provides a natural subject for a combined experimental and theoretical study.

Anharmonic modes Reference [28] gives a detailed discussion of computer modeling of the localized modes induced by anharmonic interactions among atoms. It may suggest issues for investigation or help you address questions similar to those raised above for the localized mode. Using "born" you should be able to develop a convincing experimental demonstration that the critical parameter in defining the nature of the anharmonic mode is the ratio $K'A^2/K$ where K' is the quartic coupling constant, K the quadratic constant, and A the displacement amplitude.

4.B "born" – the program

"born" is a lattice dynamics simulation of a one-dimensional chain of atoms. The simulated crystal can be pure or have a mass defect, and the nearest neighbor potential can be purely harmonic or contain an additional quartic component.

4.B.1 *Dynamics of a one-dimensional atomic chain*

"born" models the motion of N atoms placed at positions na, $n = 0,\ldots,(N-1)$ in a one-dimensional chain with lattice constant a. For simplicity, in "born" we make the two assumptions, that only neighboring atoms interact and that the interaction is harmonic with a possible additional quartic component. The potential energy $V(t)$ of such a chain of N atoms with displacements $u_n(t)$ from their equilibrium position

$$V(t) = \sum_{n=0}^{N-1} \left\{ \tfrac{1}{2} K \left[u_n(t) - u_{n+1}(t) \right]^2 + \tfrac{1}{4} K' \left[u_n(t) - u_{n+1}(t) \right]^4 \right\}. \quad (4.11)$$

The harmonic spring constant K is set to 1 in the "born" program and the QUARTIC SPRING CONSTANT K' is an adjustable parameter. The motion of each atom n of mass M_n is described by Newton's second law with the force $F_n = -\partial V / \partial u_n$,

$$\begin{aligned} M_n \ddot{u}_n(t) = &- K \left[2u_n(t) - u_{n-1}(t) - u_{n+1}(t) \right] \\ &- K' \left\{ \left[u_n(t) - u_{n-1}(t) \right]^3 + \left[u_n(t) - u_{n+1}(t) \right]^3 \right\}. \end{aligned} \quad (4.12)$$

For the (simple) case in which the masses of all the atoms are the same and the quartic spring constant K' is zero (CRYSTAL TYPE PURE), we can easily solve (4.12):

$$u_n(t) \sim e^{i[qna - \omega(q)t]}, \quad (4.13)$$

with the dispersion relation $\omega(q)$ specified in Eq. (4.2) on page 64. Much of the interesting physics, however, takes place when we put an impurity atom of different mass into the chain (CRYSTAL TYPE IMPURE), or have an anharmonic interaction (CRYSTAL TYPE ANHARMONIC). "born" uses a numerical solution to the equations of motion (4.12) which "works" even in these more general cases.

4.B.2 Leap-frog algorithm

Given the initial displacement and velocity of each atom in the chain, "born" solves numerically the set of N coupled equations of motion (4.12) by first converting this set of second order differential equations into two sets of first order difference equations and then solving these difference equations with the *leap-frog algorithm* described below.

The two first order difference equations for the amplitude $u_n(t)$ and the velocity $v_n(t) = du_n(t)/dt$ are

$$\Delta u_n(t) = \Delta t \, v_n(t) \quad (4.14)$$

and

$$\begin{aligned} \Delta v_n(t) = -\frac{\Delta t}{M} \Big(&K \left[2u_n(t) - u_{n-1}(t) - u_{n+1}(t) \right] \\ &- K' \left\{ \left[u_n(t) - u_{n-1}(t) \right]^3 + \left[u_n(t) - u_{n+1}(t) \right]^3 \right\} \Big). \end{aligned} \quad (4.15)$$

Given $u_n(t)$ and $v_n(t)$ at time t, we can determine through Eqs. (4.14) and (4.15) $u_n(t + \Delta t) = u_n(t) + \Delta u_n(t)$ and $v_n(t + \Delta t) = v_n(t) + \Delta v_n(t)$

at some later time $t + \Delta t$. With this simple numerical integration method the error accumulates very quickly, and soon the numerical solutions have little to do with the *true* solutions of the differential equations (4.12). The trick to increase the numerical stability of a difference equation while keeping the time step Δt reasonably large is to update the displacement of each atom by only half a time step,

$$u_n(t + \tfrac{1}{2}\Delta t) = u_n(t) + \tfrac{1}{2}\Delta t \, v_n(t), \qquad (4.16)$$

then to update the velocity by a full time step,

$$v_n(t + \Delta t) = v_n(t) + \Delta v_n(t + \tfrac{1}{2}\Delta t), \qquad (4.17)$$

and lastly to update the displacement again by half a time step,

$$u_n(t + \Delta t) = u_n(t + \tfrac{1}{2}\Delta t) + \tfrac{1}{2}\Delta t \, v_n(t + \Delta t). \qquad (4.18)$$

This two-step scheme, used by "born", is known as a leap-frog algorithm.

"born" chooses a time step which is a $1/50$ of the period of one harmonic oscillation (see Eq. (4.2)) for a wave vector $q = \pi/a$, which gives a time step $\Delta t = 0.02(\pi\sqrt{M/K})$. Such a time step leads to sufficient numerical stability for the physically interesting region of the parameters.

4.B.3   Boundary conditions

The displacement and velocity of any atom *inside* the chain (atom 1 to atom $(N-2)$) follow Eqs. (4.14) and (4.15). The dynamics of the first and last atoms (atoms 0 and $(N-1)$), however, depend on the BOUNDARY conditions which can be either FIXED or PERIODIC. For FIXED boundary conditions the displacement and velocity of the 0th and $(N-1)$th atoms are zero at any time t. For the PERIODIC choice the change in displacement Δu and change in velocity Δv of atoms 0 and $(N-1)$ are still given by Eqs. (4.14) and (4.15), but with the $u_{-1}(t)$ and $u_N(t)$ appearing in the equations identified with $u_{N-1}(t)$ and $u_0(t)$ respectively.

4.B.4   Initial conditions

"born" allows the specification of the initial displacements and velocities of the atoms in the chain through the five different choices for INITIAL. The initial values for u_n and v_n below are derived from the harmonic solution (4.13) and the dispersion relation (4.2). (Note: for a TRAVELING and STANDING wave "born" requires the WAVE VECTORS to be consistent with the chosen BOUNDARY conditions. For PERIODIC BOUNDARY conditions the WAVE VECTOR q must be a multiple of $2\pi/Na$ while the q values for FIXED BOUNDARY conditions are restricted to be multiples of $\pi/(N-1)a$. "born" automatically chooses the closest consistent wave vector to the wave vector specified on the slider and sets the slider to that value.)

Traveling wave For a TRAVELING WAVE, the initial displacements are taken to be proportional to the imaginary part of (4.13),

$$u_n(0) = A \sin(qna), \tag{4.19}$$

where q and A are the chosen WAVE VECTOR (in units of π/a) and AMPLITUDE, respectively. We have chosen $u_n(0)$ to be proportional to the imaginary part rather than the real part of (4.13) in order to satisfy the FIXED BOUNDARY condition at $n = 0$. The initial velocities are then proportional to the time derivative of $u_n(t)$ in Eq. (4.13) at time $t = 0$,

$$v_n(0) = -2\sqrt{K/M} \left| \sin(\tfrac{1}{2}qa) \right| A \cos(qna). \tag{4.20}$$

Standing wave For a STANDING WAVE, the initial displacement of the nth atom is given by the same relation (4.19) as for the TRAVELING WAVE. The initial velocity of each atom is zero.

Traveling packet For a TRAVELING WAVE PACKET, the displacement of each atom $u_n(t)$ at small times t can be taken to be proportional to the real part of the harmonic solution (4.13) times a Gaussian envelope that travels with the group velocity $v_g = d\omega/dq$,

$$u_n(t) = A e^{-\frac{[(n-n_0)a - v_g t]^2}{2\sigma_0^2}} \cos[q(n - n_0)a - \omega t], \tag{4.21}$$

with the wave packet of PACKET WIDTH σ_0 centered at atom n_0 (PEAK POSITION n_0) at time 0. Note that Eq. (4.21) is a valid approximation for the solution of the equation of motion (4.12) only for small times t – it doesn't, for example, account for the dispersive spread of the wave packet. The initial displacement of the nth atom is then given by

$$u_n(0) = A e^{-\frac{(n-n_0)^2 a^2}{2\sigma_0^2}} \cos\left[q(n - n_0)a\right]. \tag{4.22}$$

(We have chosen here the real part of (4.13) rather than the imaginary part so that the displacement is maximal at the center of the wave packet.)

The initial velocities (see also Exercise 4.26) are the time derivatives $\dot{u}_n(t)$ of the displacements $u_n(t)$ in Eq. (4.21) at time $t = 0$,

$$v_n(0) = A\sqrt{\frac{K}{M}}\, e^{-\frac{(n-n_0)^2 a^2}{2\sigma_0^2}} \left\{ \frac{(n - n_0)a^2}{\sigma_0^2} \cos[q(n - n_0)a] \cos(\tfrac{1}{2}qa) \right.$$

$$\left. + 2 \sin[q(n - n_0)a] \sin(\tfrac{1}{2}qa) \right\}. \tag{4.23}$$

Standing packet For a STANDING WAVE PACKET, the initial displacement of the nth atom is given by the same relation (4.22) as for the TRAVELING WAVE PACKET. The initial velocity of each atom is zero.

Random (thermal) motion For the RANDOM motion, the random initial displacements of the atoms are distributed uniformly between $-A$ and A. The initial velocity of each atom is zero.

4.B.5 Displayed quantities

The upper display gives the positions of the N atoms at time t, along with a grid with a spacing of ten atom distances. (The NUMBER OF ATOMS N can be changed in the CONFIGURE menu.) For the INITIAL choices of TRAVELING and STANDING WAVE, the continuous function $u(x) = A\sin(qx)$ can be shown as well.

The graph shows, for the DISPLAY choice SINGLE, the position versus time of the center atom in the chain. Choose TRIPLE or QUINTUPLE to see in addition the positions versus time of the two or four neighbors to the right of that center atom. The color coding in the graph is as follows: blue for the center atom, and green, red, magenta, and purple for the next four neighbors to the right. The graph is only updated every GRAPH UPDATE INCREMENTS×SPEED time steps.

The numerical computation (i.e., the leap-frog algorithm) of the atom positions involves little numerical effort, and the speed at which "born" runs is limited mostly by the display. Therefore the display of the atom positions is only updated every SPEEDth time step. The positions of the central atom and its four neighbors are stored in the graph also every SPEEDth time step. A large SPEED therefore not only increases the speed at which the simulations runs but it also decreases the resolution in the graph. The numerical accuracy, however, which is determined by the time step $\Delta t = 0.02(\pi\sqrt{M/K}) = \text{const.}$, is *not* affected by the choice of speed. If you want to observe the behavior of the atoms at very large times (see, for example, Exercise 4.40), RUN with a large SPEED until you reach the times of interest, and then reduce the SPEED. You will save yourself not only a lot of time but also your computer a lot of memory (see below).

4.B.6 Bugs, problems, and solutions

Because the size of the time step in "born" is fixed, the leap-frog algorithm will become unstable for some choices of parameters such as very light IMPURITY MASSES or very large QUARTIC SPRING CONSTANTS.

The initial conditions for the TRAVELING PACKET and the STANDING PACKET are not consistent with either choice of BOUNDARY conditions. If the packet is placed too close to the end of the chain, or if the PACKET WIDTH is taken too large, there will be obvious distortions in the packet shape.

If "born" runs many time steps, the amount of data displayed in the graph can be very large, and the computer may run out of memory. Less memory is used if "born" runs at a high SPEED since the data is stored for the graph only every SPEEDth time step.

5

"debye" – Lattice dynamics and heat capacity

Contents

5.1 Introduction

"debye" continues the discussion started in "born". The pictures developed in "born" gave insights into the normal mode spectrum of solids, but gave no hint how these finally connect to measurable properties. What are the dispersion relations for a real three-dimensional solid? How do calculated results compare with experimental values, accessible via inelastic neutron diffraction? What is the amplitude of the atomic motion at a given temperature, measurable via the Debye–Waller factor in x-ray scattering? How much energy must be added to a crystal to change its temperature by a specified amount, the heat capacity?

"debye" explores the lattice dynamics, in the harmonic approximation, of face centered cubic (fcc) structures with nearest neighbor interactions. The power of the computer is used not to produce illustrative simulations but to calculate observable properties of a three-dimensional model system. We start with experimentally determined macroscopic mechanical properties of three materials, aluminum, copper, and lead: either velocities of propagation of sound or else elastic constants determined from sound velocities. We use these parameters to deduce a value for the nearest neighbor interaction constant. This constant is used by "debye" to calculate the frequency of an elastic wave (or phonon) propagating with an arbitrary wave vector. The frequencies may be compared with values determined experimentally in neutron diffraction experiments.

Knowledge of the normal mode frequencies allows calculation of the number of modes with frequencies in a specified narrow range, the density of modes. The density of modes is reflected in results of tunneling experiments and in the heat capacity of the crystals. "debye" allows comparison of the numerically generated heat capacity results with experimental data and also with the predictions of the Debye model.

5.2 Brillouin zone

The equations of motion for the atoms in a three-dimensional crystal can be satisfied by plane wave solutions, just as was done in one dimension. Restricting attention to a single atom per primitive unit cell, for any given wave vector $\mathbf{q}$ there are three possible polarizations of the propagating wave and three corresponding allowed frequencies. Our first task in understanding the effects of thermal motion is to deduce the dispersion relations for these waves, $\omega_p(\mathbf{q})$, where p is a polarization index running from 1 to 3. In "born" we found that we did not need to find an ω for arbitrary values of q, but only for q values in the limited range $-\pi/a < q \leq \pi/a$, a cell centered on the origin of the reciprocal lattice and called the first *Brillouin zone*.

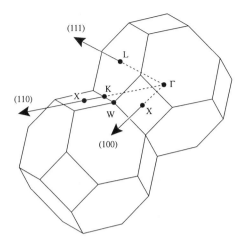

Figure 5.1: First Brillouin zone and neighboring cell for fcc crystals. (Adapted from *Solid State Physics* by Neil W. Ashcroft and N. David Mermin, copyright 1976 by Saunders College Publishing, reproduced by permission of the publisher.)

To describe the dispersion relations for the three-dimensional crystal we again need only to specify the frequencies ω corresponding to wave vectors **q** lying in the first Brillouin zone. In three dimensions, this zone is a volume of the reciprocal space centered at the origin and bounded by the perpendicular bisecting planes of the low index (small (hkl)) reciprocal lattice vectors. It may also be regarded as the locus of points closer to the origin than to any other point of the reciprocal lattice. Figure 5.1 illustrates the first Brillouin zone for fcc crystals, along with one repeated image of it. The origin of reciprocal space is conventionally denoted by Γ, while X, L, K, and W are assigned to various symmetry points on the surface of the zone.

Exercise 5.1 (M) *Consider the fcc structure with a cubic cell with lattice constant a and atoms at $[l\,m\,n]$ but only for l, m, and n such that $l + m + n =$ even. Show that the lattice reciprocal to the fcc lattice is a body centered cubic (bcc) lattice with a cubic cell with a lattice constant $4\pi/a$ (not $2\pi/a$).*

For this problem we conventionally label vectors in reciprocal space in terms of a simple cubic reciprocal lattice with lattice constant $2\pi/a$. Using this convention, the bcc reciprocal lattice points are specified by the cubic indices (hkl), but with the restriction that h, k, and l are either all even or all odd.

Exercise 5.2 (M) *Show that the set of vectors $(h\,k\,l)$ on a cubic lattice of lattice constant $2\pi/a$, with the restriction that h, k, and l are either* all even or all odd, is the bcc lattice with a cubic lattice constant $4\pi/a$.

Exercise 5.3 (C) *Draw a sketch of this bcc reciprocal lattice. Using the simple cubic indexing, the shortest reciprocal lattice points are* $(2\,0\,0)$, $(1\,1\,1)$ *and the symmetry related vectors, e.g.,* $(0\,2\,0)$, $(1\,\bar{1}\,1)$, *etc. Convince yourself that the planes defining the Brillouin zone in Figure 5.1 are the perpendicular bisecting planes of these two sets of reciprocal lattice vectors. Where in Figure 5.1 would the plane bisecting the* $(1\,1\,0)$ *reciprocal lattice vector lie?*

Exercise 5.4 (M) *What are the coordinates in reciprocal space of the points in Figure 5.1 denoted* Γ, X, *and* L; *i.e., the origin and the intersections of the* $(1\,0\,0)$ *and* $(1\,1\,1)$ *axes respectively with the surfaces of the first Brillouin zone?*

Exercise 5.5 (M**) *What are the coordinates in reciprocal space of the point denoted* K, *the intersection of the* $(1\,1\,0)$-*axis with the surface of the first Brillouin zone? Note that the K point is not the bisector of the* $(2\,2\,0)$ *reciprocal lattice vector. Rather it is the common intersection of two of the* $(1\,1\,1)$ *bisectors with the* $(1\,1\,0)$ *direction. It requires a little more thought than do the X and L points. According to Figure 5.1 you can reach an X point either by going along* $(1\,0\,0)$ *to the zone boundary, or by going beyond the zone boundary along a* $(1\,1\,0)$ *direction. Use the translational and rotational symmetry of the dispersion relation* $\omega(\mathbf{q})$ *to show that* ω *will be the same at these two points.*

It is difficult to illustrate the results of the calculation of the dispersion relation in a multidimensional plot. Instead imagine some line in the Brillouin zone, e.g., a line along the $(1\,0\,0)$ direction from the origin or Γ point to the X point, and plotting the ωs as a function of $|\mathbf{q}|$ as it moves along one or another of these lines. PRESET 1 shows such a plot for the $(1\,0\,0)$ direction. We need to see how "debye" generates the data for this plot.

1

5.3   Dispersion relations

5.3.1   Symmetry directions

The model is the simplest imaginable: an fcc array of atoms of mass M, with nearest neighbors connected by springs of FORCE CONSTANT $B/2$. Note the unconventional factor of 2 in the definition of B: see page 108 after Eq. (5.11). "debye" is able to calculate the three frequencies ω_p for an arbitrary choice of wave vector $\mathbf{q}$. The method is explained in Section 5.B. For special, high symmetry directions of $\mathbf{q}$ it is straightforward to perform

Table 5.1: The dispersion relations for $\mathbf{q}$ along the three symmetry directions. M is the mass of the atom and B the force parameter of the nearest neighbor interaction?

$\mathbf{q}$	ω_L^2	ω_T^2	$\omega_{T'}^2$
$(1\,1\,0)$	$\frac{2B}{M}\sin^2\left(\frac{qa}{4\sqrt{2}}\right) + \frac{2B}{M}\sin^2\left(\frac{qa}{2\sqrt{2}}\right)$	$\frac{2B}{M}\sin^2\left(\frac{qa}{4\sqrt{2}}\right)$	$\frac{4B}{M}\sin^2\left(\frac{qa}{4\sqrt{2}}\right)$
$(1\,0\,0)$	$\frac{4B}{M}\sin^2\left(\frac{qa}{4}\right)$		$\frac{2B}{M}\sin^2\left(\frac{qa}{4}\right)$
$(1\,1\,1)$	$\frac{4B}{M}\sin^2\left(\frac{qa}{2\sqrt{3}}\right)$		$\frac{B}{M}\sin^2\left(\frac{qa}{2\sqrt{3}}\right)$

the calculation with a less formal procedure, and to see much of the physics.

Exercise 5.6 (M*) *Find the dispersion relations for the longitudinal and transverse phonon modes, with $\mathbf{q}$ along $(1\,0\,0)$, for an fcc structure with nearest neighbor springs with force constant $B/2$. Remember that for a plane wave with $\mathbf{q}$ along $(1\,0\,0)$, all of the atoms in each plane perpendicular to the x-axis move together, either in the x-direction for the longitudinal wave, or the y- or z-direction for the transverse waves. Thus, for all of the atoms in each plane, a single coordinate serves to define their displacements. Compare your result qualitatively with the curves in PRESET 1. Is the ratio of the longitudinal to the transverse frequency for a given q the value you would predict?*

ZOOM the full vertical axis to lock out the auto-scaling. Then test the changes in the graphs produced by changing the MASS and FORCE CONSTANT. The dispersion relations in this model are given in Table 5.1 for our three favorite symmetry directions.

Exercise 5.7 (C) *For waves propagating along $(1\,0\,0)$, the two transverse polarizations have the same frequency. Now change the choice of $\mathbf{q}$ direction, at the bottom of the left hand display, to $(1\,1\,0)$. Explain why the transverse modes are degenerate for propagation along $(1\,0\,0)$ but not for $(1\,1\,0)$. (Hint: think about the relative motion of the atoms in two successive $(1\,1\,0)$ planes, the planes perpendicular to $\mathbf{q}$, for the two polarizations with displacements parallel to $(1\,\bar{1}\,0)$ and to $(0\,0\,1)$.)*

You can watch the gradual change in going from $(1\,0\,0)$ to $(1\,1\,0)$ in the following way. Use the sliders to read in a $\mathbf{q}$ direction $(20\,0\,0)$ and you'll see the $(1\,0\,0)$ plot again. Now increase the second index by clicking, one unit at a time. You see the degeneracy gradually lifted as you leave the $(1\,0\,0)$ direction.

Exercise 5.8 (C*) *Do you expect the transverse modes to be degenerate for* **q** *along* (111)? *Construct an argument to justify your guess? It's not as straightforward as for the* (1 0 0) *case. See what the plot in "debye" shows. Try starting with* (20 20 20) *and reducing the last index a unit at a time to see the transition to* (1 1 0).

A convenient and conventional display of the dispersion relations is seen by choosing SYMMETRY DIRECTIONS from the SHOW menu above the left hand display. Now you see the dispersion relations for each of the three symmetry directions as well as those for the line connecting the X and K points.

Exercise 5.9 (C) *Override the auto-scaling by* ZOOMING *the full vertical axis and use the slider to change the* FORCE PARAMETER. *Does the shape of the dispersion curves change or only the vertical scale factor? Explain why?*

Note the crossing of the longitudinal and one of the transverse branches for **q** in the (110) direction, and the connection of these branches with the curves for the other directions. You see that there is some ambiguity in what should be labeled transverse and what should be labeled longitudinal. Only in the symmetry directions are the modes strictly longitudinal and transverse: for directions of **q** without symmetry the modes are of mixed character.

5.3.2   Force parameter

To use "debye" to calculate dispersion relations for real materials requires an appropriate choice of the FORCE PARAMETER B. One semiempirical approach is to work from knowledge of the elastic constants or from known velocities of sounds of the material of interest. Table 5.2 gives values from the literature for some of these data, along with the atomic masses and lattice constants of the three materials, Al, Cu, and Pb. To make explicit connection with the microscopic model, recall that the low frequency plane wave solutions in the $q \to 0$ limit correspond to sound or acoustic waves.

Exercise 5.10 (M*) *Use Table 5.1 to obtain the three different velocities of sound for waves propagating in the* (110) *direction in terms of the constant B. (Hint: do not work out the derivative $d\omega/dq$ and then take the limit $q \to 0$. Instead, first evaluate the expression for $\omega(q)$ to lowest order in q in the limit $q \to 0$ and then take the derivative.) This, and results for the other symmetry directions, give a number of relations for B in terms of the sound velocities.*

Table 5.2: Material constants for Al, Cu, and Pb. The velocities of sound are given for the longitudinal (L) and two transverse (T, T') waves propagating along the $(1\,1\,0)$ directions.

	Al	Cu	Pb
Lattice constant (Å)	4.05	3.61	4.95
Mass number	27.0	63.5	207.2
C_{11} (GPa)	119		54
C_{12} (GPa)	62		44
C_{44} (GPa)	31		18
v_L (10^3 m/s)		5.07	2.43
v_T (10^3 m/s)		1.69	0.66
$v_{T'}$ (10^3 m/s)		3.00	1.26

Exercise 5.11 (M*) *The velocities of sound are intimately related to the elastic constants of the material via the equations of Table 5.3. Now find relations that give the elastic constants in terms of the force parameter B, or B in terms of the elastic constants.*

Exercise 5.12 (C) *Compare the material constants given in Table 5.2 for Pb and Al. Without doing any detailed calculation, predict which metal will have the higher velocity of sound? Explain your choice!*

Exercise 5.13 (M*) *Find a single value for the* FORCE PARAMETER B *that gives a plausible fit to the three different (110) velocities of sound for Cu. Since the velocities are determined by various combinations of three elastic constants, don't be surprised that the simple model, with only a single free parameter, does not give a good fit to all three velocities. Choose a sensible compromise. The discrepancies among the three values gives one measure of the expected accuracy of the model. Caution: it's easy to end up with factors of 2π in the wrong place and to waste a lot of time. Your B value should be in the range 10–100 N/m.*

5.3.3   Comparison with neutron data

Open PRESET 1, but change the ARBITRARY DIRECTION to $(1\,1\,0)$ and the FORCE PARAMETER B to the one you have chosen in the preceding exercise. With a little help from "debye", your quantitative prediction of the dispersion relation for Cu for wave vectors in the $(1\,1\,0)$ direction is then displayed. How can we find out how good a job we did?

Table 5.3: Relations among the density, the velocities of sound and the elastic constants.

$\mathbf{q}$	$\rho v_L^2 =$	$\rho v_T^2 =$	$\rho v_{T'}^2 =$
$(1\,1\,0)$	$\frac{1}{2}(C_{11} + C_{12}) + C_{44}$	$\frac{1}{2}(C_{11} - C_{12})$	C_{44}
$(1\,0\,0)$	C_{11}	C_{44}	
$(1\,1\,1)$	$\frac{1}{3}(C_{11} + 2C_{12} + 4C_{44})$	$\frac{1}{3}(C_{11} - C_{12} + C_{44})$	

Inelastic neutron scattering gives a direct measure of the dispersion curves, $\omega(\mathbf{q})$. We saw half of the physics in "laue". A lattice wave of wave vector $\mathbf{q}$ gives satellite diffraction peaks separated from the Bragg peaks by $\mathbf{q}$. The part of the story not mentioned in "laue" is that the neutrons in this scattering event will have lost or gained an amount of energy equal to $\hbar\omega(\mathbf{q})$, see reference [25]. Thus, energy analysis of the scattered beam gives the dispersion relation quite directly.

Exercise 5.14 (C) *Select* COPPER *from the* COMPARE WITH *menu to see the data from neutron scattering experiments [30] which determine the dispersion curves experimentally. You should have plausible agreement with the initial slopes of the curves because it was the velocities of sound which were fit to get a value for B. If the agreement is poor, or if you did not do Exercise 5.13, then adjust the* FORCE PARAMETER *B to get a good fit to the initial slopes. If you did Exercise 5.13, how does the size of this adjustment compare with the differences in the Bs you needed to fit the different velocities?*

Exercise 5.15 (C) *Select* SYMMETRY DIRECTIONS *from the* SHOW *menu above the left hand plot. Is there a reasonable fit to the dispersion relations in the other directions in* $\mathbf{q}$*-space? Can you improve the overall agreement with adjustments in the* FORCE PARAMETER *B? Note down the best value of B for use later.*

Exercise 5.16 (C) *Before you get too impressed with your newly found power to calculate dispersion relations, try again for Pb. To save you time, use* PRESET 2 *which includes parameters appropriate to Pb. Can you improve the fit with other values of the* FORCE PARAMETER *B?*

2

Now, are you upset at not being able to get a better fit for Pb? In fact, you should be pleased that with a single parameter you come anywhere close! To give you comfort, the fits in Figure 5.2 were obtained with a model which takes account of interactions with not just the nearest neighbors, but rather out to eighth nearest neighbors [27]! The fit is

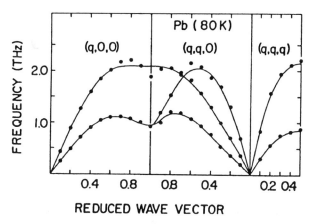

Figure 5.2: Born–von Karman fit with 26 parameters of the dispersion curves for Pb (from reference [27] with permission from Elsevier Science, Ltd). The solid lines are the calculated values, the points the neutron data.

made not with one fitting parameter but with 26! They do pretty well, but for the longitudinal mode in the (100) direction they're still in trouble. (With 26 parameters you should be able to fit to several elephants.)

We're seeing a phenomenon often encountered in condensed matter physics: the roughest approximation works remarkably well while attempts at further improvement are frustratingly unsuccessful. The forces among the atoms in a crystal are simply not describable in terms of short range two-body atom–atom interactions; attempts to fit using such models are doomed to frustration.

There's another example in PRESET 3 which gives the comparison for   # 3 Al. Lucky again!?

5.4   Density of states

5.4.1   From dispersion relations to density of states

How do we go from the dispersion relations to specific material properties? Many depend upon the distribution of normal mode frequencies, not the detailed nature of the modes. The normal modes are determined from the plane wave solutions of the infinite medium by applying periodic boundary conditions. We have already seen this idea in "born" with the condition that the allowed wave vectors for a wave on a chain of N atoms were $q = 2\pi n/Na$, with n an integer. A similar result holds in three dimensions.

Exercise 5.17 (M) *Consider a crystal defined by a unit cell with primitive lattice vectors $\mathbf{a}_1$, $\mathbf{a}_2$, and $\mathbf{a}_3$. Suppose the crystal to be a parallelepiped with edges of length N_1a_1, N_2a_2, and N_3a_3 which are parallel*

to the corresponding primitive lattice vectors. Verify that the plane wave
solutions will satisfy the periodic boundary conditions if **q** is chosen to be

$$q = \left(\frac{n_1}{N_1}\right) b_1 + \left(\frac{n_2}{N_2}\right) b_2 + \left(\frac{n_3}{N_3}\right) b_3, \qquad (5.1)$$

with the n_i integers.

For each **q** value in the grid defined in Eq. (5.1) we have, with a single
atom per primitive cell, three possible normal mode solutions, one for
each of three polarizations. To avoid duplication, **q**s are restricted to a
single unit cell of reciprocal space, usually the first Brillouin zone. As
in "born" the motion of a normal mode with **q** chosen outside the first
Brillouin zone is physically indistinguishable from one associated with a
q′ within the first zone with **q** − **q**′ = (reciprocal lattice vector).

We need to know how many modes have frequencies in a specified range
ω to $(\omega + d\omega)$. We define this number of modes to be $g(\omega)d\omega$ and call
the distribution function, $g(\omega)$, the *density of modes*. The term *density*
here refers to the number of modes per unit frequency interval, *not* per
unit volume. It may, depending upon the normalization used, refer to
modes per unit frequency *and* per unit volume, but the important idea is
the idea of a *density in frequency*. In the quantum picture, the different
modes correspond to different possible states for phonons, the mode of
frequency ω corresponding to an allowed phonon state of energy $E = \hbar\omega$.
Thus the density of modes $g(\omega)$ is frequently referred to as a *density of
states for phonons* (DOS) for phonons.

"debye" finds the density of modes or density of phonon states essentially in the following way. (See Section 5.B to see just what it does.) The
program picks a **q** value at random from the grid defined by Eq. (5.1), calculates the three phonon frequencies associated with that **q**, and assigns
them to appropriate *bins* of a histogram, with different bins corresponding
to different frequencies. It repeats this process many times. The NUMBER
OF MC POINTS (MC = Monte–Carlo) from the CONFIGURE dialog box
gives the number of **q** points sampled when you CALCULATE, and in the
main panel you may set the NUMBER OF BINS in the histogram. PRESET
4   4 has sorted 5000 samples of three frequencies each into 30 bins.

Exercise 5.18 (M**) *"debye" does not follow the rule above that* **q**
*values should be chosen from within the first Brillouin zone. Instead, it
chooses* **q**s *randomly from the volume defined by* $0 < q_x, q_y, q_z \leq 2\pi/a$
*which is one octant of the cubic cell within which the first Brillouin zone
is contained. Show that this procedure will give the proper DOS. (Hint:
use the fact that the dispersion relation* $\omega(\mathbf{q})$ *is periodic in* **q**, *as well as
invariant under the symmetry operations of the cube, to map different
volumes in* **q**-*space onto each other.)*

5.4.2 Resolution versus noise*

PRESET 4, after you click CALCULATE, shows the DOS for Pb using the value for the FORCE PARAMETER B from PRESET 2. You may calculate the DOS for any set of parameters listed on the sliders simply by clicking on CALCULATE but remember that it takes some time to do the calculations, and you will lose any current DOS data. The right-hand display of "debye" shows a plot of the DOS $g(\omega)$ with various peaks and valleys, but a coarse grained structure, imposed by the choice in PRESET 4 of too large a bin size, is also evident.

Exercise 5.19 (C) *Improve the resolution by increasing the* NUMBER OF BINS *to 200 and* CALCULATE *again. Certainly the new plot has finer resolution and a lot more structure. Is all of the structure real?* COPY GRAPH, CALCULATE *again with the same parameters, and* STEAL DATA *to superpose data from the two runs. Now what do you conclude about the significance of the fine structure?*

Double the number of sampled $\mathbf{q}$ *values by clicking* DOUBLE MC STEPS. *Is the change evident? Repeat until either you are satisfied that most of the features you see are real or you get tired of waiting. Remember that each doubling takes twice as long as the preceding one! The total accumulated number of MC steps is shown in the upper right hand corner of the DOS plot.*

Unfortunately, reducing the bin size to get better resolution in frequency decreases the number of samples per bin. The penalty is the larger fractional fluctuations of the numbers of samples in each bin: with n samples in a bin, the rms fractional fluctuation, from run to run, of the number in the bin is $1/\sqrt{n}$. Note that the total number of frequency samples equals $3N_{MC}$ since each $\mathbf{q}$ value gives three frequencies.

Exercise 5.20 (M)** *If you know the* NUMBER OF BINS *and the total number of samples you should be able to estimate the number of points contributing to any bin in the histogram. (Hint: imagine a rectangular distribution extending over the full frequency range of the DOS graph and with the same area as the calculated DOS. Knowing how many samples and how many bins there are, you should be able to find the number of samples per bin for the rectangle and thus calibrate the vertical scale.) Is the noise in the DOS consistent with this estimate, using the* $1/\sqrt{n}$ *argument?*

The problem here is a recurring one, not only in numerical physics but in much of experimental physics: trade-off between resolution and signal to noise ratio. A particularly painful aspect is that a two-fold increase in signal to noise ratio requires a four-fold increase in observing time.

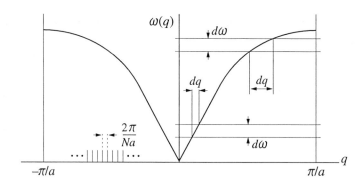

Figure 5.3: Influence of $d\omega/dq$ on the DOS.

Exercise 5.21 (M*) *Give the argument to justify the statement that it requires four times the observing time to double the signal to noise ratio.*

  Warning: "debye" saves the data generated by the CALCULATE and the DOUBLE MC STEPS buttons until the CALCULATE button is again pressed. The same data set is used for all three of the available right hand graphical displays. Those data and the graph are *not* updated when parameters are changed. If a parameter change is made, the number of accumulated Monte–Carlo steps indicated in the upper right corner of the graph changes to red to warn that the display is probably *not* appropriate to the listed parameters. Updating the data with CALCULATE, of course, deletes any existing data.

5.4.3   Van Hove singularities*

If you DOUBLE MC STEPS enough times you should be convinced that the discontinuities in slope, the cusps, in the $g(\omega)$ are real. These *Van Hove singularities* are characteristic of DOS functions for phonons, electrons, and other excitations such as magnons and excitons. They occur at the frequencies (or energies) of the extrema and saddle points of the dispersion relations $\omega(\mathbf{q})$.

Exercise 5.22 (M) *Identify as many as you can of the frequencies of the features in the $g(\omega)$ plot on the right with the maxima, minima, and saddle points of dispersion relations shown in the graph on the left: e.g., which singularity in the DOS is to be associated with the maximum of the transverse mode in the (111) direction, etc. (Using the (x,y)-readout of the mouse cursor gives the easiest way to get quantitative values for graphical coordinates.) Print a copy of the DOS graph and note on it the frequencies of the singularities and their identifications to use later.*

Figure 5.3 shows the dispersion relation for a one-dimensional chain. Two equal frequency ranges $d\omega$ are shown, one at low ω and the other at high. The DOS at each frequency is proportional to the number of allowed modes or states in the frequency range $d\omega$. Remember that the values of q permitted by the periodic boundary conditions are on a fine grid uniformly spaced in q at intervals of $2\pi/Na$. The DOS in frequency requires converting the known number of states per unit range of wave vector into the number of states per unit range of frequency.

Exercise 5.23 (C) *Argue from Figure 5.3 that the DOS will be larger at the higher frequency than at the lower: i.e., a small slope of the $\omega(q)$ curve implies a high DOS.*

Exercise 5.24 (M*) *Show quantitatively that the dispersion relation $\omega(q) = 2\sqrt{K/M}\sin(qa/2)$ for the one-dimensional chain of atoms results in a DOS $g(\omega)$ given by*

$$g(\omega) = \frac{2N}{\pi} \frac{1}{\sqrt{\omega_c^2 - \omega^2}} \qquad \text{with} \qquad \omega_c = 2\sqrt{\frac{K}{M}}. \qquad (5.2)$$

Be sure to express the result in terms of ω, not q, and don't forget to include both the right hand and the left hand pieces of the dispersion curve. (Hint: require that the number of states within a range dq be the same as the number of states within a corresponding range of $d\omega$.)

Note that the DOS becomes infinite at the *cut-off frequency* ω_c. Can you see why this is so from a qualitative argument? This is an example of a Van Hove singularity for a one-dimensional system. You might be concerned that the divergence must be physically unrealistic. As long as it is an integrable singularity this poses no problem.

Exercise 5.25 (M*) *Show that the singularity for the one-dimensional DOS $g(\omega)$ is integrable and verify that the total number of modes, given by $(\int g(\omega)d\omega)$, is N as it should be.*

You will note that the singularities in the DOS computed by "debye" are milder than that of the one-dimensional case: discontinuities or infinities in the slopes of $g(\omega)$ rather than a divergence of $g(\omega)$ itself.

Exercise 5.26 (M**) *Consider a maximum in a three-dimensional dispersion relation for which $\omega(\mathbf{q})$ has the Taylor series expansion*

$$\omega(\mathbf{q}) = \omega_0 - A(q_x^2 + q_y^2 + q_z^2) = \omega_0 - A|\mathbf{q}|^2. \qquad (5.3)$$

Show that the DOS near this maximum of the dispersion relation may be expressed as $g(\omega) = (L^3/4\pi^2 A^{3/2})\sqrt{\omega_0 - \omega}$.

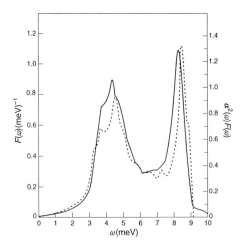

Figure 5.4: Density of modes for Pb (from reference [29] with permission from Oxford University Press.) The dashed curve, $F(\omega)$, is computed from neutron scattering. The solid curve, $\alpha^2(\omega)F(\omega)$, is from tunneling experiments.

5.4.4   Superconducting tunneling**

We saw that the calculated dispersion relations could be compared with neutron scattering data. Having worked out the DOS, are there experiments with which to compare the DOS curves? The superconducting tunneling experiment gives a fairly direct measure of the shape of the phonon DOS curve. A device is constructed using two different superconductors separated by a thin insulating layer, typically an oxide of one of the metals. The tunnel current in these devices includes both a purely electronic contribution, and an additional contribution from processes in which the electron, as it tunnels from one metal to the other, also excites a phonon in one of the metals. This process will be possible only if the electron has sufficient energy to excite the phonon. Thus the tunnel current as a function of voltage gives information about the number of phonon modes that are available to participate in these *inelastic tunneling* events. The solid curve in Figure 5.4 gives, apart from a proportionality factor, the density of phonon states in Pb determined in this fashion. The dashed curve is computed, much in the spirit of what "debye" has done, from dispersion curves determined by neutron scattering.

Exercise 5.27 (M*) *Compare the "debye" graph of the calculated DOS for Pb with the results of the tunneling experiment shown in Figure 5.4. You will not be able to match successfully the frequencies of the cusps in your DOS with those of these tunneling data. Recall how poorly the calculation fitted the neutron data. You should, however, be able to use the identifications you made in Exercise 5.22 to label the singularities*

evident in Figure 5.4. A semiempirical scheme for developing a more
sophisticated model than the one in "debye" is to use fits to tunneling
data as the basis for adjustment of the force constants in a model with a
larger number of free parameters.

5.4.5 Debye density of states

Now, this numerical evaluation of the DOS takes a substantial computing
effort, and required considerably more effort in the 1920s before workstations were invented. The Einstein and Debye models were developed to
give a basic understanding of the physics of the heat capacity, without
involving a detailed calculation of all of the normal mode frequencies. In
the *Einstein model*, the atom motion was modeled as harmonic motion
of each atom about its lattice site without regard to the motion of its
neighbors. The resulting DOS is a delta function at the single frequency
assigned to this motion, which is the same for all of the atoms. The *Debye
model*, as well as the numerical one we've been discussing, is based on a
knowledge of the velocities of sound, the slopes of the phonon dispersion
curves at low frequency. Its simplest version assumes the same linear dis-
persion relation $\omega(\mathbf{q}) = \mathbf{v_s}|\mathbf{q}|$, for all three polarizations, up to a cut-off
frequency ω_D defined to give the correct total number of modes. If you
click the DEBYE toggle button between the graphs, the DOS in the Debye
model, appropriate to the chosen lattice constant a and velocity of sound
v_s, will be added to the graph. The velocity of sound v_s is used *only* as
a fitting parameter for the Debye model and is unrelated in the "debye"
program to choices made for the atomic MASS and FORCE PARAMETER.
The v_s SLIDER may be changed without corrupting the set of numerical
data.

Exercise 5.28 (C) *Adjust the v_s SLIDER to get a good match to the
numerical DOS for small values of q. This is easier to do if you display
the $g(\omega)$s on a LOG/LOG plot using the LINEAR–LOG/LOG toggle above the
graph. Verify from the LOG/LOG plots that the limiting behavior of $g(\omega)$
for both curves is ω^2. (Don't be concerned if the LOG/LOG plot seems out
of control at the low frequency end. The fractional uncertainties become
very large for the lowest frequency bins, and the density for a bin with
zero samples is plotted arbitrarily as 1.)*

   Though we can get a good fit to the numerical results at low T, the
Debye approximation ignores a lot of the complex structure of the true
DOS. It would clearly have little relevance, for example, to a discussion
of the tunneling data just presented. Let's see whether it's adequate for
describing the temperature dependence of the heat capacity.

5.5   Thermal energy

Knowledge of the DOS allows us to work out the internal energy associated with the thermal motion of the atoms, and from this we can obtain the heat capacity. Key to this development is the relationship, from the *Planck distribution*, giving the average energy $\mathcal{E}(\omega,T)$ associated with a normal mode of frequency ω in equilibrium with a heat bath of temperature T,

$$\mathcal{E}(\omega,T) = \hbar\omega \left(\frac{1}{e^{\hbar\omega/k_B T} - 1} + \frac{1}{2} \right) \equiv \mathcal{E}_T(\omega,T) + \frac{1}{2}\hbar\omega. \qquad (5.4)$$

Our concern is only with $\mathcal{E}_T(\omega,T)$, the thermal contribution to the energy of the mode: we set aside the zero point energy.

Exercise 5.29 (M) *Verify that $\mathcal{E}_T(\omega,T) \to kT$, independent of ω, at high temperature as implied by the classical equipartition theorem. Show that the thermal contribution $\mathcal{E}_T(\omega,T)$ to the mode energy becomes exponentially small at low temperature. What are the criteria for T that define the high and low temperature limits, i.e., high and low compared with what?*

To get the internal energy of the full crystal, we must sum the contributions from all of the modes. The internal energy dU associated with the modes in the frequency range $d\omega$ is the product of the thermal energy associated with each mode, and the number of modes in the frequency range, $\mathcal{E}_T(\omega,T)g(\omega)d\omega$. Open PRESET 5, click on CALCULATE, and DOUBLE MC STEPS a few times to get good statistics. The right hand plot shows a family of ten plots, for different values of T, of the product $\mathcal{E}_T(\omega,T)g(\omega)$. For a given T, adding up all of the contributions, the area under the plot, gives the total internal thermal energy $U(T)$.

5

Exercise 5.30 (C) *Explain qualitatively the behavior of the family of curves. It may be helpful to put side by side, using COPY GRAPH, graphs for a low T_{max} and for a high T_{max}: possibly 10 K and 100 K. Explain why, at the high temperatures, the curves for different Ts have the same shape, except for a vertical scale factor, as the DOS, which can be accessed from the right hand SHOW menu. Why are the curves in the low temperature family so distorted from the shape of the DOS curve?*

Exercise 5.31 (M*) *Use the toggles to replace the SIMULATION data with the DEBYE data and choose a moderate maximum temperature (30 K is convenient) to make a similar set of curves from the Debye DOS. For each temperature T find the frequency $\omega_{max}(T)$ at which the energy density has a maximum and make a plot of $\omega_{max}(T)$ versus T. Does your plot fit the Wien displacement law [19] which relates the peak of the* *spectral intensity of black-body radiation to the source temperature by* $\omega_{\max}(T) = AT$, *where A is a constant? Why does this relation fail for the internal energy of a solid if you go to too high a temperature, but not if you're speaking of black-body radiation? Don't forget to* ZOOM *to see the lower temperature plots.*

Exercise 5.32 (M**) *Construct a formal scaling argument based on the forms of* $\mathcal{E}_T(\omega, T)$ *and the Debye* $g(\omega)$ *to validate the Wien displacement law for low temperature and to show at what temperature it should begin to fail.*

5.6   Heat capacity

We've noted that the area under the curves in the thermal energy plots gives the thermal contribution to the internal energy of the crystal.

Exercise 5.33 (C) *Explain why it is appropriate to calculate the heat capacity as the area between two successive curves of the internal energy graph divided by the difference in the two associated temperatures.*

5.6.1   Lead

You have just explained in Exercise 5.33 how to calculate the heat capacity. "debye" has followed your instructions and in the right hand graph in PRESET 6 plots the heat capacity as determined from the SIMULATION.   # 6
The formal statement may be written as

$$C_V(T) = \frac{1}{V}\frac{\partial U}{\partial T} = \frac{1}{V}\frac{\partial}{\partial T}\int d\omega\, g(\omega)\,\mathcal{E}_T(\omega, T)$$

$$= \frac{1}{V}\frac{\partial}{\partial T}\int d\omega\, g(\omega)\,\frac{\hbar\omega}{e^{\hbar\omega/k_B T} - 1}. \qquad (5.5)$$

"debye" has chosen to calculate the heat capacity per unit volume, hence the normalization by $1/V$. In the graph the SIMULATION is COMPARED WITH data for LEAD [31] and a curve derived from the DEBYE model. A velocity v_s was chosen which gave a good low frequency match of the Debye DOS to the SIMULATION DOS.

Exercise 5.34 (C) *Compare the predictions of the two models for the heat capacity with the experimental data. Do the predictions give good fits? You will get a more severe test if you display the results using a* LOG/LOG *graph. Note that by adjusting the sound velocity* v_s *in the Debye model you can get a good fit either with the experimental data or with the simulation, though the choices of velocity differ according to the type of plot you happen to choose for comparison. If you wanted to spend* the time, you could bring the numerical result into better agreement with the experiment by making small changes in B. (*Don't try it unless you're prepared to wait for the lengthy calculation of the DOS!*)

Exercise 5.35 (M) *Determine the temperature dependence of the* HEAT CAPACITY *in the low temperature limit, using the* LOG/LOG *choice for the* AXES *of the graph.*

Exercise 5.36 (M**) *Without explicit evaluation of the integral for the heat capacity in Eq. (5.5), derive the temperature dependence of the heat capacity in both the high and low temperature limits. For the low temperature limit, use the knowledge that the DOS $g(\omega)$ varies as ω^2 at small ω. (Hint: introduce dimensionless variables within the integral so that the integral simply becomes an undetermined constant of proportionality in the expression for the heat capacity. This can be done only in the low and high temperature limits.)*

Exercise 5.37 (C*) *Explain why the Debye model, for which the DOS is so far from reality, can give, with a suitable choice of the velocity, a prediction for the heat capacity which is so nearly identical to that of the numerical model?*

Exercise 5.38 (M**) *How does the velocity of sound v_s which gives a good heat capacity fit for the Debye model compare with the measured sound velocities in Pb from Table 5.2? Compare v_s with both the average of the measured velocities $\langle v_M \rangle$ and the inverse cube root of the average inverse velocity cubed $\langle v_M^{-3} \rangle^{-1/3}$. Although in principle this should be an average over all directions, using just the three velocities for the $(1\,1\,0)$ direction should give a rough value. Give a theoretical rationale for the second, apparently obscure, choice. (Hint: focus your attention on what determines the low frequency limit of the DOS.)*

Exercise 5.39 (M*) *Suppose you have two (imaginary) fcc crystals in which the* FORCE PARAMETERS B *are identical, in which the atoms have the same* MASSES, *but in which the* LATTICE CONSTANTS *differ by a factor of 2.*

1. *What will be the ratio of the high temperature heat capacities per unit volume and which will be greater?*
2. *What will be the ratio of the low temperature heat capacities per unit volume and which will be greater?*
3. *What will be the ratio of the temperatures at which the heat capacities reach half of their high temperature limit and which will be higher?*

*Use "debye" to check your predictions. (*COPY GRAPH *will be useful to preserve one plot while making the second.)*

Exercise 5.40 (M*) *Repeat Exercise 5.39 for the case in which the two crystals have the same* FORCE PARAMETER *B, the same* LATTICE CONSTANT, *but atomic* MASSES *which differ by a factor of 2.*

Exercise 5.41 (M*) *Repeat Exercise 5.39 for the case in which the two crystals have the same* LATTICE CONSTANT, *the same atomic* MASS, *but* FORCE PARAMETERS *B which differ by a factor of 2.*

Exercise 5.42 (C) *Predict (without calculations!) from the constants given in Table 5.2, which of Pb and Cu will have the higher Debye temperature. Is this prediction consistent with the experimental* HEAT CAPACITY *data?*

5.6.2 Copper and aluminum

Data are included in PRESETS 7 and 8 for Cu and Al. The presets compare the Debye model with the experimental results. If you wish to see the numerical results as well, CALCULATE after opening the PRESET to have the DOS and heat capacity updated to the new parameter set. You may also wish to DOUBLE MC STEPS once or twice to improve the statistics. Use the LOG/LOG graph to COMPARE the heat capacities from experiments on COPPER and ALUMINUM with the DEBYE model. The departure of the heat capacity from the T^3 law at the lowest temperatures is striking for both of these metals.

7,8

Exercise 5.43 (M*) *Determine the power law of the lowest temperature experimental data for aluminum. Ignore the point at the lowest temperature.*

What's going on? "debye" has considered only the energy associated with the motion of the atoms, or the *phonon heat capacity*. But Pb, Cu, and Al are metals with electrons which are, in some sense, free to move throughout the material. Isn't there a contribution to the internal energy, and hence heat capacity, from the kinetic energy of motion of the electrons? Indeed there is, and this is the source of the extra heat capacity seen for Cu and Al. Theory [1, 6] predicts a linear temperature dependence for the *electronic heat capacity*.

Now the puzzle is the absence of such a contribution from the conduction electrons in the Pb. For Pb we appeal to the effect of the superconducting transition at about 7 K, which suppresses the electronic heat capacity leaving the phonon term the dominant contribution at all temperatures. A final mystery is the point at the lowest temperature, 1.1 K, for Al. This point, just below the superconducting transition for Al at 1.2 K, shows a contribution from the superconducting heat capacity anomaly. Data extending to lower T would show the heat capacity returning rapidly to the extrapolation of the phonon T^3 term. The corresponding anomaly for Pb is too small to be observed in the data shown.

5.7   Summary

The term *phonon* pervades much of the solid state physics literature. Many properties of solid systems are strongly influenced by thermal motion of the atoms, by the deviations from perfect periodic order. "debye" works through a simple model for real systems to illustrate some of the language and methods used to treat the physics of the thermal motion.

Detailed solution of the equations of motions gives the dispersion relations $\omega_p(\mathbf{q})$ for the different branches, results which can be compared with experimental neutron scattering data. To address many issues, it is sufficient to know the number of normal modes with frequencies in a given frequency range, the density of modes or density of phonon states, the DOS. "debye" uses a Monte–Carlo technique to determine the DOS which, for a limited set of materials, may be compared with the DOS as determined from superconducting tunneling experiments. Before the advent of the neutron scattering technique, the classic experiment for testing models of the thermal motion in solids was the heat capacity experiment. The heat capacity is calculated straightforwardly once the DOS is known.

The numerical technique illustrated by "debye" is cumbersome, however, for making rough theoretical estimates in a variety of contexts. The Debye model, which uses an extrapolation of continuum elastic theory to high frequencies to approximate the dispersion relation, is often adequate. As "debye" has shown, the Debye theory explains the heat capacity very well. For other properties, ones which are more sensitive than the heat capacity to details of the DOS function, the model is, of course, less useful.

Finally, in comparing experimental heat capacity data with the predictions of "debye", an anomaly was observed at the lowest temperatures, which the theory was at a loss to explain. Its resolution requires the development of models with which to describe the electronic degrees of freedom of the crystal.

5.A   Deeper exploration

7,8   *Electronic heat capacity*   PRESETS 7 and 8 show a heat capacity at low temperature which exceeds the prediction for the phonon heat capacity. Analyze carefully experimental data taken from the graphs to determine the temperature dependence of the excess and its magnitude at some temperature. Compare your results with expectations [1, 6] for both aluminum and copper.

Debye model In this chapter the velocity of sound parameter v_s has been treated more or less as a fitting parameter for the Debye model. Use the information from Tables 5.2 and 5.3 to calculate velocities in all three symmetry directions. Deduce from these the value of v_s which should be used in the Debye model. In addition to the question of the appropriate power of v_s to average, see Exercise 5.38, there is the question of appropriate weighting of different directions in **q**. How does your choice of v_s compare with the velocity for which "debye" gives the best low temperature fit to the data?

Black-body radiation Read up on the nature of black-body radiation [19] and use the low temperature limit of the Debye model, as simulated by "debye" to illustrate important ideas.

5.B "debye" – the program

"debye" computes the phonon dispersion relation, the phonon density of states, the heat capacity, and the spectral density of the internal energy for fcc crystals. The model underlying the algorithm can be chosen to be either a lattice dynamics model or the Debye model.

5.B.1 Lattice dynamics model

The lattice dynamics model of "debye" is based on a harmonic interatomic potential with one free parameter, the force constant of the nearest neighbor interactions. We give here a short summary of this model.[1]

Take as a starting point an fcc crystal in which every atom undergoes small oscillations around its equilibrium position. Given a harmonic interatomic potential, we seek plane-wave harmonic (sinusoidal) solutions to the equations of motion of the atoms. The sinusoidal time dependence is determined by the phonon frequency ω which is a function of the phonon wave vector **q**. As we will see below, the phonon frequency ω of each mode is closely related to the eigenvalues of the *dynamical matrix* $\mathcal{D}(\mathbf{q})$, which in turn depends on the atomic arrangement of the crystal and the interatomic potential.

The total harmonic potential of a three-dimensional crystal can be written in terms of the displacement vector **u** of each atom from its equilibrium position **R** as

$$U^{\text{harm}} = \tfrac{1}{2} \sum_{\mathbf{RR}'} \mathbf{u}(\mathbf{R}) \cdot \mathcal{D}(\mathbf{R} - \mathbf{R}') \cdot \mathbf{u}(\mathbf{R}'), \tag{5.6}$$

---

[1]For a more comprehensive description see Ashcroft & Mermin [1], Chapter 22 and especially Problem 22.5.

where the sum extends over all pairs of atoms located at $\mathbf{R}$ and $\mathbf{R}'$. The quadratic matrix $\mathcal{D}(\mathbf{R} - \mathbf{R}')$ can be specified in terms of the interatomic pair potential $\Phi(\mathbf{R} - \mathbf{R}')$.

From the potential (5.6) it then follows that the motion of each atom of mass M is determined by

$$M\ddot{\mathbf{u}}(\mathbf{R}) = -\sum_{\mathbf{R}'} \mathcal{D}(\mathbf{R} - \mathbf{R}') \cdot \mathbf{u}(\mathbf{R}'). \tag{5.7}$$

The equation of motion (5.7) has solutions of the form

$$\mathbf{u}(\mathbf{R}, t) = \boldsymbol{\epsilon}_p e^{i[\mathbf{q}\cdot\mathbf{R} - \omega_p(\mathbf{q})t]}, \tag{5.8}$$

where $\boldsymbol{\epsilon}_p$ ($p = 1, 2, 3$) is one of the three polarization vectors and $\omega_p(\mathbf{q})$ is the phonon frequency associated with polarization p and wave vector $\mathbf{q}$. As it turns out, the polarization vectors and their associated phonon frequencies can be determined from the dynamical matrix $\mathcal{D}(\mathbf{q})$,

$$\mathcal{D}(\mathbf{q}) = \sum_{\mathbf{R}} \mathcal{D}(\mathbf{R}) e^{-i\mathbf{q}\cdot\mathbf{R}}. \tag{5.9}$$

The three polarization vectors $\boldsymbol{\epsilon}_p$ are the three real eigenvectors of $\mathcal{D}(\mathbf{q})$ and the angular frequency ω_p is given in terms of the corresponding eigenvalue λ_p,

$$\omega_p(\mathbf{q}) = \sqrt{\lambda_p(\mathbf{q})/M}. \tag{5.10}$$

If we assume that the interatomic pair potential Φ contributes to the sum (5.6) only for nearest neighbors then (5.9) reduces to a sum over the 12 nearest neighbors $\mathbf{R}$ in an fcc crystal,

$$\mathcal{D}(\mathbf{q}) = B \sum_{\mathbf{R}} \sin^2(\tfrac{1}{2}\mathbf{q} \cdot \mathbf{R}) \, \hat{\mathbf{R}}\hat{\mathbf{R}}, \tag{5.11}$$

where $\hat{\mathbf{R}}\hat{\mathbf{R}}$ is the dyadic (the outer product) $(\hat{\mathbf{R}}\hat{\mathbf{R}})_{ij} = \hat{\mathbf{R}}_i\hat{\mathbf{R}}_j$ of the unit vectors $\hat{\mathbf{R}} = \mathbf{R}/R$. The force parameter B is given in terms of the interatomic potential Φ as $B = 2\Phi''(d)$, where d is the equilibrium nearest neighbor distance ($d = a/\sqrt{2}$) for an fcc crystal.

5.B.2 Lattice dynamics simulation

The (normalized) phonon DOS $g(\omega)$ is proportional to the (unnormalized) number $N(\omega)$ of phonon frequencies ω in the frequency interval ω to $\omega + \Delta\omega$, with $\Delta\omega$ defined below. To calculate $N(\omega)$ we have to sample all of reciprocal space evenly (or, see below, a part of it which is equivalent to the whole space), determine the dynamical matrix $\mathcal{D}$ in Eq. (5.11) for each of the sampled wave vectors $\mathbf{q}$, and then compute the three phonon frequencies $\omega_p(\mathbf{q})$ from the three eigenvalues according to Eq. (5.10). To determine the eigenvalues λ_p of the dynamical matrix $\mathcal{D}(q)$, for any phonon wave vector $\mathbf{q}$, "debye" computes the characteristic polynomial of $\mathcal{D}(q)$ and finds its three real roots.[2]

Instead of sampling all of reciprocal space, "debye" chooses the positive octant $(0 < q_x, q_y, q_z < 2\pi/a)$ of a cubic (bcc) unit cell centered at the origin. Sampling $\mathbf{q}$ in the positive octant is equivalent to sampling it in all of reciprocal space (see also Exercise 5.18): it suffices to pick a unit cell because of translational symmetry. Because of the cubic symmetry of the bcc reciprocal lattice, only one octant of the unit cell is needed. The remaining octants have dispersion relations which can be inferred from those of the chosen octant using the cubic symmetry, and thus have the same phonon spectrum associated with them.

A simple Monte–Carlo method is then used to compute the (unnormalized) number of phonon frequencies $N(\omega)$ in a given frequency interval:

1. To determine the frequency range 0–$\omega_{\max}$, "debye" samples 100 random $\mathbf{q}$-vectors in the positive octant of the unit cell centered at the origin, computes for each $\mathbf{q}$ the three associated phonon frequencies ω_p, and finds from these 300 values the maximum, $\max(\omega_p)$. The frequency $\omega_{\max} = 1.25 \max(\omega_p)$ is then used as an estimate for an upper limit of the frequency range. Any frequency above $\omega_{\max}$ is disregarded in the simulation described in step 3.

2. "debye" then divides the frequency range 0–$\omega_{\max}$ into NUMBER OF BINS bins of size $\Delta\omega = \omega_{\max}/\text{NUMBER OF BINS}$.

3. "debye" samples a new set of NUMBER OF MC POINTS (N_{MC}) random $\mathbf{q}$-vectors in the positive octant of the unit cell, computes the associated phonon frequencies ω_p, and sorts these phonon frequencies into the appropriate bins. The number of phonon frequencies in each bin, $N(\omega)$, corresponds to the number of phonon frequencies in the interval ω to $\omega + \Delta\omega$. Any ω larger than $\omega_{\max}$ is disregarded.

Once the number $N(\omega)$ of frequencies in the interval ω to $\omega + \Delta\omega$ is computed, the DOS is given by

$$g(\omega) = \frac{N(\omega)}{N_{MC}\Delta\omega}\frac{4}{a^3}V. \tag{5.12}$$

Here a is the lattice constant of the fcc crystal whose atomic density, $n = N/V$, with four atoms per cubic unit cell is given by $n = 4/a^3$. (It is straightforward to check that this DOS (5.12) satisfies the normalization condition $\int_0^\infty d\omega g(\omega) = 3N$, where $3N$ is the number of phonon modes in a crystal with N atoms.)

---

[2] Note that $\mathcal{D}$ in Eq. (5.11) is a real and symmetric matrix. Therefore it can be diagonalized and has indeed three real eigenvalues.

5.B.3 Debye model

Instead of calculating the DOS with the lattice dynamics simulation described above, we can make a crude approximation (known as the Debye model) by assuming the same linear dispersion relation for each of the three polarization directions, $\omega(\mathbf{q}) = v_s q$. Here v_s is the Debye velocity of sound.[3] This approximation results in a particularly simple DOS,

$$g_d(\omega) = \begin{cases} (3V/2\pi^2)\,(\omega^2/v_s^3) & \text{if } \omega < \omega_d, \\ 0 & \text{otherwise,} \end{cases} \tag{5.13}$$

where the factor of 3 arises from the three directions of polarization. Normalization requires that the number of modes in a crystal with N atoms is $3N$, $\int_0^{\omega_d} d\omega\, g_d(\omega) = 3N$, which determines the cut-off frequency ω_d. In an fcc crystal with density $n = N/V = 4/a^3$ we thus find

$$\omega_d = \sqrt[3]{24\pi^2}\, v_s/a. \tag{5.14}$$

5.B.4 Displayed quantities

The left hand display in "debye" shows the dispersion relation either along the three SYMMETRY DIRECTIONS $(1\,0\,0)$, $(1\,1\,0)$, and $(1\,1\,1)$, or along an ARBITRARY DIRECTION which can be specified on the panel. The right hand display offers the choice of the DENSITY OF STATES, the spectral density of the INTERNAL ENERGY, or the HEAT CAPACITY. Quantities computed from the lattice dynamics SIMULATION are displayed in blue, quantities computed from the DEBYE model are displayed in red, and experimental neutron diffraction data as a series of points. The wave vector for the experimental data is scaled to whatever lattice constant happens to be chosen in the panel: i.e., the units for the wave vector axis are appropriately thought of as π/a rather than $1/\text{Å}$. The neutron data are available only for the $(1\,0\,0)$, $(1\,1\,0)$, and $(1\,1\,1)$ directions.

The computation of the dispersion relation requires little numerical effort, and "debye" updates the dispersion relation immediately whenever any relevant quantity (force parameter B, LATTICE CONSTANT, or MASS) has changed. Good statistics for the DENSITY OF MODES, HEAT CAPACITY, and INTERNAL ENERGY based on the lattice dynamics SIMULATION, however, take considerable computational effort and time, and

---

[3]If there are two transverse modes with velocity v_T and $v_{T'}$, and one longitudinal mode with velocity v_L then the Debye velocity of sound is given as

$$1/v_s^3 = (1/v_T^3 + 1/v_{T'}^3 + 1/v_L^3)/3.$$

these quantities are only updated when the CALCULATE button is clicked or the SPACEBAR is pressed. The right hand graph shows the number of MC STEPS in black as long as the panel settings correspond to the current display. The color changes to red as a warning flag, when any relevant physical parameter has been changed but the DOS has not yet been updated with a CALCULATE. If the velocity of sound v_s is changed, however, the DENSITY OF MODES, HEAT CAPACITY, and INTERNAL ENERGY based on the (simpler) DEBYE model are updated immediately, and the simulation results remain valid since they do not depend on the choice of the velocity of sound.

Dispersion relation The dispersion relation $\omega_p(\mathbf{q})$ in "debye" is computed according to the lattice dynamics model, Eq. (5.10), described in Section 5.B.2. Given a direction $\hat{\mathbf{q}}$ in reciprocal space, "debye" determines $\omega_p(\mathbf{q})$ for 100 evenly spaced values of the wave vector $\mathbf{q} = \mu\hat{\mathbf{q}}$, with $\mu = 0$, ..., $2\sqrt{2}\pi/a$.[4] The dispersion relation can be compared with data from neutron scattering experiments [30] for Al (green), Cu (orange), and Pb (black).

Density of states The DENSITY OF STATES for the Monte–Carlo SIMULATION of the lattice dynamics model is computed according to Eq. (5.12), as described in Section 5.B.2, and displayed in blue. The NUMBER OF BINS determines the frequency intervals in which the computed frequencies ω are sorted, and the NUMBER OF MC POINTS (see the CONFIGURE menu) determines how many $\mathbf{q}$-vectors are sampled during the initial Monte–Carlo calculation. The accuracy of the DOS (its statistics) can be improved by pressing DOUBLE MC STEPS. This will run the Monte–Carlo algorithm again with the same parameters, adding as many new points as already accumulated in previous runs. (If you DOUBLE MC STEPS after any relevant parameter such as the force parameter B has changed since the last CALCULATE, you will be warned by the red MC STEPS display that the results are not meaningful.) The total accumulated number of MC STEPS is shown in the right hand graph. The DENSITY OF STATES for the DEBYE model is computed according to Eq. (5.13) with the cut-off frequency given by Eq. (5.14). It is displayed in red.

---

[4]The maximum value $2\sqrt{2}\pi/a$ is the distance from the Γ point (the origin of reciprocal space) to the X point in the *second* Brillouin zone (see Exercise 5.4). This is according to the convention that in the (1 1 0) direction the dispersion relation is specified not only within the first Brillouin zone (from the Γ to the K point) but also in parts of the second Brillouin zone (up to the X point). The maximum distance from the origin within the first Brillouin zone is only $\sqrt{5}\pi/a$ (from the Γ to the W point). (Recall from Exercise 5.1 that the bcc unit cell in reciprocal space has a cubic side length $4\pi/a$ if the fcc unit cell in real space is taken to have a side length a.)

Internal energy   The thermal contribution to the INTERNAL ENERGY per unit frequency interval (energy density) of a crystal is given by

$$U(\omega, T) = g(\omega)\, \mathcal{E}_T(\omega, T) = g(\omega)\, \hbar\omega\, \frac{1}{e^{\hbar\omega/kT} - 1}, \qquad (5.15)$$

see Eq. (5.4). "debye" computes the spectral density of the internal energy $U(\omega, T)$ for both the lattice dynamics SIMULATION (displayed in blue) and the DEBYE model (displayed in red). That is, the DOS $g(\omega)$ in Eq. (5.15) is taken either from Eq. (5.12) or from Eq. (5.13). The internal energy is calculated at NUMBER OF BINS frequencies in the interval 0–ω_{max} for 10 different temperatures equally spaced from $0.1\, T_{max}$ to T_{max}. The frequency ω_{max} is determined as described in Section 5.B.2 and the maximum temperature TMAX can be specified.

Heat capacity   The total thermal energy of a crystal is the integral (or sum) over all frequencies ω of the internal energy density $U(\omega, T)$ in Eq. (5.15), $U(T) = \int_0^\infty d\omega U(\omega, T)$. The heat capacity per unit volume C_V then follows from the temperature derivative of the internal energy $U(T)$, as given in Eq. (5.5). "debye" computes the heat capacity for both the lattice dynamics model (displayed in blue) and the Debye model (red): i.e., the DOS in Eq. (5.5) is given by either Eq. (5.12) or Eq. (5.13). The integral over ω in Eq. (5.5) is converted into a sum over NUMBER OF BINS values of ω. The heat capacity is then plotted at 100 equally spaced temperatures T from $0.01\, T_{max}$ to T_{max}. These heat capacities can be compared with data [30, 31] for Al (green), Cu (orange), or Pb (black).

5.B.5   Bugs, problems, and solutions

Poor statistics at small frequencies ω can give meaningless fluctuations in the LOG/LOG plots at the lowest temperatures. Also, if the DOS, internal energy, or heat capacity has the value zero at some temperature the value will be plotted as 1 in the LOG/LOG plots.

6

"drude" – Dynamics of the classical free electron gas

Contents

6.1   Introduction

The simplest description of electrical conduction in metals and semiconductors is that given by the Drude model. The electrical current is assumed to be carried by a classical gas of electrons which do not interact with one another, but which have their velocities randomized by scattering events characterized by a mean time between collisions τ. Initially proposed as a description of metallic conduction, the Drude model was only partially successful, but it still provides the language we usually use in speaking of electron transport in semiconductors.

"drude" simulates this model for a small number of electrons (typically 32) moving in two dimensions. An electric field, either DC or AC at a frequency that can be specified, may be applied in the plane of the electron motion and a magnetic field may be applied perpendicular to that plane. Each electron has a probability dt/τ of scattering in the time interval dt. In the scattering event, memory of the velocity before the scattering is completely lost, the position of the electron is unchanged, and a new velocity assigned which is consistent with a Maxwellian distribution at temperature T. The motion of the individual electrons during the time between collisions is computed numerically from Newton's law, using as initial conditions the velocity and position of each electron after its most recent scattering event.

6.2   No applied fields

6.2.1   Velocity and real space

We're used to thinking of motion in terms of trajectories of particles in real space, as displayed in the left hand panel of "drude". In problems in solid state physics we usually direct our interest to velocity or momentum (later reciprocal) space. In electrical conduction it is the average *velocity* of the electrons that is the issue; the density of electrons in real space remains uniform and tells us nothing. RUN "drude" using PRESET 1 to become acquainted with the two displays.

1

The display on the right, the more important of the two, is of velocity space. The dots give the velocities of the individual electrons, the large red dot is the origin and the large green dot the average velocity of the electrons. On the left is the real-space display. Initially the electrons are in a cluster near the origin of real space to enable the visualization of the spreading of the electrons due to diffusion and of the average motion of a selected set of electrons. A few electrons are colored to aid in correlating the trajectories of particular electrons in the two spaces. Several successive positions of the electrons are displayed to give a sense of the speed and direction of the moving points in both spaces.

Exercise 6.1 (C) *Why do the electrons move smoothly in real space but remain stationary in velocity space, except for occasional jumps from one point to another? Pick out one electron (by color) and check that the change in direction and speed of its motion in real space correspond correctly to its change in location in velocity space.*

One essential feature of the Drude model is the randomization of the velocities with each scattering event. "drude" performs this by distributing the velocities after the scattering event according to a Maxwellian distribution with temperature T. Watch the velocity display as you change the TEMPERATURE with the slider.

Exercise 6.2 (M) *Click on* SHOW GRAPH *over the velocity display. This graph gives the mean square* DEVIATION *of the electron velocities from their mean (of zero for this case).* RUN *briefly at a few* TEMPERATURES *from 10 K to 1000 K and determine $\langle v^2 \rangle$ for each temperature from the graph. Make a log/log plot of your data and determine the power n in the relation $\langle v^2 \rangle \propto T^n$.*

The Maxwellian probability distribution $P(\mathbf{v})$ for finding the electron at velocity $\mathbf{v} = v_x \hat{\mathbf{x}} + v_y \hat{\mathbf{y}}$ is

$$P(v_x, v_y) = \frac{m}{2\pi k_B T} e^{-m \frac{v_x^2 + v_y^2}{2k_B T}}. \tag{6.1}$$

The rms value of either component of the velocity is $\sqrt{kT/m}$. Are your data consistent with this relation?

Exercise 6.3 (M) *Verification of the T dependence is one thing: but is the numerical coefficient correct? (Values of the fundamental constants are available from the SSS* HELP.)

6.2.2 Diffusion

Though we've suggested that velocity space is more important for most problems than real space, the real-space display does give a convenient demonstration of diffusion. A critical parameter in the Drude model is the *scattering time* τ. Alternatively we often speak of a *mean free path* λ, the distance an electron typically moves during one scattering time. If you watch the real-space simulation, you can estimate λ roughly as the typical distance between kinks in the real-space trajectories.

Exercise 6.4 (M) *From the real-space display, estimate the mean free path λ for the parameters of* PRESET 1. *How does the ratio λ/τ compare* with the thermal velocity $v_T \equiv \sqrt{2kT/m}$? (Note that the numerical coefficient 2 is appropriate for our two-dimensional model. More generally it is the dimensionality of the space in which the particles move.) Drag first the SCATTERING TIME τ and then the TEMPERATURE T SLIDER to check out the qualitative dependence of λ on these two parameters.

A consequence of the random electron velocities is that electrons which start together wander off in all directions and gradually disperse. This dispersion of the electrons with time is described by the diffusion equation and can be of importance in devices in which the spatial distribution of electrons is important and where rapid time response is an issue.

Exercise 6.5 (C) *With* PRESET 1, *use the* SHOW GRAPH *toggle over the left hand window to open a graph which gives the mean square deviations of the electron x- and y-positions from their means. As the system* RUNS, *the diffusive motion of the electrons gradually increases their dispersion in space. On average this mean square deviation increases linearly with time.* COPY GRAPH *to save the result and* RUN *again with a* SCATTERING TIME τ *chosen to be larger by a factor of 3.* STEAL DATA *for comparison with the first. Reduce the* TEMPERATURE *by a factor of 3 and repeat again. Why does the rate of dispersion of the electrons increase with increasing T and with increasing τ?*

If D is the diffusion constant, an approximate solution of the diffusion equation in two dimensions gives the spatial distribution of the electrons at time t as

$$P(x, y; t) = \frac{1}{2\pi\sigma^2} e^{-\frac{x^2+y^2}{2\sigma^2}}, \qquad (6.2)$$

with $\sigma^2 = (\sigma_0^2 + 2Dt)$ and σ_0, the rms width of the initial distribution, equal to about 0.17×10^{-12} m^2. (There is no significance to our particular choice of σ_0.) In the Drude model, the diffusion constant D is related to the model parameters by $D = v_T^2\tau/2 = v_T\lambda/2 = \lambda^2/2\tau$. (The numerical factor of $\frac{1}{2}$ would be $\frac{1}{3}$ or 1 in a three- or one-dimensional system. Also remember that, in two dimensions, $v_T^2 = 2\langle v_x^2 \rangle = 2\langle v_y^2 \rangle$.)

Exercise 6.6 (M) *Determine the slope of a graph of the mean square deviation of position versus time and deduce the diffusion constant from that slope. Does it agree with the prediction $D = v_T^2\tau/2$ of the Drude model?*

6.3   Electric field (DC)

\# 2   PRESET 2 focuses on a single electron with an electric field applied in the positive x-direction and a SCATTERING TIME chosen large enough that no scattering occurs until after the electron has left the real-space screen. The electron in velocity space now moves as it accelerates in the electric field; and in real space there is the motion under constant acceleration that is so familiar from freshman physics.

Shorten the SCATTERING TIME τ to 3 ps and watch the displays. The effect of the scattering becomes evident. Our problem is to describe the motion of a large number of electrons, which are all in this sort of disjoint motion: it is illustrated in PRESET 3. The impression is of a complex behavior, with electrons going every which way.

3

6.3.1 Drift velocity

Our concern, fortunately, is rarely with the behavior of individual electrons, but rather with the ensemble's average behavior which can be quite simply described. The average velocity and position are indicated by the large green dots in the displays, and the corresponding x- and y- components are plotted in the two graphs. (Click the two SHOW GRAPH toggles if the graphs are not already open.)

Exercise 6.7 (C) *Change the* SCATTERING TIME τ *to* 10^3 *ps in* PRESET 3 *and verify from the graphs that the average velocity and position behave appropriately in response to the field: i.e., with what power of t do they vary. Now shorten τ back to 3 ps. The average velocity soon settles to a constant, though quite noisy, value called the drift velocity v_D, while the average displacement x_D is proportional to t. Explain why the dependences of velocity and position on time have changed from t and t^2 to t^0 and t.*

Exercise 6.8 (M) *Find experimentally, by varying the magnitudes of the electric field and the scattering time, the dependence of the drift velocity on each of these parameters.*

In the *Haynes–Shockley experiment* [46, p. 54–56] a very short pulse of electrons is injected into a semiconductor. Suppose the electrons have a mass equal to the free electron mass (we'll see later why this need not be so), a SCATTERING TIME of $\tau = 3 \times 10^{-12}$ s, and are subject to an electric field $E = 3 \times 10^4$ V/m. The electrons drift in the electric field as described by the Drude model and the pulse becomes diffuse because of the random component of the electrons' motions.

Exercise 6.9 (M**) RUN *"drude" for about 400 ps with* PRESET 3 *to simulate this experiment. Watch the behavior of the electrons in real space. For short times all you see is the broadening of the distribution of electrons because of the diffusion. At much later times, the pulse has* drifted a distance from the origin which is large compared with its width. Compare your observations of the simulation display with each of the two position graphs: the AVERAGE and the DEVIATIONS.

Compare the rms deviation of the electron positions from their average with the value of the average displacement for the times, 50 and 300 ps. Explain why diffusion dominates at short times and drift at long times. Because of the small number of electrons, these graphs will be far from ideal: rough estimates are adequate.

Predict the *crossover time*, i.e., the time at which the drift distance is comparable with the diffusive width of the distribution. The Haynes–Shockley experiment provides the most direct measurement of the mobility $\mu \equiv e\tau/m$ for minority charge carriers in a semiconductor.

6.3.2   Approach to steady state

We have seen how the steady state drift velocity of the electrons depends upon the field and the scattering time. Next consider the problem of the approach to this steady state and of the return to the equilibrium, zero current, state after the field has been turned off.

#4   **Exercise 6.10 (C)** RUN PRESET 4 *long enough for the system to reach a steady state, and then* STOP. *Change the* E_x *slider to zero and* RUN *again briefly (without* INITIALIZING*) to allow the average velocity to return to zero.* COPY GRAPH, RUN AGAIN, *and* STEAL DATA *a couple of times to help distinguish meaningful signal from the noisy fluctuations. Why does the average value not change instantaneously to the appropriate steady state value?*

Exercise 6.11 (M) *What physically determines the time for the initial approach of the average velocity to its steady state value,* v_D? *And the time for its subsequent decay to zero?*

Exercise 6.12 (M) *Now reopen* PRESET 4 *and record a graph, including just the initial transient.* COPY GRAPH *these data, and repeat the process several times for successively shorter values of the* SCATTERING TIME *in steps of* $\sqrt{10} \approx 3$. *For each successive* RUN, REINITIALIZE, *and then* STEAL DATA *to develop a superposition of the results for the several values of* τ. *How do the initial slopes of the plots depend upon* τ? *How do the limiting velocities depend upon* τ?

Exercise 6.13 (M*) *Show how the results above follow from analytic solutions of the Drude equation for the average velocity,*

$$\frac{d\mathbf{v}}{dt} = -\frac{e\mathbf{E}}{m} - \frac{\mathbf{v}}{\tau}, \tag{6.3}$$

including the transient as well as steady state response. Note that we will refer to the average electron velocity as v or $\mathbf{v}$ with no subscript.

6.4 Noise*

The graphs you have seen for the average velocity versus time were quite noisy, annoyingly so! Let's explore the origin of that noise. It is *not* merely an artifact of the simulation but reflects sources of noise in real systems. Remember that the current density in an electronic conductor is related to the average electron velocity by the relation $\mathbf{j} = -ne\mathbf{v}$, where n is the electron density and e the magnitude of the electronic charge. Noise in the drift velocity implies noise in the current.

6.4.1   Johnson noise*

While RUNNING PRESET 1 but with the velocity graph switched to AVER-   # 1
AGE, with no applied field, the graph reminds us that there is noise even in absence of current flow. This is because of the random motions of the electrons after scattering. Though the long time average of $\mathbf{v}_D$ will be zero, there are fluctuations about this average.

Exercise 6.14 (M*) *Estimate the magnitude of the fluctuations of the average velocity. What happens to this magnitude as you change the* TEMPERATURE *by a factor of 10? Switch the* GRAPH *to* DEVIATIONS *to see the mean square deviations of the individual electron velocities from the mean. How is this mean square deviation related to the fluctuations in the average velocity? Does this check quantitatively?*

The Johnson noise in a resistor may be characterized by the relation $\langle \delta V^2 \rangle = 4Rk_BT\Delta\nu$. Let's use the "drude" model to relate the noise in the average velocity to the macroscopic conductance of the sample.

Exercise 6.15 (M*) *Forget completely, for the moment, the real-space display. Think of the resistor constructed from our two-dimensional gas of length L and width W containing N electrons, hence a (two-dimensional) density of electrons $n = N/WL$. Establish the following relations:*

1. *the sample conductance (inverse resistance) is $G = Ne^2\tau/mL^2$;*
2. *the sample current is $I = Nev_D/L$;*
3. *the mean square fluctuations in current are $\langle \delta I^2 \rangle = Gk_BT/\tau$.*

Exercise 6.16 (M**) *Establish the connection between this result for the current noise and the familiar expression for the Johnson voltage noise power in bandwidth $\Delta\nu$, $\langle \delta V^2 \rangle = 4Rk_BT\Delta\nu$. (Hint: you will need to make some assumptions about the Fourier spectrum of the noise. For*

*example, you might consider it to be a white spectrum, with a high fre-
quency cut-off related somehow to the physics of the conduction process.)*

This source of noise is variously referred to as *thermal*, *Nyquist*, or
Johnson noise. The aim of good electronics design is to have system
sensitivity limited by thermal noise, not by excess noise from the electronic
devices.

6.4.2   Statistical noise*

We have noted the presence of the Johnson noise in the absence of an
applied field. If a current is driven through the system, the noise will
typically be larger.

Exercise 6.17 (M*) RUN PRESET 1 *with the* SPEED *changed to 10, for
200 ps; then type in an* ELECTRIC FIELD *of* 2.5×10^4 *V/m and* RUN *for
another 200 ps. Repeat, increasing the field by a factor of 2 each time
up to* 20×10^4 *V/m. Note both the increasing noise with increasing
field for the x-component of the velocity and the contrast between the
fluctuations in the x- and y-components. From your data determine how
the magnitude of the additional velocity-dependent noise depends on the
average velocity v?*

Exercise 6.18 (M**) *To explore the field and temperature dependence
of the noise in more detail, take additional data from "drude" to make a
plot of the mean square fluctuations in the average velocity versus electric field for several different values of the temperature. (Be sure to go
to large enough values of field that the average displacement of the electrons in velocity space is quite apparent.) The "noise in the noise" is
large: be content with rough estimates. Interpret the slopes and intercepts of these plots in terms of the thermal noise and the additional
velocity dependent noise. Deduce the temperature and field dependence
of the velocity-dependent noise.*

This additional current noise arises from the use by "drude" of a small
number of electrons in the simulation. Consider the zero temperature case
in which the electrons are returned to zero velocity after each collision. At
any time you can see from the velocity-space display that the electrons
have a variety of velocities from zero to the order of a few times the
average velocity $eE\tau/m$. The rms variation of the velocities among the
electrons is of the order of the average velocity, and the variation of the
average velocity for the N electrons is $\langle 1/\sqrt{N} \rangle$ smaller.

Exercise 6.19 (M*) *Verify that this picture is consistent with the magnitude measured for the velocity-dependent contribution to the noise.*

Compare the magnitude of the noise in the current with the value of the current itself.

Exercise 6.20 (M**) *In the* DEVIATION *plot for the velocities with the electric field at* 10×10^4 *V/m and a* TEMPERATURE *of 3 K, there are occasional "events" in which the mean square deviation drops by a factor of 3 or more in a very short time. What happens to the* AVERAGE *velocity in the same time interval? Give a semiquantitative explanation of such events in terms of the scattering of one, or possibly two, electrons which have survived, without scattering, an anomalously long time before the event. (A single point, far outside of a distribution, can give a very large contribution to the mean square deviation.)*

The statistical noise seen in "drude" is related to the *shot noise* in electronic devices, though it differs in detail from the usual shot noise. Shot noise is *not* an issue in real metals because of the strong correlations in the motions of the electrons induced by their Coulomb interaction. The Coulomb interaction suppresses the statistical fluctuations in both the charge density and the current in a fashion related to the suppression of shot noise in diodes operating under conditions of space charge limited current.

In using "drude" it is frequently desirable to reduce the noise level when making quantitative measurements. The thermal noise can be effectively eliminated by reducing the temperature. Unfortunately, the statistical noise can only be reduced by using the CONFIGURE menu to increase the NUMBER OF ELECTRONS. Although that sounds easy, you must remember that there will be the usual penalty to pay: a slower simulation speed for better signal to noise.

Exercise 6.21 (M*) *How big is that penalty? Assume that doubling the number of electrons doubles the computing time. By how much will the computing time be increased if the number of electrons is increased enough to reduce the noise by a factor of 10?*

6.5   Magnetic field

A magnetic field can have a profound influence on the properties of electrically conducting solids, and the use of magnetic field experiments has been essential in elucidating details of electronic properties of solids. We shall see in Chapter 9 that electrons in solids behave in a bizarre fashion. Sometimes they respond as if their mass were very different from the free electron mass. Sometimes they move in open rather than closed orbits in a magnetic field. It is primarily experiments with magnetic fields that have led to our current understanding of the strange dynamics of electrons in solids. First we'll use "drude" to check out the behavior of conventional electrons in a magnetic field, with no electric field present.

6.5.1   Circular orbits

5   In PRESET 5 the ELECTRIC FIELD **E** is turned off and a MAGNETIC FIELD $B_z\hat{\mathbf{z}}$ is applied instead, perpendicular to the plane in which the electrons move. The motion of the individual electrons is governed by Newton's law in which the force is the Lorentz force. Since we are frequently interested only in the electron velocities, it is most useful to treat the problem in terms of the two first order differential equations

$$\frac{d\mathbf{v}}{dt} = -\frac{e}{m}\,\mathbf{v} \times \mathbf{B} \quad \text{and} \quad \frac{d\mathbf{r}}{dt} = \mathbf{v}. \tag{6.4}$$

RUNNING the program shows the circular trajectories of the electron motion in both real and velocity space. These trajectories are referred to as *cyclotron orbits*. The SCATTERING TIME has been set quite long so only occasionally will you see the motion interrupted by a scattering event.

Exercise 6.22 (C) *First watch the real-space display. The electrons wander away from the vicinity of the origin and many appear to be lost off screen. Wait a little while and they return to congregate again near the origin. Explain what's going on.*

Exercise 6.23 (C) *Check out qualitatively in both spaces how the orbit size and the frequency of circulation of the electrons depend upon the magnitude of the MAGNETIC FIELD. Be sure you are able to explain each dependence you find.*

Exercise 6.24 (C) *Explain why all of the orbits in velocity space have a common center, but in real space have different centers.*

Exercise 6.25 (M*) *Find the solution to Eq. 6.4 for both the position and velocity of an electron moving in a 1 T magnetic field **B** in the z-direction. Choose the initial position of the electron to be at the origin and its initial velocity to be in the y-direction with magnitude 10^5 m/s. Sketch the two trajectories, in real space and in velocity space, for the first $\frac{3}{4}$ of a cycle of the circular motion, indicating which end of the arc is at the initial time. In "drude", follow a colored electron in both spaces, stopping the simulation frequently, to verify the relation between the phases of the motion in the two spaces.*

Exercise 6.26 (M*) *The radii of the trajectories in real space and in velocity space are determined by the initial speed of the electron and the* magnetic field. *Predict the relationship between the radius in one space and that in the other? Verify it by observation of "drude". A modified version of this relation will reappear in Chapter 9 for a more complicated problem in which the orbits are not circular.*

6.5.2 Cyclotron frequency

You may have noted that all of the electrons seem to be moving in lock step. Whatever their initial positions and velocities, they take the same time (called the cyclotron period) to complete the full orbit.

Exercise 6.27 (M) *What must be the relation between the orbit radius and the electron speed in order for the slower electrons to complete their orbits in the same time as needed by the faster electrons?*

Exercise 6.28 (M) *Show that the angular frequency of the cyclotron motion, the cyclotron frequency, is given by $\omega_c = eB/m$. Measure the cyclotron frequency of the electrons in "drude" and show that they have the free electron mass.*

The anomalous cyclotron frequencies of charge carriers in semiconductors and metals are the most direct evidence for dynamical masses of electrons in solids which are quite different from the free electron mass.

6.6 Crossed fields

In real life we cannot just look into the solid and watch the trajectories of electrons. We learn about the trajectories by seeing how the application of a magnetic field alters the response of a material to an electric field, possibly a DC field, possibly a field oscillating at frequencies in the infrared, microwave, or radio-frequency regions of the spectrum. We focus first on the dynamics of individual electrons. Then we take a look at the average response, the behavior of the average velocity.

6.6.1 Velocity space orbits

PRESET 6 is similar to the preceding PRESET, but with an electric field in the positive x-direction. For the moment focus your attention on velocity space. Superficially the display resembles the case of zero electric field, a set of concentric circular trajectories. With a closer look, however, we see that the circles are no longer centered on the origin but on a point displaced downward from the origin.   # 6

Recall that the electrons (between scattering events) are in motion governed by Newton's second law with the force given by

$$\mathbf{F} = -e\,(\mathbf{E} + \mathbf{v} \times \mathbf{B}).\tag{6.5}$$

Exercise 6.29 (M*) *From the expression for the force due to the combined electric and magnetic fields, predict the value of the velocity of the point which forms the center of the circular orbits in velocity space? (Hint: an electron at this velocity, if stationary in velocity space, has no acceleration, and hence no net force due to the fields. What does that imply about the velocity? What would be the motion of such an electron in real space?)*

Exercise 6.30 (C) *With the* SCATTERING TIME *set to 10 ps, the green dot denoting the average velocity of the electrons stays near the common center of the orbits. Reduce the* SCATTERING TIME *to 1 ps. The individual orbit segments are still centered on the same point as before. Why, with the shorter scattering time, is the average or drift velocity so very different from the velocity of the orbit centers?*

Exercise 6.31 (C*) *Vary the parameter values to determine qualitatively the effect of:*

1. *the* ELECTRIC FIELD *on the position of the orbit centers;*
2. *the* MAGNETIC FIELD *on the position of the orbit centers;*
3. *the* SCATTERING TIME *on the position of the orbit centers;*
4. *the* SCATTERING TIME *(as it is shortened) on the position of the average velocity (red dot).*

6.6.2   Real-space orbits

In contrast to the simplicity of the velocity-space trajectories, in real space a variety of qualitatively different orbits is seen with PRESET 6. Increase the SCATTERING TIME τ to 30 ps to remove most of the scattering, and INITIALIZE whenever you need to return the electrons in real space to the screen.

Exercise 6.32 (C) *Occasionally you may see one electron that moves in nearly a straight line. When you do,* STOP *and identify that same electron in velocity space. Find its velocity with the mouse cursor and compare with the velocity you computed in Exercise 6.29.*

Exercise 6.33 (M*) *Figure 6.1 qualitatively illustrates segments from several orbits in real space, though rotated in the figure by 90° in order to fit better onto the page. Let's call the straight line orbits type A, the wavy orbits type B, the cusped orbits type C, and the looped orbits type D.*

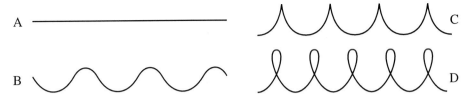

Figure 6.1: Possible real space trajectories.

*Suppose for each, the initial velocity had zero x-component. What is the equality, or inequality, for the y-component of the initial velocity, in terms of E and B, which defines in which class the orbit will fall. Assume the **E**-field is in the positive x-direction and the **B**-field in the z-direction out of the plane of the screen.*

Exercise 6.34 (M**) *Repeat Exercise 6.33, except find the more general conditions on the initial vector **v** that define the various classes of orbits. (That is, don't assume $v_x = 0$ initially.)*

6.6.3 *Drift velocity and Hall effect**

At first sight the task of finding the average velocity for the complex of orbits in PRESET 7 seems daunting. Yet if we watch the green dots   # 7 while running they seem not to do much, except for small variations due to statistical noise. (Much of the thermal noise has been removed by reducing the TEMPERATURE to 30 K.) The average velocity stays roughly fixed and the average position moves slowly and uniformly to the lower left. We see little reminder in the averages of the complicated behavior of the individual electron orbits.

Exercise 6.35 (C) *Change the τ SLIDER slowly and observe the effect on the average velocity. Be sure to cover the range of τ for which the green dot moves from the negative v_x- to the negative v_y-directions. Does the center of the circular **v**-space orbits vary as you do this? Why not? Repeat by moving the B SLIDER.*

Exercise 6.36 (M*) *Using "drude" find a couple of pairs of the PARAMETERS B and τ which leave the direction from the origin to the green dot unchanged, e.g., at 30° or 45° from the horizontal. Feel free to change the magnitude of the ELECTRIC FIELD to keep the drift velocity at a convenient value. Be sure to cover a substantial range (e.g., ×10) of these parameters. What empirical relation can you find relating the two members of each pair? (You may find it convenient to use the slopes of the position graphs rather than the averages of the velocity graphs. Why is the one so much less noisy than the other?)*

Exercise 6.37 (M*) *Use* PRESET 7 *with the* SCATTERING TIME *set to 3 ps and* RUN *for at least 200 ps. In the velocity plots you see lots of noise. Do you see any correlation between the noise in the two components of the velocity? What is the apparent delay time in this correlation?*

Exercise 6.38 (M**) *Amplify the preceding exercise by exploring the noise correlations for a range of* SCATTERING TIMES. *Which is the determining factor in the delay time in the correlations, the scattering time or the cyclotron period?*

The influence of the magnetic field on electrical transport was first studied through the *magnetoresistance* and *Hall effect* experiments. Let's use "drude" to help visualize these experiments. In the presence of a transverse magnetic field, we have just seen (the green dot) that the drift velocity, and hence electrical current, is *not* parallel to the electric field. The angle between the electric field and the resultant current density is called the *Hall angle* ϕ. In terms of the drift velocities, for an electric field in the positive x-direction, $\phi = \tan^{-1} v_{Dy}/v_{Dx}$.

Exercise 6.39 (M*) *Apply the steady state condition to the Drude equation for the average electron velocity,*

$$d\mathbf{v}/dt = (-e/m)(\mathbf{E} + \mathbf{v} \times \mathbf{B}) - \mathbf{v}/\tau, \tag{6.6}$$

to show that the magnitude of the Hall angle is given by

$$\phi = \tan^{-1} \omega_c \tau \quad \text{with} \quad \omega_c \equiv eB/m. \tag{6.7}$$

Assume, to correspond to "drude", that the **B**-*field is in the z-direction, and the* **E**-*field in the x-direction.*

Exercise 6.40 RUN *"drude" using* PRESET 7 *and use either average values of the velocity components from the velocity graph or the slopes of the average position graphs to determine the Hall angle. Repeat for a variety of values of both τ and B and plot the Hall angle as a function of the product $B\tau$. Are your measurements of ϕ consistent with Eq. (6.7)?*

Exercise 6.41 (M*) *In the usual Hall effect experiment, the experiment defines the x-component of the field and imposes the boundary condition that the current be in the x-direction. The y-component of the field required to keep the current in the x-direction is then evidenced as the Hall voltage. For an $\omega_c \tau \approx 1$, and a moderate E_x, adjust E_y to give the drift velocity, and hence current, parallel to $\hat{\mathbf{x}}$. Deduce the Hall angle from the values of the x- and y-components of the* **E**-*field. Check this Hall angle against the plot of the Exercise 6.40.*

Consider the geometry of Exercise 6.41, with E_y adjusted to make the y-component of the drift velocity zero. The nominal resistance of this sample would be proportional to E_x/v_x and could be calculated from that ratio with appropriate geometrical constants, along with the electron charge e and concentration n. From the complexity of the individual electron orbits, one would expect the resistance of such a sample to depend upon the magnitude of the magnetic field. For real materials such a dependence is found and is referred to as a *magnetoresistance*. What happens in the Drude model?

Exercise 6.42 (M**) *Note down the value of the x-component of v_D when, as in the preceding exercise, E_y is chosen to give zero for the y-component of the drift velocity. Now turn off the magnetic field, set E_y to zero, and leave E_x unchanged. What is the magnitude of v_D? Is it significantly different from the value with the magnetic field present? In the Drude model there is zero magnetoresistance. Can you give a simple argument why? (Hint: work from Eq. (6.6).)*

Exercise 6.43 (C**) *In Exercise 6.42, the noise in both velocity channels was about the same in the presence of the magnetic field. Why, with E_y and B_z turned off but E_x unchanged, is the noise in the two channels so different?*

The lack of magnetoresistance is one of many shortcomings of the Drude model. A more dramatic failure concerns the sign of the Hall coefficient: in our language, the sign of the Hall angle. In "drude" we always find the electron drift velocity in the third quadrant if the electric field is in the positive x-direction. The corresponding current density $-nev$ is always deflected to the left of the applied field. This corresponds to the behavior of many metals and semiconductors, but by no means all. Numerous examples exist for which the current density is deflected to the right of the total electric field: i.e., the Hall angle has the wrong sign.

Exercise 6.44 (C) *Draw an appropriate sketch to demonstrate that the sign of the Hall angle is reversed if the carriers are positively instead of negatively charged.*

A serious problem in solid state physics for a number of years was the inability to explain this anomaly in the sign of the Hall coefficient. The development of the band theory of electron states in a periodic potential, introduced in Chapter 8, finally gave a most elegant solution. We will return to the issue in Chapters 9 and 10.

6.6.4   Tensor conductivity**

We've just seen that in the presence of a magnetic field the current density and electric field are not parallel to one another. In the large $\omega_c \tau$ limit we have the interesting case of $\mathbf{j}$ and $\mathbf{E}$ being nearly perpendicular to one another, the Hall angle is nearly $90°$. In order to describe situations with linear response but $\mathbf{j}$ and $\mathbf{E}$ not parallel, the familiar electrical resistivity ρ and conductivity σ must be treated as second rank tensors and the two alternative forms of Ohm's law become

$$\mathbf{j} = \sigma \cdot \mathbf{E} \quad \text{and} \quad \mathbf{E} = \rho \cdot \mathbf{j}. \tag{6.8}$$

Exercise 6.45 (M**) *For a magnetic field in the z-direction show from the steady state condition for the drift velocity that the conductivity tensor may be written*

$$\sigma = \begin{pmatrix} \sigma_{xx} & \sigma_{xy} & \sigma_{xz} \\ \sigma_{yx} & \sigma_{yy} & \sigma_{yz} \\ \sigma_{zx} & \sigma_{zy} & \sigma_{zz} \end{pmatrix} = \frac{ne^2\tau}{m} \begin{pmatrix} \dfrac{1}{1+\omega_c^2\tau^2} & \dfrac{-\omega_c\tau}{1+\omega_c^2\tau^2} & 0 \\ \dfrac{\omega_c\tau}{1+\omega_c^2\tau^2} & \dfrac{1}{1+\omega_c^2\tau^2} & 0 \\ 0 & 0 & 1 \end{pmatrix} \tag{6.9}$$

with $\omega_c \equiv eB/m$. Find the inverse of this tensor, the resistivity tensor $\rho = \sigma^{-1}$. Show explicitly the non-intuitive result that the xx- and yy-components of both ρ and σ go to zero as τ becomes large.

Additional physical insight may be gained by examination of the *conductivity losses* in the weak and strong B-field limits. The Ohm's law losses of a material are proportional to $\mathbf{E} \cdot \mathbf{j}$ or, for an imposed field $\mathbf{E} = E\hat{\mathbf{x}}$, are given by $P = \sigma_{xx}E^2$ where P is the density of power dissipation. The tensor form above for the conductivity shows that the losses are proportional to τ for small $\omega_c\tau$, but proportional to τ^{-1} for large $\omega_c\tau$.

Exercise 6.46 (C**) *Give a physical argument for each of these limiting τ dependences. (Hint: consider the work done by the electric field on an individual electron as it moves in the crossed fields in the two limits.)*

Exercise 6.47 (M*) RUN PRESET 7 *again, but now watch what happens to v_y as the scattering time τ is made very long. What is the limiting value of v_y? How does it compare with the prediction from Eq. (6.9)? (Hint: you can obtain the drift velocity for the "drude" simulation from the current density by the relation $\mathbf{v}_D = \mathbf{j}/ne$. Eq. (6.9) will give the current density in terms of the electric field.)*

The off-diagonal component of the resistivity tensor is called the *Hall resistivity* and has the limiting value at high magnetic field $\rho_{yx} = B/ne$, independent of τ. At very low temperatures and in two-dimensional systems this Hall resistivity becomes quantized with remarkable precision. The phenomenon is referred to as the *quantized Hall effect* and the Hall resistivity in the two-dimensional electron gas provides the current international standard of resistance.

6.7 AC electric field

The discussion until now has focused on the response of the electron gas to time-independent fields, i.e., to the steady state response to DC fields. The SCATTERING TIME τ enters principally in limiting the magnitude of the response. If the applied fields are time-dependent, then τ plays a more significant role which depends upon the relative values of τ and the time characteristic of the changing fields. We next address solutions of the basic Drude equation (6.6) for the case of a time-dependent electric field.

6.7.1 Zero magnetic field

PRESET 8 applies an oscillating electric field with an angular frequency   # 8
10^{11} rad/s, a microwave frequency.

Exercise 6.48 (C*) *Without* INITIALIZING *the program between successive parameter changes,* RUN *for several oscillation periods at each of the* SCATTERING TIMES: *1, 2, 4, 8, 16, 64, and 128 ps. Explain qualitatively why the amplitude of the oscillations in the velocity, or position, first increases with increasing τ and then becomes independent of τ as τ becomes even larger.*

Exercise 6.49 (M*) *Show formally how this dependence on τ follows from the Drude equation (6.6). (Hint: assume there is a sinusoidal time dependence of the electric field and find the amplitude of the resultant sinusoidal v_D response. Using complex notation gives substantial simplification.)*

Exercise 6.50 (M**) *By* ZOOMING *in to a small time interval on the graph you can determine the phase of the velocity response relative to the phase of the driving field. Check this out in the two limits $\omega\tau \ll 1$ and $\omega\tau \gg 1$ and show the results to be consistent with the theoretical prediction from the Drude model.*

Exercise 6.51 (M**) *Convert the results of the calculation of the velocity response of the Drude model to an expression for the complex conductivity $\sigma(\omega) = \sigma'(\omega) + i\sigma''(\omega)$, where $\sigma'(\omega)$ represents the in-phase response and $\sigma''(\omega)$ the out-of-phase response. Explain qualitatively why the real part of the conductivity increases with τ for small $\omega\tau$, but decreases with increasing τ for large $\omega\tau$. (Hint: think about what determines the rate of transfer of energy from the field to the electrons and how that is influenced by the magnitude of the parameter $\omega\tau$.)*

6.7.2   Cyclotron resonance*

One of the surprising features of electron dynamics in solids is an apparent mass different from that of the free electron. The most compelling experiment demonstrating this concept is the cyclotron resonance experiment. We've noted already that electrons in a magnetic field circulate in orbits with an angular frequency, $\omega_c = eB/m$, which is independent of the electron velocity. We might expect a system of electrons to show a resonant response to an electric field which oscillates at or near the cyclotron frequency $\omega_c = eB/m$, giving an experimental way to measure the dynamic mass of the carriers.

Have "drude" simulate the response of a single electron to an AC electric field in the presence of a perpendicular DC magnetic field using PRE-SET 9. The frequency of the AC field is tuned slightly off (can you determine how much?) the cyclotron resonance frequency. To interpret the VELOCITY graph, note that there are at least two issues: the phase of the response relative to that of the driving field and the periodic variation in the envelope of the oscillations.

9

Exercise 6.52 (C*) *Compare the phase of the x-component of the electron velocity with the phase of the driving field at a time shortly before the envelope is a minimum. How does the phase difference compare with the relative phase shortly after the minimum. What does this allow you to say about the transfer of energy between the field and the electron? (Hint: how is the transfer of energy related to the field and the electron's velocity?)*

Exercise 6.53 (C*) *What determines the beat period in the envelope of the oscillations in the velocity response graph? Change OMEGA to the calculated resonance value and watch what happens. If you don't soon lose the electron altogether, you've got the wrong frequency for the resonance. The most likely error is a factor of 2π.*

Exercise 6.54 (C*) *Leave OMEGA at the resonance frequency, increase the TEMPERATURE to 300 K and INITIALIZE frequently while running. (You don't need to STOP in order to start again.) Sometimes you see the electron spiraling out, but sometimes it spirals in before it spirals out. How do you explain that?*

10

Exercise 6.55 (M*) *Change to PRESET 10 to see the response of a collection of electrons. Using the GRAPH again, measure the amplitude of the response for several values of OMEGA and estimate the Q of this cyclotron resonance. (The Q of the resonance is $Q \equiv \omega_c/\Delta\omega$ with ω_c the cyclotron frequency and $\Delta\omega$ the deviation of the frequency from ω_c when* *the amplitude of the response is reduced to 70% of its peak, on-resonance, value.) How is the Q related to the parameter $\omega_c \tau$?*

Exercise 6.56 (C*) *In* PRESET 10 *the driving field is made left circularly polarized. Note the amplitude of the steady state response which is fairly well established after 200 ps. Now change the* PHASE *of the y-component of field from* $-\frac{1}{2}\pi$ *to* $\frac{1}{2}\pi$ *which gives a right circularly polarized field. What is the amplitude of response now? Explain with a physical argument why there is resonant response for one sense of circular polarization, but not for the other. (Hint: look at the nature of the energy transfer between the field and the electrons as the field cycles through each of the four quadrants.)*

Exercise 6.57 (M**) *Starting from the Drude equation, derive an expression for the time dependence in steady state (neglect initial transients) of the average velocity for a driving field* $\mathbf{E}(t) = E_0(\hat{\mathbf{x}} \cos \omega t + \hat{\mathbf{y}} \sin \omega t)$. *Show explicitly from your solution the dependence of the resonance on the sense of circular polarization.*

The "drude" simulation, with negatively charged electrons, shows cyclotron resonance only for a left circularly polarized field. How can it be that there are materials which show a cyclotron resonance with right circular polarization? This turns out to be the same puzzle as the wrong sign of the Hall coefficient.

6.8   Summary

Control of the electrical properties of materials, particularly semiconductors, is one of the greatest concerns of current advanced technology. The Drude picture of electrical conduction remains the basic language for understanding these properties and for constructing models for device behavior. "drude" provides a microscopic illustration of much of the relevant physics.

The basic theme throughout the discussion is the interplay of two competing processes. One is the acceleration of the electrons by applied electric and magnetic fields which drives the electron system out of equilibrium. The resulting distribution has a finite average electron velocity, implying an electric current. This drive out of equilibrium is countered by scattering processes which work to return the system toward equilibrium. Though the picture of the many trajectories of the different electrons appears very complex, the Drude model provides a simple phenomenological description, Eq. (6.6), in terms of the average electron velocity v.

"drude" illustrates the rich variety of situations for which this model has something to say, including both AC and DC response. The Drude model is the basis of discussions of metallic and semiconductor response to electric fields at frequencies as high as the infra-red. With the addition of a static magnetic field, experiments are able to probe in exquisite detail features related to the electronic states of the conduction electrons in metals. Cyclotron resonance is only one of a variety of such experiments.

Although the Drude model is useful, we have noted some serious short-comings. It gives us no hint as to why there are materials with *both* signs of Hall constant; or why there are systems that behave as if the mass of the electrons is less than 10% of, or more than 10 times, the free electron mass; or why some materials show a cyclotron resonance response to the "wrong" circular polarization of the electric field.

6.A   Deeper exploration

Lawn sprinkler   This is more fun than physics, but gives a sensitive way to tune precisely to the cyclotron resonance condition. Open PRESET 10 and change the SPEED to 35. Next reduce the TEMPERATURE to 1 K and watch what happens as it RUNS. Increase or decrease the MAGNETIC FIELD by a factor of a few percent. Explore and explain.

Hall effect   In Exercise 6.41 we looked at the essential physics of the usual Hall effect experiment, but in terms of the drift velocities and electric fields instead of currents and voltages. Express the results of the simulation in terms of a typical experiment on a two-dimensional electron gas by supposing the N electrons distributed over a sample of dimensions W by L. Show how the electron density N/WL and the electron mobility $e\tau/m$ could be deduced from measurement of the current and the voltages along and transverse to the sample. Take measurements with "drude" and check the numerical agreement.

Cyclotron resonance   Explore the simulation of the cyclotron resonance in more detail. Possible issues include analysis of the transient response, study of the phase relation between the driving field and the electron velocity as a function of detuning from resonance, and the influence of the relaxation time τ on the width of the resonance.

6.B   "drude" – the program

"drude" simulates, in two dimensions, the free motion of Drude electrons in electric and magnetic fields. In contrast with "sommerfeld" and "peierls", the electrons obey classical statistics, and their velocities are given by the Maxwell distribution.

6.B.1 Drude model

Dynamics of the Drude model Each Drude electron is taken to move independently of all other electrons and its motion is determined by Newton's second law, with the force given by the Lorentz force,

$$\mathbf{F}(t) = m\,\mathbf{a}(t) = q\,[\mathbf{E}(t) + \mathbf{v}(t) \times \mathbf{B}(t)]. \tag{6.10}$$

Here $q = -e$ is the charge of an electron, and $\mathbf{B}$ and $\mathbf{E}$ are the applied magnetic and electric fields. With the $\mathbf{E}$-field and the electron velocity restricted to lie in the xy-plane and the $\mathbf{B}$-field restricted to be perpendicular to that plane, the equation of motion (6.10) reduces to two coupled differential equations,

$$\left.\begin{aligned}
\ddot{x}(t) &= \frac{q}{m}\,[E_x\cos(\omega t) + B_z\,\dot{y}(t)] \\
\ddot{y}(t) &= \frac{q}{m}\,[E_y\,\cos(\omega t + \phi) - B_z\,\dot{x}(t)].
\end{aligned}\right\} \tag{6.11}$$

Here we have made the additional assumptions that the magnetic field is static, $B_z(t) = B_z$, and that the components of the electric field in x- and y-direction vary sinusoidally. The amplitudes E_x, E_y, and B_z, the angular frequency ω, and the phase ϕ are adjustable parameters in "drude".

Scattering in the Drude model   Thermal equilibrium in the Drude model is maintained through scattering of the electrons, with no assumption made about the origin of that scattering. One simply takes each electron to scatter with a probability $p = \Delta t/\tau$ during the time interval Δt, where τ is the SCATTERING TIME and $\Delta t \ll \tau$. The position of an electron is unaltered by the scattering while the new velocity points in some random direction with a magnitude (speed) appropriate to the temperature of the system.

6.B.2 "drude" simulation

The equations of motion (6.11) are a set of linear, coupled, second order differential equations which can be solved analytically. We nevertheless choose in "drude" to solve Eqs. (6.11) numerically, using a Runge–Kutta algorithm [65] described below. The numerical solution has two advantages: first, given a numerical integration routine (which we needed for other simulations as well), it is much easier to implement in the program. Second, the divergence of the analytic solution, as $1/(\omega_c^2 - \omega^2)$ with $\omega_c = qB_z/m$, forces the use of different formal solutions on and off resonance, which further complicates the analytic approach . (The speed at which "drude" runs is not much affected by the use of the more computationally intensive numerical integration since the display time is the limiting factor.)

Numerical integration   The equations of motion (6.11) are rewritten as a set of four linear, coupled, first order differential equations. "drude" uses a fifth order Runge–Kutta algorithm to solve these equations for $\mathbf{r}(t)$ and $\mathbf{v}(t)$. A Runge–Kutta algorithm can be viewed as an advanced Euler method. The Euler method uses the value and derivative of a function at time t_0 to estimate its value at some later time t_n. A Runge–Kutta method evaluates the derivative typically at several intermediate (trial) times t_i which results in a better estimate of the value of the function at the final time t_n. Although this method requires several intermediate computations, it turns out to be worth the effort: under "normal" circumstances the time step $t_n - t_0$ can be chosen much larger than in the Euler method, maintaining the same numerical accuracy with increased stability. With a time step Δt, the error in the Euler method is of the order Δt^2, while it is of the order of Δt^6 in our Runge–Kutta method.

"drude" algorithm   The electron dynamics and scattering in "drude" are implemented in the following way:

1. *Initial distribution of positions and velocities:* Initially, the positions of the electrons are randomly distributed in a circular region of radius 10^4 Å. (Note that the x- and y-positions of the electrons do not enter into the equations of motion (6.11) which makes our choice of radius one of convenience. A radius 10^4 Å guarantees that the electrons are initially separated enough to be distinguished and remain within the displayed region for some time when the simulation is running.) The initial velocities are distributed according to the Maxwell velocity distribution with the TEMPERATURE T specified as one of the input parameters. The probability $P(v_i)$ for a velocity component v_i, $i = x, y$, to lie in the interval v_i to $v_i + dv_i$ is given by

$$P(v_i)dv_i = \frac{1}{\sqrt{2\pi}\sigma}e^{-v_i^2/2\sigma^2}\,dv_i, \tag{6.12}$$

   with the standard deviation $\sigma = \sqrt{k_B T/m}$. This velocity distribution guarantees that the electrons are initially in thermal equilibrium. "drude" uses the routines `gaussian` and `ranmar` to compute random numbers with the Gaussian distribution (6.12).

2. *Advancing* $\mathbf{r}(t)$ *and* $\mathbf{v}(t)$ *by one time step:* Given the positions and velocities of all electrons at time t, each electron's position and velocity are computed at time $t + \Delta t$, using the routine `rkf` to

integrate Eqs. (6.11). (See below for finding the "right" time step Δt.)

3. *Scattering:* After each time step Δt, each electron scatters with a probability $p = \Delta t/\tau$. If the electron scatters, its new velocity components will be assigned a random value according to the Maxwell distribution (6.12). If the electron does not scatter its velocity is given by the velocity calculated through the Runge–Kutta integration in step 2. The positions and velocities of the electrons are displayed (a few displays may be skipped, see below) and "drude" returns to step 2 to advance the positions and velocities by yet another time step.

Computation of the time step Two issues arise in choosing the "right" time step Δt for the algorithm above: first, the condition $\Delta t \ll \tau$ must be satisfied, and second, the speed at which "drude" runs should be roughly proportional to the SPEED chosen.[1] "drude" picks Δt to be less than $\frac{1}{5}\tau$ and, with the assumption that the speed of "drude" is limited mostly by the graphics display speed, skips a number N_{skip} of displays to speed up the simulation. In particular, the number of displays skipped, N_{skip}, and the exact time step Δt are determined with respect to a typical time scale $t_{norm} = 0.2$ ps,

$$N_{skip} = \text{round}\left(\frac{\text{SPEED} \times t_{norm}}{0.2 \times \tau}\right), \text{ and } \Delta t = \frac{\text{SPEED} \times t_{norm}}{N_{skip}}. \quad (6.13)$$

The smallest value N_{skip} can take is set to 1. (The function round() rounds its argument to the closest integer.) Note that for large SPEEDS, however, the computational time will be the limiting factor of the simulation and increasing the speed will not result in a faster display speed.

6.B.3 Displayed quantities

Displays The positions of the N electrons (the NUMBER OF ELECTRONS N can be specified in the CONFIGURE menu) are shown in the left hand (real-space) display, and the corresponding velocities in the right hand (reciprocal-space) display. Each electron is displayed as a *worm* whose five dots represent the position or velocity of that electron at the last five displayed time steps. Most electrons are shown in white while some are colored so that it is easier to follow their motion. The average position and velocity of the electrons are shown as a larger green dot, and the origin of either space as a larger red dot. The electric field $\mathbf{E}$ is indicated

---

[1] Numerical stability is *not* an issue since the Runge–Kutta integrator chooses its own (much smaller) time step for the numerical integration of Eqs. (6.11) according to some error limits specified.

by a red bar in the velocity display. Pointing with the mouse cursor to any electron will show its xy-position or velocity at the bottom of the corresponding display.

Graphs   Toggle the SHOW GRAPH button below either display to get a graph showing the AVERAGE electron position (velocity) or the corresponding mean square DEVIATION versus time. Note that these averages and mean square deviations are averages and deviations over the number of electrons, *not* over time. For example, the x-AVERAGE is given by the sum over the N electron positions x_i

$$\langle x \rangle = \frac{1}{N} \sum_{i=1}^{N} x_i, \tag{6.14}$$

and the mean square x-DEVIATION by

$$\sigma_x^2 = \frac{1}{N} \sum_{i=1}^{N} (x_i - \langle x \rangle)^2. \tag{6.15}$$

The x-component of the electric field (with an arbitrary scale) is also shown in the graphs. The graphs are updated only every GRAPH UPDATE INCREMENTS (initially 10) time steps. This number can be changed in the CONFIGURE menu.

7

"sommerfeld" – Dynamics of quantum free electrons

Contents

7.1 Introduction

We should not be surprised to find shortcomings with the Drude model. Electrons often do *not* behave as classical particles, and surely scattering from the ion cores must somehow be important. The Sommerfeld model addresses the first issue by including the effects of the Pauli exclusion principle. However, as with the Drude model, the electrons are still assumed to move in a constant potential and to move independently of one another. The effect of the ion potentials is addressed in following three chapters.

The consequences of the exclusion principle for electrical transport are not dramatic. However, the model suggests a qualitatively new point of view, which is important when considering the implications of band structure. Rather than working with average properties of the electrons, e.g., velocity or energy, we focus on the behavior of the electrons of highest energy, the ones with energies near the Fermi energy. It is the scattering rates, velocities, and accelerations of these electrons alone which determine the electrical transport. Electrons of lower energies, because of the exclusion principle, effectively play no role.

"sommerfeld" simulates the response of a two-dimensional, degenerate, free-electron gas to applied DC electric and magnetic fields. Realistic simulation of the metallic system is made difficult by the strong inequalities $mv_F \gg k_B T/v_F \gg eE\tau$ typically satisfied for real metals. The Fermi momentum is much greater than the $k_B T/m$ "fuzz" of the momenta at the Fermi surface, which in turn is much greater than the momentum accumulated by electrons between successive collisions. In order to make visible the essential qualitative features of the quantum free-electron gas, "sommerfeld" violates these inequalities, thereby introducing substantial non-linearities in its response to applied fields. An interesting, though non-essential, issue also considered by "sommerfeld" is the distinction between elastic and inelastic scattering and the implications for the heating of the electron gas by the applied electric field.

"sommerfeld" is the only program in the SSS with a name longer than eight letters, a length that causes problems for some computer systems. All file names related to the "sommerfeld" program are abbreviated to SOMMER. For example, to open "sommerfeld" on a Unix system, you must type SOMMER, not SOMMERFELD.

7.2 Ground state

If we assume a constant potential and neglect the electron–electron interactions, it is straightforward to find the quantum mechanical one-electron energy states for the problem of an electron confined to a volume V. The ground state of the *many-electron* system is then defined by assigning the electrons, one by one, to these energy states, in order of increasing energy.

7.2.1 One-electron states

Useful infinite medium solutions of the one-electron Schrödinger equation are plane waves with the dispersion relation $\mathcal{E}(\mathbf{k}) = |\hbar\mathbf{k}|^2/2m$. As in the case of phonons, the finite system is usually defined by periodic boundary conditions (PBC). If the PBC are applied to a cubic box of edge length L, the allowed values of $\mathbf{k}$ lie on a grid in reciprocal space with spacing $2\pi/L$.

Exercise 7.1 (M) *Consider a cubic volume of edge length L. Verify that plane wave solutions $\psi(\mathbf{k}) = e^{i\mathbf{k}\cdot\mathbf{r}}/L^{3/2}$ to the Schrödinger equation will satisfy the PBC if $\mathbf{k}$ is chosen to be*

$$\mathbf{k} = \left(\frac{2\pi n_1}{L}\right)\hat{\mathbf{x}} + \left(\frac{2\pi n_2}{L}\right)\hat{\mathbf{y}} + \left(\frac{2\pi n_3}{L}\right)\hat{\mathbf{z}}, \tag{7.1}$$

with the n_i integers.

"sommerfeld" defines a two-dimensional grid of points in reciprocal or wave vector space by relation (7.1) and specifies the initial state of the system in terms of which of these allowed $\mathbf{k}$-states are occupied. The occupied states in $\mathbf{k}$-space are the white dots in the "sommerfeld" display. In real systems, because of the spin of the electron, each $\mathbf{k}$-state may be occupied by two electrons, one with spin up, and one with spin down. "sommerfeld" pretends the electrons have no spin and limits the occupation of any $\mathbf{k}$-state to a single electron.

This $\mathbf{k}$-space display is intimately connected to the velocity space in "drude". As with phonons, we make the connection between the wave and particle languages by constructing localized electron wave packets made up of linear combinations of plane wave solutions. These packets move with a group velocity

$$\mathbf{v}(\mathbf{k}) = \frac{1}{\hbar}\nabla_{\mathbf{k}}\mathcal{E}(\mathbf{k}) = \frac{\hbar\mathbf{k}}{m}. \tag{7.2}$$

Relation (7.2) connects the velocity space used by "drude" with the reciprocal space or space of wave vectors used for the "sommerfeld" display. For the free-electron model, the spaces are connected by appropriate scale factors of Planck's constant $\hbar$ and the electron mass m. In the Sommerfeld or Drude models, with electrons moving in a uniform potential, you are free to work with any of the three variables, $\mathbf{v}$, momentum, or $\mathbf{k}$. Be forewarned, however, that for electrons moving in a *periodic* potential the distinction becomes critical because the electron's velocity is no longer necessarily proportional to its wave vector: $\mathbf{v}(\mathbf{k}) \neq \hbar\mathbf{k}/m$.

7.2.2   Fermi sphere

In the Sommerfeld model the ground state of the N-electron system is obtained by filling the N one-electron states of lowest energy, each with a single electron. PRESET 1 shows the occupied states in **k**-space for the ground state of a few electrons in a very small box. The BOX SIZE may be set from the CONFIGURE dialog box.

1

Exercise 7.2 (M) *Deduce the dimensions of the two-dimensional "box" to which these electrons are confined. (Hint: measure the spacing of the grid of allowed **k** values.) How can you tell from the "sommerfeld" display that it is a square and not a rectangular box? (You may use the CONFIGURE dialog box if you wish to change the box size.)*

The maximum energy of the filled states $\mathcal{E}_F$ is called the *Fermi energy*. The surface (line in two dimensions) in **k**-space dividing the filled states from the empty states is called the *Fermi surface*. Because of the simplicity of the dispersion relation, $\mathcal{E}(\mathbf{k}) = |\hbar\mathbf{k}|^2/2m$, the Fermi surface for free electrons is a sphere (in "sommerfeld", a circle). Its radius, the *Fermi wave vector*, is denoted by k_F.

Exercise 7.3 (M*) *Deduce the relation, appropriate to the real (spin included) two-dimensional electron gas, between the Fermi energy and the electron concentration $n = N/V$. Here, N is the number of electrons, and V the confining volume (area in this two-dimensional problem). (Hint: find first the Fermi wave vector, knowing that the Fermi circle must contain $(N/2)$ **k**-states. The factor of 2 is because each **k**-state can be occupied by a spin up and a spin down electron.)*

Exercise 7.4 (M) *Verify that the Fermi energy for the two-dimensional electron gas is linearly proportional to the electron concentration by counting the electrons within the Fermi circle in "sommerfeld" for different values of the FERMI ENERGY. The area of the confining "box" is $A = L^2$ where $L = 60$ Å as given in the CONFIGURE dialog box. Is the proportionality constant equal to the predicted value $2\pi\hbar^2/m$? (This constant differs by a factor of 2 from the result of the Exercise 7.3 because "sommerfeld" uses spinless electrons, only one per **k**-state.)*

The number of allowed states per energy interval is the density of states (DOS) for electrons. Exercises 7.3 and 7.4 show the DOS to be independent of energy for the two-dimensional electron gas. In one dimension, however, it diverges as $1/\sqrt{\mathcal{E}}$ and in three dimensions it varies as $\sqrt{\mathcal{E}}$ where $\mathcal{E}$ is the kinetic energy of the electron.

7.2.3 Thermal energies

Exercise 7.5 (M) *For the* FERMI ENERGY *of* PRESET 1*, what is the Fermi velocity, $v_F = \hbar k_F/m$? How does it compare with a typical thermal velocity in the Drude model? What is the ratio of the Fermi energy of the electron gas in* PRESET 1 *to the thermal energy of the Drude model at room temperature?*

This ratio of energies is a measure of the importance of the exclusion principle in determining the properties of the electron gas. In a metal at room temperature, basically all of the kinetic energy of the electron gas is that forced by the exclusion principle.

The states of the electron gas corresponding to finite temperature differ from the picture described here because some states not far below the Fermi energy will be found to be empty, and some states above it will be occupied. The range in energy near the Fermi energy over which the states go from fully occupied to completely empty is of the order of $k_B T$. If you RUN PRESET 1 you will see a little action at the Fermi surface as electrons are scattered from below to above the Fermi energy.

Exercise 7.6 (M*) *Modern technology permits fabrication of semiconductor structures (heterojunctions and quantum wells) in which a two-dimensional electron gas can be produced. A typical electron concentration is 5×10^{11} cm^{-2} and an interesting temperature for experimentation is of the order of 0.1 K. Is this system better modeled as a zero temperature Sommerfeld gas or a finite temperature Drude gas? What would the answer be for a room temperature experiment?*

7.3 Electric field

7.3.1 Exclusion principle

The equilibrium descriptions of the Drude and Sommerfeld models are markedly different. Surprisingly these differences do not influence the response of the electrons to electric and magnetic fields in a major way. Select PRESET 2 and RUN to see the effect of an electric field in the x-direction for the case of a long relaxation time τ_i. (Ignore the red line in the graph and leave the parameter τ_e at a large value until Section 7.5. The time τ_i corresponds roughly to the τ in "drude".) Just as in "drude", the distribution of electrons simply drifts to the left at a constant rate. In "drude" that rate is the acceleration, $d\langle \mathbf{v} \rangle/dt = e\mathbf{E}/m$.

2

Exercise 7.7 (M) *In "sommerfeld" we speak of the drift at a constant rate in* **k***-space: a drifting Fermi sphere. Show that this drift rate is $d\langle \mathbf{k} \rangle/dt = e\mathbf{E}/\hbar$. For a free electron model the velocity and the wave*

vector languages are the same except for scaling by the factor of $\hbar/m$; but for electrons moving in a periodic potential, we will see later that it is only the **k**-space language which remains simple. A velocity-space picture is of little use.

Will the exclusion principle somehow restrict the ability of an electric field to accelerate the electrons? In response to the electric field, an individual electron changes its **k**-state. But can it be accelerated to a different value of **k** if that "new" **k**-state is already occupied? A little thought shows that if all of the electrons are accelerated at the same rate, then the "new" state is emptied just in time to be available for the electron to move into it. For a more sophisticated argument, imagine a small volume of phase space, filled with electrons to the density permitted by the exclusion principle, and enclosed by a surface S. This surface evolves in time according to the equations of motion, and *Liouville's theorem* insures that the volume enclosed by the surface remains constant. Thus, at a later time, the same number of electrons may be accommodated within S and the exclusion principle places no constraint on the motion.

INITIALIZE and RUN briefly to get a displaced Fermi sea. Then reduce TAU-I (τ_i) to 1 ps and RUN again to see the effect of the scattering in limiting the displacement of the Fermi sphere to a small value, roughly $e\mathbf{E}\tau_i/\hbar$. Now the issue of the exclusion principle comes up again. Does it inhibit the scattering? Yes indeed it does: an electron at an energy many k_BT below the Fermi energy cannot be scattered. All other states of energy within k_BT are occupied, and available scattering processes can, at most, add only the order of k_BT to the electron's energy. The exclusion principle does inhibit the scattering of electrons deep *beneath* the Fermi surface. The inhibition is evident on watching the display. An electron on the right, within the Fermi surface, marches quietly across the screen until it reaches the left hand side of the the Fermi surface. Only then does it become vulnerable to scattering.

For electrons well *above* the Fermi surface there is no inhibition. For scattering of electrons *near* the Fermi surface, the inhibition is a factor of $\frac{1}{2}$. If the energy of the final state of a scattering event is at the Fermi level, then the probability (at or near equilibrium) that the state is full is $\frac{1}{2}$ and the exclusion principle will reduce the scattering by $\frac{1}{2}$. The scattering time τ_i enters the "sommerfeld" program as the scattering time that would be relevant in the absence of the exclusion principle. For small deviations from equilibrium, the relaxation to equilibrium will be characterized by a time $2\tau_i$ because of the exclusion principle suppression of scattering near the Fermi surface.

What are the consequences for the electrical conductivity? Apart from the factor of $\frac{1}{2}$ in the scattering rate, nothing. Consider the Fermi distribution, initially displaced a small amount to the left. Left to RUN without an applied field, in a time roughly τ_i most of the electrons near the Fermi surface have scattered randomly around the Fermi surface. This single scattering of the electrons near the Fermi surface gives a distribution well on its way towards the equilibrium one. The fact that the low energy electrons have not scattered is irrelevant to the relaxation of the full distribution. It is sufficient to scatter just the ones near the Fermi energy. If we *approximate* the rate of scattering of these electrons as $1/2\tau_i$ then the average or drift velocity of the full electron distribution is $v_D = \langle v \rangle = 2eE\tau_i/m$, the same as the result for the Drude model if the τs are suitably redefined. We will continue to distinguish between τ, the *resistivity relaxation time*, i.e., the τ which enters the expression for the drift velocity, and τ_i, the parameter which is used by "sommerfeld" to characterize the inelastic scattering. Near equilibrium, we expect $\tau = 2\tau_i$.

Exercise 7.8 (C) *"sommerfeld" draws a graph of the average k_x and k_y as a function of time. To see the transient response of the system, start PRESET 1 with the ELECTRIC FIELD E equal to zero and RUN briefly. Change E_x to 10^6 V/m and RUN again. Finally, when the steady state has been reached, turn off the ELECTRIC FIELD and watch the recovery to zero current. Verify that both the approach to steady state and the recovery to equilibrium are governed roughly by the inelastic scattering time τ_i.* # 1

Exercise 7.9 (C) *Change the inelastic scattering time τ_i SLIDER by factors of 2 in the presence of the electric field to see the effect on the average wave vector $\langle \mathbf{k} \rangle$.*

In the formal development of the transport coefficients, the effect of the electric field is treated as a small perturbation. In "sommerfeld" we have exaggerated the effect of the field in order to make its effects evident. We might ask whether, in real-life situations, the perturbation of the distribution function by the electric field is really small.

Exercise 7.10 (M) *Consider a copper wire in a typical household circuit carrying 10 A in a 1 mm^2 cross section. Increasing the current much above this value melts the wire! The resistivity of copper at room temperature is 1.6 $\mu\Omega$ cm and the electron concentration 8.5×10^{22} cm^{-3}.*

1. *What is the drift velocity v_D of the electrons?*
2. *Calculate the ratio of this drift velocity to the much larger Fermi velocity, $v_F \equiv \hbar k_F/m$?*
3. *Picture the distribution of electrons in this wire as a displaced Fermi sphere in $\mathbf{k}$-space. By what fraction of the radius of the sphere is the sphere displaced?*

4. What is the difference in kinetic energies of electrons moving with velocity $(v_F + v_D)$ and with velocity $(v_F - v_D)$?

5. How does this difference compare with $k_B T$ at room temperature?

6. What are the mean free time and mean free path between collisions for electrons at the Fermi surface?

7. What is the electric field required to drive this current? (Compare this with the absurdly high values of the fields used in "sommerfeld", which are needed in order to see the shift of the Fermi sea on the computer screen.)

8. It is possible, with care, to prepare copper with a conductivity, at $T = 4$ K, which is 10^4 times that at room temperature. What is the electron mean free path in this material?

7.3.2   Displaced Fermi sea

For a Sommerfeld gas at $T = 0$ K, it is helpful to think in two different ways of the current $j = nev_D = ne^2 E\tau/m$. One is that there are n electrons per cm^3, each moving with a time averaged velocity equal to the drift velocity $-eE\tau/m$. The other is to note that for most of the states within the displaced Fermi sea, the contribution to the current of an occupied state with positive wave vector (momentum) just cancels the contribution from the state with the opposite wave vector (momentum). The only non-canceling contributions come from those states in the thin skin of thickness $2eE\tau/\hbar$ which are occupied on one side of the Fermi sea but empty on the other.

Figure 7.1 shows an equilibrium, undisplaced, Fermi circle drawn in heavy black. A second circle defines the steady state distribution after it is displaced to the left by $eE\tau/\hbar$. The vertical hatching shows occupied states of the displaced Fermi circle lying outside the equilibrium Fermi circle. The white crescent shows the area of states of the equilibrium distribution which have been left empty as the distribution displaces to the left. The crescent with the horizontal hatching is related to the white crescent by inversion in the origin of **k**-space. The dotted area, containing the vast majority of the occupied states, is symmetrically disposed with respect to the origin: for any state in the dotted region with wave vector **k** there is another with wave vector $-$**k** which gives an opposite contribution to the electric current density. Adding up the contributions of all the states in the dotted area gives zero current. Remembering that $eE\tau/\hbar \ll k_F$, we realize that the current, in this interpretation, is being carried by a tiny minority of the total number of electrons, namely those in the two crescent regions. Though only a few of the electrons are involved, each has a velocity of the order of the Fermi velocity which is orders of

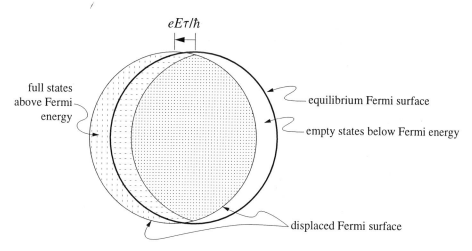

Figure 7.1: Displaced Fermi surface.

magnitude greater than the drift velocity. There are two pictures then: either all the electrons contribute to the current, each with a small drift velocity to the left; or else a few electrons each contribute a large Fermi velocity systematically directed to the left.

Exercise 7.11 (M*) *Show that the fraction of electrons in the vertically hatched crescent outside the equilibrium Fermi circle in "sommerfeld" is given by*

$$\text{(area of crescent)}/\text{(area of Fermi circle)} = 2eE\tau/\pi\hbar k_F. \qquad (7.3)$$

Show that the average velocity of the electrons in the crescent is then $\pi\hbar k_F/4m \approx v_F$.

Combine these two results to show that the average velocity for all the electrons is the familiar $\langle v \rangle = eE\tau/m$. (Hint: recall that $eE\tau/\hbar \ll k_F$ and be sure to include the contribution from both crescents.)

Exercise 7.12 (M**) *Repeat this argument for the three-dimensional case where the crescents become skullcaps.*

The two pictures are, of course, just two ways of keeping track of the same physics. Practice thinking in the *second* of these two pictures. It carries over simply to metals where band structure effects are important; and where the *first* picture becomes useless. The two are equivalent *only* for a limited number of special cases, ones which can be mapped onto this free-electron model. The second picture also reflects a general feature of many of the electronic properties of metals: all the action is at the Fermi surface and it is only the properties of the electrons at the Fermi surface (in this case their velocities and scattering times) which determine the physical behavior of interest. The electrons deep below the Fermi surface are unresponsive and irrelevant.

7.3.3   Distribution function**

The ground state of the Sommerfeld model was described by assigning the probability 1 for the occupancy of energy states below the Fermi energy, and 0 for states above the Fermi energy. For equilibrium at a finite temperature the sharp transition from 1 to 0 is fuzzed out over an energy range of the order of $k_B T$ near the Fermi energy: the equilibrium is described by the Fermi–Dirac distribution function,

$$f_0(\mathbf{k}) = f_0[\mathcal{E}(\mathbf{k})] = \frac{1}{e^{[\mathcal{E}(\mathbf{k})-\mu]/k_B T} + 1}. \tag{7.4}$$

A formal discussion of the behavior illustrated by "sommerfeld" is developed by finding the non-equilibrium distribution function $f(\mathbf{k})$ which gives, in the presence of applied fields, the occupation probability of the state with wave vector $\mathbf{k}$ (momentum $\hbar\mathbf{k}$ or velocity $\mathbf{v} = \hbar\mathbf{k}/m$).

We can ask how the equilibrium distribution $f_0(\mathbf{k})$ is modified when an electric field is applied. The animation in "sommerfeld" of the motion of the electrons in momentum space shows us what happens. When the field is applied, each electron moves in momentum space at a rate $d(\hbar\mathbf{k})/dt = -e\mathbf{E}$. In the absence of scattering, the distribution (in $\mathbf{k}$) at a later time $t + \Delta t$ is the same as at t, but displaced in $\mathbf{k}$-space by the amount $(-e\mathbf{E}\Delta t/\hbar)$.

Exercise 7.13 (M**) *Show that the time dependence of $f(\mathbf{k})$ induced by the applied field may be expressed formally by*

$$\left.\frac{\partial f(\mathbf{k}, t)}{\partial t}\right|_{\text{field}} = \left(\frac{-e\mathbf{E}}{\hbar}\right) \cdot [-\nabla_\mathbf{k} f(\mathbf{k}, t)]. \tag{7.5}$$

Check, recalling the behavior illustrated by "sommerfeld", that the change in f with time in response to the electric field is indeed large only where the derivative of f with respect to the component of $\mathbf{k}$ parallel to $\mathbf{E}$ is large; and see if you can even make the sign come out right.

Now, if f is not at its equilibrium value f_0, then the collision processes tend to restore it to f_0. A commonly used approximation is to introduce a relaxation time τ and to model the effect of scattering with a contribution to the time dependence of the distribution function of the form

$$\left.\frac{\partial f(\mathbf{k}, t)}{\partial t}\right|_{\text{scattering}} = \frac{f_0(\mathbf{k}) - f(\mathbf{k}, t)}{\tau}. \tag{7.6}$$

 This τ is effectively that used in "drude", and related to but *not* equal to the τ_i of "sommerfeld". The distribution function f relaxes towards its equilibrium value at a rate $1/\tau$. If both field and collisions are present a steady state is reached in which the total time derivative of f is 0: the two terms must add to 0. The resulting equation,

$$\frac{df(\mathbf{k}, t)}{dt} = \frac{e\mathbf{E}}{\hbar} \cdot \nabla_{\mathbf{k}} f(\mathbf{k}, t) + \frac{f_0(\mathbf{k}) - f(\mathbf{k}, t)}{\tau} = 0, \qquad (7.7)$$

is called the Boltzmann transport equation; or rather a special case of it.

Exercise 7.14 (M**) *In steady state we require that the distribution $f(\mathbf{k}, t)$ be independent of time. Show that the steady state distribution function $f_{ss}(\mathbf{k})$ given by*

$$f_{ss}(\mathbf{k}) = f_0(k) + \frac{e\tau}{\hbar} \frac{df_0(k)}{dk} \hat{\mathbf{k}} \cdot \mathbf{E}, \qquad (7.8)$$

where $k = |\mathbf{k}|$, satisfies that requirement to first order in $E = |\mathbf{E}|$. (Hint: because these are the lowest order terms in an expansion in powers of the applied field E we may replace f by f_0 if it is multiplied by E. Also note that the equilibrium distribution depends only on the magnitude of $\mathbf{k}$, not its direction. Finally, think a little about the meaning of $\nabla_{\mathbf{k}} k$.)

Exercise 7.15 (M*) *Show that the steady state distribution given by Exercise 7.14 corresponds to a simple displacement of the equilibrium distribution function by the amount $\delta\mathbf{k} = -e\mathbf{E}\tau/\hbar$. Show that the magnitude of this displacement in $\mathbf{k}$ corresponds to the magnitude of the drift velocity of the Drude model.*

Knowing the probability $f_{ss}(\mathbf{k})$ that a state of given $\mathbf{k}$ is occupied and that the electron with wave vector $\mathbf{k}$ contributes $-e\hbar\mathbf{k}/m$ to the electrical current we can add up all the contributions to get the current density

$$\mathbf{j} = \int N(\mathbf{k}) \left(\frac{-e\hbar\mathbf{k}}{m}\right) \frac{e\tau}{\hbar} \frac{df_0(k)}{dk} \hat{\mathbf{k}} \cdot \mathbf{E} \, d^3\mathbf{k}, \qquad (7.9)$$

where $N(\mathbf{k}) = 1/4\pi^3$ is the uniform DOS in $\mathbf{k}$-space for a sample of unit volume.

Exercise 7.16 (M**) *Finish the calculation to show that the current density is finally given by the Drude relation,*

$$\mathbf{j} = \frac{ne^2\tau}{m} \mathbf{E}. \qquad (7.10)$$

This may seem a lot of work just to retrieve the simple Drude result (7.10). It confirms formally that, in this relaxation time approximation, the result for the current density is independent of the form of the equilibrium distribution $f_0(k)$. The classical Drude model and the Sommerfeld model, with dramatically different views of the nature of $f_0(k)$, give the same result. More importantly, the formal approach provides a base on which to build a theory for the transport in the more complicated situation that we meet in Chapter 10.

7.4   Crossed E- and B-fields

3   How does all this work out if a magnetic field is added perpendicular to the plane in which the electrons move? Try it with PRESET 3. The individual electron orbits are the same as in "drude". Can you find a state which remains fixed in k-space under the combined action of the two fields? Again, Liouville's theorem insures that the exclusion principle is irrelevant to the discussion of the individual electron trajectories in k-space. Instead of watching the individual trajectories, try to get a sense of the displaced Fermi circle and how its position depends upon the magnitude of the **B**-field.

Exercise 7.17 (C) *In what direction is the Fermi circle displaced in the limit of a large* MAGNETIC FIELD? *And in the limit of a small* MAGNETIC FIELD? *(The center of the displaced Fermi circle, apart from fluctuations caused by the small number of electrons, is best determined from the graph of the average x- and y-components of* **k**.)

Exercise 7.18 (M*) *What is the criterion which distinguishes between the limits of a small and a large magnetic field?*

Exercise 7.19 (M) *Use the graph to monitor the average x- and y-components of the wave vector. Sketch the qualitative behavior of each as a function of* MAGNETIC FIELD. *(It may be useful to use different values of the* ELECTRIC FIELD E *to keep the values of* $\langle k \rangle$ *small but readable. In making the graph, assume* $\langle k \rangle \propto E$ *and normalize all of the* $\langle k \rangle$ *values to a common value of* E.)

Exercise 7.20 (M**) *Generalize the Boltzmann equation (7.7) to include the effects of the magnetic field. Show that the distribution function* $f_0(\mathbf{k})$, *shifted in k-space by* $(m/\hbar)$ *times the Drude drift velocity, satisfies this Boltzmann equation. Compare this with the empirical result of Exercise 7.19.*

As in the response to the electric field alone, the quantitative prediction for the crossed field response is the same in the Drude and Sommerfeld models. The difference is not in the quantitative prediction but in our visualization of the current as due to a displacement in wave vector space of the distribution of occupied states, the Fermi sea. Again, the new point of view is essential to our later understanding of the response of electrons whose dynamics is influenced by their interaction with the periodic crystalline potential.

Not surprisingly, the results for the Hall effect are unaltered by the introduction of the exclusion principle. The Sommerfeld model does *not* solve the riddle of the frequently wrong sign of the Hall effect. It does, however, provide answers to other questions, not addressed in this set of simulations, such as the rationale for the linear temperature dependence of the electronic heat capacity, noted in "debye", and the gross inadequacy of the Drude theory in explaining the magnitude of the thermoelectric effect.

7.5   Heating of the electron gas**

An interesting, though subtle, question arises in the study of the electrical conductivity of metals at low temperatures, $T \approx 1$ K and below. In all but the purest of metals, in this temperature range the scattering of electrons is predominantly *elastic scattering* by impurities. This scattering can randomize the directions of the electron momenta and thus destroy any directional drift of the electrons, but no energy can be transferred from the electron system to the other degrees of freedom. For both the Drude and Sommerfeld models, in the forms discussed in Chapters 6 and 7, the scattering has been assumed to be *inelastic*. The scattering events return the system towards thermal equilibrium, removing *both* any net momentum in the electron distribution and energy added to the electrons by their acceleration by the electric field.

The neglect of the consequences of the elasticity of the dominant scattering mechanism seems hardly warranted for the case of real metals at low temperature where the scattering is almost entirely elastic; yet the naive picture appears to be successful. Only recently, for metal systems, have experiments [41] revealed non-linearities of the electron response which result from the limited ability of the electrons to get rid of the energy added by the electric field. Let's explore this issue. However, be aware that the parameter values used here have little relevance to real life! These absurd values are required in order to make the qualitative effects evident in the display.

"sommerfeld" models the scattering in an *ad hoc* fashion as described in Appendix 7.B.2. Because of the discreteness of the grid in **k**-space, we cannot force a scattering event to be precisely elastic. "sommerfeld" keeps track of the inherent energy gains and losses in the elastic scattering events and allows an elastic scattering only if it tends to maintain a zero net gain or loss. The inelastic scattering events either involve a loss of a small energy increment $\Delta\mathcal{E}$, which can be specified in the CONFIGURE dialog box, or are quasi-elastic scattering. The model illustrates qualitative ideas but is not to be taken literally.

The red curve in the graph in "sommerfeld" shows the excess of the total kinetic energy of the electron system over the energy of the ground state. It is a convenient measure of the heating of the electron gas and is measured here in units of the Fermi energy per electron. PRESET 4 illustrates the problem we're addressing: the elastic scattering time τ_e has the value 1 ps we have previously used for τ_i while τ_i is set very high. What happens to the electron distribution as the program RUNS? It seems clear from the excess energy plot that the system will never reach a steady state condition.

Exercise 7.21 (C*) *Compare the behaviors for the two limiting cases in which the scattering is purely elastic and purely inelastic. In* PRESET *4 the scattering has been chosen to be entirely elastic.* RUN *for about 20 ps and* COPY GRAPH. *Then* INITIALIZE *and* RUN *again but now with $\tau_i = 1$ ps and $\tau_e = 10^4$ ps: i.e., inelastic instead of elastic scattering. Compare the behavior of the electron heating for the two cases. And compare the values of the average wave vectors for the time interval. Explain why the heating curves are so different while the values of the steady state average k values, which are proportional to the drift velocity, are essentially the same. Also note for later comparison the steady state value of the excess energy, when the inelastic scattering is dominant.*

Exercise 7.22 (M*) *Now repeat Exercise 7.21 but with the* ELECTRIC FIELD *increased by a factor of 2. Do the heating effects vary linearly or quadratically with the magnitude of the electric field? Which would you expect? To make effects clearly visible, "sommerfeld" drives the system far beyond its* linear *response regime. Don't expect things to work very quantitatively. (Hint: how does "Joule, or Ohm's law, heating" depend upon the magnitude of E?)*

Exercise 7.23 (M*) *To contrast the effects of the two relaxation times, carry out the following sequence. With* SLIDERS *set at $\tau_i = 5$ ps, $\tau_e = 1$ ps, and $E = 10^6$ V/m,* RUN *until a steady state has been established for both the average velocity and the excess energy.* STOP *but do* not INITIALIZE. *Set the* ELECTRIC FIELD *to zero and run again until equilibrium is established. What is the characteristic time for relaxation of the averaged wave vector $\langle k \rangle$? And for the excess energy? Interpret these equilibration times in terms of the "sommerfeld" model.*

In real metals the inequality $\tau_i \gg \tau_e$ is typical at low temperatures. The simulation results suggest that electron heating should be an important issue in the theory of conductivity at low temperature, yet in many treatments the issue is not even raised. It turns out that a relatively small amount of inelastic scattering is easily able to maintain the metal near equilibrium for most experimental situations.

Exercise 7.24 (M*) *Check the value of the steady state excess energy for the parameters of Exercise 7.23. Then reduce the elastic scattering time to $\tau_e = 0.3$ ps and repeat. Explain why decreasing the elastic scattering time, with τ_i fixed, reduces the magnitude of the heating.*

Exercise 7.25 (M)** *Return to* PRESET *4, but with $\tau_i = 3$ ps and $\tau_e = 10^4$ ps.* RUN *until the energy increase has leveled off to its noisy steady state. Now apply a* MAGNETIC FIELD *$B = 5$ T and explain why application of the magnetic field reduces the magnitude of the electron heating.*

The heating effects, as illustrated in "sommerfeld", are an extreme exaggeration of heating phenomena in real metals. The same ideas may be translated into consideration of the electron temperature for a more realistic Sommerfeld metal.

Exercise 7.26 (M)** *Show, for a free-electron metal with a conductivity σ and DOS at the Fermi surface $N(\mathcal{E}_F)$, that the rate of heating of the electron system is given by*

$$\frac{dT}{dt} = \frac{\sigma E^2}{(\pi^2/3)k_B^2 T N(\mathcal{E}_F)} = \frac{e^2 \tau E^2 v_F^2}{\pi^2 k_B^2 T}. \tag{7.11}$$

(Hint: you may wish to look in a textbook for an expression for the electronic heat capacity.) The scattering time here is taken as the inverse of the sum of the two rates, $\tau = \tau_i \tau_e/(\tau_i + \tau_e)$. Argue from this result that the fractional change in electron temperature $\Delta T/T$ is given, for small values of $\Delta T/T$, by

$$\Delta T/T = 3(eE\lambda_D/\pi k_B T)^2, \text{ where } \lambda_D \equiv \sqrt{v_F^2 \tau_i \tau/3} = \sqrt{D\tau_i}. \tag{7.12}$$

The inelastic diffusion length λ_D in Eq. (7.12) is the typical distance an electron diffuses between inelastic scattering events. Plugging in numbers for a typical low temperature experiment shows that it is difficult at $T \approx 1$ K to apply a sufficient field to heat the electron gas relative to the phonons. At lower temperatures it becomes an issue to be considered. In semiconductors, even at room temperature, it is possible to apply much larger electric fields than in metals and electron heating effects can play a quite significant part in device behavior [46], sometimes advantageous, more often detrimental.

7.6   Summary

The Sommerfeld model improves upon the Drude model as a picture for the electronic properties of metals. The critical issue is the recognition of the quantum nature of the electrons. It is more useful to think in terms of their wave vectors **k** than their velocities. The restrictions of the Pauli exclusion principle give the picture in **k**-space of the Fermi sphere of occupied states, rather than the diffuse thermal distribution of the Drude model. The response to electric and magnetic fields in the Sommerfeld model is essentially the same as in the Drude model, though our conceptualization is rather different. In particular, it is important to focus attention on the properties of those electrons with energies near the Fermi energy. More striking contrasts with the Drude model appear in discussion of specific heat and thermoelectric effects.

7.A   Deeper exploration

Electron heating   Investigate with some care the non-linear dependence of the average velocity of the electrons on the magnitude of the electric field, an analog of the experiment of Roukes [41]. Set the elastic scattering time τ_e to a large value, leave τ_i at 1 ps and vary the electric field E_x. (Heating effects are evident in "sommerfeld" even when the scattering is entirely inelastic.) Summarize the experimental results from "sommerfeld". Predict the non-linear behavior of the system using a model based on the inelastic-scattering algorithm described in Section 7.B.2. Do *not* expect any quantitative connection with the work of Roukes, but the qualitative ideas of that work are applicable.

Temperature   In "sommerfeld" the parameter $\Delta\mathcal{E}$ plays a role similar to temperature. It gives a smearing out of the transition in energy from occupied to unoccupied states over a range of about $\Delta\mathcal{E}$, and the rate of energy transfer in the inelastic scattering is proportional to $\Delta\mathcal{E}$. Explore and understand the changes in behavior of "sommerfeld" as $\Delta\mathcal{E}$ is varied, using the CONFIGURE dialog box, over a substantial range of values.

7.B   "sommerfeld" – the program

"sommerfeld" simulates, in two dimensions, the motion of free electrons in static electric and magnetic fields. The PBC give rise to quantized **k**-states, and the Pauli exclusion principle allows at most one (spinless) electron per state. Elastic scattering of the electrons provides momentum relaxation, and inelastic scattering provides both momentum relaxation and a means of cooling the electron gas.

7.B.1   Sommerfeld model

"sommerfeld" lets the (quantized) **k**-states evolve deterministically in time. Any state can be occupied by at most one electron, and each scattering event exchanges an occupied with an empty state. In contrast to the "drude" model, we follow the time evolution of *states* rather than *electrons*, and let the electrons scatter amongst the states. This gives a convenient way to insure compliance with the exclusion principle.

k-*states*   The PBC restrict the initial electron wave vectors **k** in the Sommerfeld model to the quantized set (7.1). With the simplified model of spinless electrons, each of these quantized **k**-states can be occupied by at most one electron. The potential in "sommerfeld" is taken to be zero: i.e., the energy $\mathcal{E}$ of each electron of mass m is

$$\mathcal{E}(\mathbf{k}) = \frac{\hbar^2 \mathbf{k}^2}{2m}. \tag{7.13}$$

In the ground state only those states with energy less than the Fermi energy $\mathcal{E}_F$ are occupied by electrons; all other states remain empty.

Dynamics in the Sommerfeld model   Each state in "sommerfeld" moves independently of all other states and according to Newton's equation, with the force given by the Lorentz force,

$$m \frac{d\mathbf{v}}{dt} = \mathbf{F}(t) = -e \left[\mathbf{E} + \mathbf{v}(t) \times \mathbf{B} \right]. \tag{7.14}$$

Here **E** and **B** are the applied electric and magnetic fields.

In "sommerfeld" we prefer to choose the wave vector **k** as the dynamical variable rather than the velocity **v**. Because of the free-electron dispersion relation (7.13), the two quantities[1] differ only by a constant, $\hbar \mathbf{k} = m\mathbf{v}$. For the two-dimensional electron gas treated by "sommerfeld", the static electric field lies within the plane of motion (xy-plane), $E_z = 0$, and the static magnetic field is perpendicular to that plane, $B_x = B_y = 0$.

---

[1] The distinction between the *crystal momentum* $\hbar\mathbf{k}$ and $m\mathbf{v}$, with **v** the *group velocity*, is crucial when the band structure is no longer the simple free-electron energy (7.13), as it will become clear in "ziman" or "peierls" from the semi-classical equations of motion (9.12) and (9.13).

Equation (7.14) can then be written in terms of k_x and k_y as

$$\left.\begin{aligned}
\dot{k}_x(t) &= \frac{-e}{\hbar}\left[E_x + \frac{\hbar}{m}k_y(t)B_z\right], \\
\dot{k}_y(t) &= \frac{-e}{\hbar}\left[E_y - \frac{\hbar}{m}k_x(t)B_z\right].
\end{aligned}\right\} \tag{7.15}$$

The equations of motion (7.15) can be solved analytically. For the case $B_z \neq 0$ we find for $k_x(t)$ and $k_y(t)$

$$\left.\begin{aligned}
k_x(t) &= A\cos(\omega_c t + \phi) + (m/\hbar)(E_y/B_z), \\
k_y(t) &= A\sin(\omega_c t + \phi) - (m/\hbar)(E_x/B_z),
\end{aligned}\right\} \tag{7.16}$$

with the cyclotron frequency $\omega_c = eB_z/m$. The two constants A and ϕ can be determined from the initial values $k_x(0)$ and $k_y(0)$.

For the case $B_z = 0$, Eqs. (7.15) are easily solved to give

$$\left.\begin{aligned}
k_x(t) &= (-e/\hbar)E_x t + k_x(0), \\
k_y(t) &= (-e/\hbar)E_y t + k_y(0).
\end{aligned}\right\} \tag{7.17}$$

Scattering in the Sommerfeld model The time evolution of states, described by Eqs. (7.16) or (7.17), is deterministic. Scattering is a statistical (random) process in which an occupied electron state becomes empty and an empty state is filled by an electron. (As in the Drude model we make no assumptions about the origin of the scattering.) "sommerfeld" allows for elastic *and* inelastic scattering. In an elastic scattering event, the energy of the electron is unaltered: the electron's wave vector remains constant in magnitude and its new direction is independent of the direction prior to the scattering. In an inelastic collision, the electron loses an amount of energy, roughly $\Delta\mathcal{E}$, where $\Delta\mathcal{E}$ can be specified in the CON-FIGURE menu. The electron's wave vector decreases in magnitude and its new direction is again chosen at random.

7.B.2 "sommerfeld" algorithm

At time t "sommerfeld" evolves all states by one time step Δt and then checks for each electron whether it scatters into a new state. The time step Δt is chosen to be smaller than one fifth of both the elastic and inelastic scattering times. (For details of how the time step is chosen, see Section 6.B.2.)

Initial **k**-*states* With the electron gas confined in a square box of linear dimension L, which can be changed in the CONFIGURE menu, it follows from the PBC that the states are placed initially on a square grid with a grid spacing $\Delta k = 2\pi/L$ and are characterized by their grid position i, j.

Initially the states below the Fermi energy $\mathcal{E}_F$ are occupied by electrons, the states above $\mathcal{E}_F$ are empty.

Evolution of **k**-*states*   According to the solutions (7.16) and (7.17) of the equations of motion (7.15), at time t the grid of states is only shifted and rotated with respect to its position and orientation at time 0. To update the grid of **k**-states at every time step Δt, it therefore suffices to keep track of the time evolution of the origin $\mathbf{k}_{0,0} = (0,0)$ and two neighboring grid-points, $\mathbf{k}_{1,0} = (\Delta k, 0)$ and $\mathbf{k}_{0,1} = (0, \Delta k)$. The wave vector at any other grid-point (i, j) is then given by

$$\mathbf{k}_{i,j}(t) = \mathbf{k}_{0,0}(t) + i\left[\mathbf{k}_{1,0}(t) - \mathbf{k}_{0,0}(t)\right] + j\left[\mathbf{k}_{0,1}(t) - \mathbf{k}_{0,0}(t)\right]. \qquad (7.18)$$

Scattering   In each time step Δt every electron scatters with the probability p. The probability p_e for elastic scattering is given by

$$p_e = \Delta t/\tau_e, \qquad (7.19)$$

with the elastic scattering time τ_e specified on the panel. Similarly, the probability p_i for inelastic scattering is given by

$$p_i = \begin{cases} \Delta t/\tau_i & \text{if } \mathcal{E} \leq \mathcal{E}_F, \\ (\mathcal{E}/\mathcal{E}_F)\,(\Delta t/\tau_i) & \text{if } \mathcal{E} > \mathcal{E}_F, \\ 1 & \text{if } (\mathcal{E}/\mathcal{E}_F)(\Delta t/\tau_i) > 1, \end{cases} \qquad (7.20)$$

with the inelastic scattering parameter τ_i specified on the panel. To speed up the cooling of electrons with energies above the Fermi energy, we have chosen an inelastic scattering probability p_i which increases with increasing energy.

The scattering algorithm for every electron and every time step can be divided into the following four steps:

1. Per time step Δt, an electron in the state **k** tries to scatter into a new state $\mathbf{k}'$ with the "probability" $p_e + p_i$. The scattering trial[2] will be elastic with the probability $p_e/(p_e + p_i)$ and inelastic with the probability $p_i/(p_e + p_i)$.

2. For elastic scattering, no energy should be lost or gained in the scattering. For inelastic scattering, the energy of the new state should be smaller by the amount $\Delta\mathcal{E}$. Ideally, given the magnitude k of the electron's wave vector prior to scattering, k' should equal k for elastic scattering and reduce to

$$k' = \sqrt{k^2 - \frac{2m}{\hbar^2}\Delta\mathcal{E}} \qquad (7.21)$$

---

[2]Note that it can happen that $p_e + p_i > 1$. In that case the electron always tries to scatter.

for inelastic scattering. Electrons with energies less than $\Delta\mathcal{E}$ can only scatter elastically. The direction of the new wave vector $\mathbf{k}'$ is chosen at random.

3. However, the states are quantized, and $\mathbf{k}'$ will be unlikely to fall on an allowed grid-point. Therefore the four grid-points defining the cell that contains $\mathbf{k}'$ are determined.

    (a) For inelastic scattering the electron attempts to scatter into the closest grid-state.

    (b) For elastic scattering we cannot allow the electrons simply to scatter into the closest empty grid-state since such a scattering algorithm would heat up the electron gas. First, because of the exclusion principle, "sommerfeld" would be more likely to find an empty state at a higher energy than at a lower one. Second, because $\mathcal{E}(\mathbf{k}) \propto |\mathbf{k}|^2$, scattering with equal probability to states $|\mathbf{k}'| = |\mathbf{k}| \pm \Delta\mathbf{k}$ would result in a systematic increase in energy. "sommerfeld" therefore keeps track of the energy imbalance of the elastically scattered electrons. If the energy imbalance is positive the electron tries to scatter into the closest of the four grid states that has a lower energy than $\mathbf{k}'$. If the energy imbalance is negative, the electron tries to scatter into the closest state with a larger energy than $\mathbf{k}'$.

    (c) Finally, the scattering is allowed *only* if that state is empty.

4. If an electron failed to scatter inelastically in step 3 because it attempted to scatter into an occupied state, then it tries to scatter elastically. This avoids the unrealistic blocking of the momentum relaxation near equilibrium when most states below the Fermi surface are occupied.

7.B.3 Displayed quantities

Electron display The left hand display shows a region of wave-vector space, 4 Å^{-1} on a side. The corners correspond to wave vectors of magnitude $k = 2\sqrt{2}$ Å^{-1}, and an energy of 30.4 eV. The wave vectors $\mathbf{k}$ of the electrons are shown as white dots. The average wave vector of all electrons is represented by a green dot, and the center $\mathbf{k} = (0,0)$ of reciprocal space by a red dot. The Fermi circle, defined by $\mathcal{E}(\mathbf{k}) = \mathcal{E}_F$, with $\mathcal{E}_F$ specified on the panel, is indicated by a red line.

The DOS in $\mathbf{k}$-space, is determined by the box size L given in the CON-FIGURE menu. Large L improves the visual impression of the simulation display, but slows the program.

Because the graphics display speed is the limiting factor in the speed of the simulation, displays can be skipped by choosing a SPEED larger than 1 to increase the simulation rate. Although this may cause the display to look jagged, the average (green dot) and the graph data are nevertheless updated every time step.

Graph The graph on the right shows the x- and y-components of the average wave vector of all electrons (the green dot in the left hand display). Additionally, the excess energy of the electrons (characterizing the electron heating), defined as

$$\mathcal{E}_{\text{excess}} = \frac{\sum_{j=1}^{N_e} \mathcal{E}_j - \sum_{j=1}^{N_e} \mathcal{E}_j^{\text{initial}}}{N_e \, \mathcal{E}_F}, \qquad (7.22)$$

is shown by a red line in the graph. Here $\mathcal{E}_j^{\text{initial}}$ is the initial energy of the electron j and N_e is the number of electrons.

8

"bloch" – Electron energy bands and states

Contents

8.1 Introduction

The Sommerfeld model, though offering useful insights, fails to provide a complete language for discussing the properties of the mobile electrons in solids. Its most dramatic shortcoming is its inability to explain why both positive and negative Hall coefficients are observed in metals. Many of the problems with the Sommerfeld picture are resolved by taking into account the interaction of the electrons with the periodic potential provided by the ion cores. What are the consequences of these interactions?

First, as a result of the periodic potential, the eigenfunctions of the one-electron problem are no longer plane waves. Nonetheless, a wave vector **k** may still be associated with each of the states, though with the same ambiguity of assignment that we saw in Chapter 4 for the case of the lattice vibrations. We speak of a *crystal momentum* or *pseudo-momentum* $\hbar\mathbf{k}$ which plays a role similar to that of the momentum in the Sommerfeld model. We must, however, learn to make a clear distinction between this crystal momentum $\hbar\mathbf{k}$ which plays the familiar role of a momentum quantum number, and the real physical momentum associated with a state which is *no longer* $\hbar\mathbf{k}$.

Second, when we determine the energy $\mathcal{E}(\mathbf{k})$ of a state of wave vector **k** we find that it no longer follows the free electron dispersion relation $\mathcal{E}(\mathbf{k}) = |\hbar\mathbf{k}|^2/2m$, but has a more complex form containing ranges of allowed energies and ranges of forbidden ones. A wide variety of properties of the electronic system may be deduced from a knowledge of $\mathcal{E}(\mathbf{k})$. Chapter 9 investigates the dramatic consequences for the electron dynamics which result from the band structure, while Chapter 10 combines ideas from Chapters 7 and 9 to illustrate the response of an ensemble of conduction electrons in a partially full band.

"bloch", though restricted to one dimension, enables exploration of many features of the electronic band structure of crystals, giving both the dispersion relation $\mathcal{E}(k)$ for a variety of potentials, and representations of the wave function for any specified energy.

8.2 Solutions in one dimension

8.2.1 Bloch's theorem

The underlying periodicity of the crystal structure imposes restrictions, summarized as Bloch's theorem, on the nature of the solutions to various equations describing the properties of crystals. In the case of the lattice vibrations, we were led to explore wave-like solutions to the classical equations of motion. In searching for solutions to the one-electron Schrödinger equation with a periodic potential, we make use of Bloch's theorem to give us specific information about the nature of the electron wave functions. Similar ideas appear when considering other properties of crystalline materials. In particular the theorem serves as a basis for the construction of wave functions describing *excitations* of the crystalline system such as magnons (or spin waves) and excitons.

There are two frequently used and equivalent statements of Bloch's theorem about the nature of the one-electron wave functions in a periodic potential:

1. For any eigenstate of the Hamiltonian there exists a **k** such that suitably chosen eigenfunctions may be written as the product of a plane wave $e^{i\mathbf{k}\cdot\mathbf{r}}$ and a cell periodic function $u(\mathbf{r})$,

$$\psi(\mathbf{r}) = e^{i\mathbf{k}\cdot\mathbf{r}}u(\mathbf{r}) \ \text{ with } \ u(\mathbf{r}) = u(\mathbf{r}+\mathbf{R}), \tag{8.1}$$

   where **R** is a translation vector of the periodic potential.

2. The eigenfunctions $\psi(\mathbf{r})$ may be chosen such that, if **R** is a translation vector of the periodic potential, then associated with $\psi(\mathbf{r})$ there is a **k** such that

$$\psi(\mathbf{r}+\mathbf{R}) = e^{i\mathbf{k}\cdot\mathbf{R}}\psi(\mathbf{r}). \tag{8.2}$$

The common theme of these statements is the ability to associate a wave vector **k** with each of the energy eigenstates of the problem of a single electron moving in a periodic potential energy.

8.2.2   Energy bands and gaps

We can easily find "exact" numerical solutions to the one-electron Schrödinger equation if we stay in one dimension and have a little help from the computer. Given a periodic potential $V(z) = V(z + a)$, we search for wave functions corresponding to a particular energy $\mathcal{E}$ which have the Bloch form. What we discover is that the wave vector **k** for that function will sometimes be pure real, and other times it will have an imaginary part.

Suppose **k** has an imaginary part. Then the plane-wave factor $e^{i\mathbf{k}\cdot\mathbf{R}}$ relating the function in one cell to that in the next will cause the function to diverge exponentially in either the positive or the negative z-direction. Hence it is inadmissible as a wave function for bulk material. We speak of energies associated with complex k as being forbidden energies for bulk states, and the range of such energies an *energy gap*. If **k** is real, then the states are extended uniformly through the sample, varying only in phase, not amplitude, on going from cell to cell. The corresponding energies lie in ranges called *energy bands* and the corresponding eigenstates are referred to as band states. Let's see how it goes!

In rather oversimplified terms, "bloch" chooses an energy $\mathcal{E}$ and numerically integrates the Schrödinger equation through a single cell. Imposing the Bloch condition, $\psi(z + a) = e^{ika}\psi(z)$ then determines the k value appropriate to the chosen $\mathcal{E}$. If this k is real, the energy is in one of the bands of allowed energies. If the k turns out to be complex, then the solution is physically forbidden and the energy must lie in an energy gap. The computer then tries another energy and repeats the process. A plot of the *allowed* values of energy versus the corresponding k values yields the dispersion relation for electrons $\mathcal{E}(k)$ as seen in PRESET 1. The choice   **# 1** of k will always be ambiguous to within an additive multiple of $2\pi/a$: "bloch" adopts the choice $-\pi/a \leq k < \pi/a$. That is, k is always chosen to lie within the first Brillouin zone of reciprocal space. This choice of convention for k is referred to as the *reduced zone scheme*. Because of the symmetry in k of $\mathcal{E}(k)$, the energy band plot in "bloch" is restricted to the right hand half of the first Brillouin zone, $0 \leq k \leq \pi/a$.

You can also ask "bloch" to show you the wave function corresponding to an allowed energy $\mathcal{E}$: either by DOUBLE-CLICKING on the dispersion curve at the corresponding energy, or by entering the desired ENERGY with the slider and clicking on CALCULATE PSI. The computer uses three-dimensional graphics to plot the complex ψ as a function of z through a single cell. Selecting PSI-SQUARED or PHASE OF PSI from the SHOW menu above the left hand graph gives additional information about the wave function for the selected energy.

8.2.3 *Finding the Bloch solution***

A simple numerical integration of the Schrödinger equation does *not* automatically give a solution of the Bloch form. The second order Schrödinger equation has two linearly independent solutions for a given $\mathcal{E}$. Only appropriate linear combinations of these will satisfy the Bloch condition. "bloch" uses the following scheme to find functions that satisfy the Bloch theorem and to allow assignment of a wave vector k for a given energy $\mathcal{E}$.

"bloch" first finds two independent functions, $\psi_1(z)$ and $\psi_2(z)$, by numerical integration of the Schrödinger equation from $z = -a/2$ to $z = a/2$, having imposed the boundary conditions on the ψs at $z = -a/2$:

$$
\left.
\begin{aligned}
\psi_1(-a/2) &= 1, & \frac{d\psi_1(z)}{dz}\bigg|_{z=-a/2} &\equiv \psi_1'(-a/2) = 0, \\
\psi_2(-a/2) &= 0, & \frac{d\psi_2(z)}{dz}\bigg|_{z=-a/2} &\equiv \psi_2'(-a/2) = \frac{1}{a}.
\end{aligned}
\right\} \tag{8.3}
$$

The most general solution for energy $\mathcal{E}$ is $\psi(z) = A\psi_1(z) + B\psi_2(z)$. Only for particular choices of A and B will this satisfy the Bloch condition.

So, "bloch" imposes the Bloch condition on both the function $\psi(z)$ and its derivative $d\psi(z)/dz$. This gives two linear homogeneous algebraic equations for the coefficients A and B.

Exercise 8.1 (M**) *Making use of the z-invariance of the Wronskian,*

$$W(z') = W(z) \equiv \psi_1(z)\psi_2'(z) - \psi_2(z)\psi_1'(z), \qquad (8.4)$$

show that the secular equation, required to give non-trivial values of A and B, gives the condition on k:

$$\cos ka = \tfrac{1}{2}\left[\psi_1(a/2) + a\psi_2'(a/2)\right] \equiv Q. \qquad (8.5)$$

Exercise 8.2 (M**) *Show that the invariance with z of the Wronskian $W(z)$ is equivalent to the conservation of particle current if you choose $\psi_2(z) = \psi_1^*(z)$.*

Now, if $|Q| \leq 1$ there are two real solutions for k, $k = \pm(1/a)\cos^{-1} Q$. These solutions remain finite on moving from the initial cell into the crystal to either positive or negative z. The energy initially assumed is an *allowed* energy, and is associated with the k value just deduced. Knowing the k appropriate to the assumed $\mathcal{E}$, "bloch" goes on to find the ratio of the coefficients A and B and finally, with a normalization condition, the Bloch function $\psi_k(z) = A\psi_1(z) + B\psi_2(z)$.

8.3   In-band states

8.3.1   Dispersion relations

There are a number of qualitative features of the band structure and of the wave functions to be explored. PRESET 1 has chosen a potential in which the atoms are represented as attractive square wells. The dispersion curve for this potential is in red and, for comparison, the curve for free electrons, $\mathcal{E}(k) = \hbar^2 k^2/2m$, in blue. The origin of energy for the free electron curve is taken as the cell average of the square well potential. (Table 8.3 on page 176 gives other forms of the potential available in "bloch".)

As the depth of the wells goes to zero, the band dispersion relation, of course, becomes identical with that for free electrons. The free electron band structure, *reduced* to the first Brillouin zone, as shown in blue by "bloch", provides the crudest possible approximation to the band structure for a crystal with a given lattice. It is constructed by displacing segments of the free electron dispersion relation, $\mathcal{E}(k) = \hbar^2 k^2/2m$, in k-space by appropriate reciprocal lattice vectors to bring them all into the first zone. It is called the *empty lattice* band structure because reduction to the first zone reflects the periodicity of the crystal lattice, even though no gaps are present. For weak potentials, the band structure is very nearly the empty lattice band structure for free electrons, but with small energy gaps at $k = 0$ and π/a.

Exercise 8.3 (C) *The three parameters, which are important in determining the qualitative behavior of the dispersion relation $\mathcal{E}(k)$, are: the well width (POTENTIAL WIDTH), its depth (POTENTIAL DEPTH), and the width of the barrier between the wells (LATTICE CONSTANT)− (POTENTIAL WIDTH). Starting from PRESET 1, vary each of these, keeping the other two fixed and determine qualitatively the effects on:*

1. *the mean positions of the bands;*
2. *the widths of the bands; and*
3. *the widths of the gaps.*

Be sure that you have set E MIN under the right hand graph to a low enough value that the plot includes all of the low lying bands. A click on the BAND LIST toggle gives a list of the bands, numbered from the lowest. (Occasionally a single band will be incorrectly listed in two segments. Watch out for two successive bands with the same index.)

8.3.2 Wave functions

A three-dimensional plot of the wave function for the ENERGY specified in the panel is shown on the left. The plot is for z ranging over a single cell. You can change your viewpoint by holding down the left hand mouse button and dragging in the display window. In the initial view you're looking along the z-axis with the real part of the wave function plotted to the right and the imaginary part plotted vertically. By changing the viewpoint to look perpendicular to the z-axis, you can see the real or imaginary parts of the wave function, or some linear combination of the two, the plots you normally see in the textbooks. Note that although the projections of the function have obvious zeros, the complex function does not, except for energies right at the edge of the band. The other two plots, PSI-SQUARED and the PHASE OF PSI, are displayed over several cells. In the phase plots the phase, in units of π, is restricted to lie in the range -1 to 1: the discontinuities of 2π in phase have no physical significance. You may find one or another of the representations more useful; but use them all since they offer different insights. Be sure you're comfortable with how to relate them to each other. Finally, the BAND INDEX to the left of the SHOW button gives the band index for the displayed wave function.

Exercise 8.4 (M*) *Use PRESET 1 with the PSI selection from the SHOW menu (the "red fan") to estimate the phase of the wave function,*

$$\phi(z) \equiv \tan^{-1}[\mathrm{Im}\psi(z)/\mathrm{Re}\psi(z)], \qquad (8.6)$$

at $z = a/2$ relative to its phase at $z = -a/2$. Be sure you get the same answer as is given by the plot of PHASE OF PSI versus z.

We will refer to this accumulation of phase on moving through the cell as the *phase shift across the cell* defined by

$$\Delta\phi(b, k) \equiv \phi(a/2) - \phi(-a/2), \tag{8.7}$$

where the phase $\phi(z)$ from Eq. (8.6) is evaluated for the state of wave vector k in the band with index b.

Exercise 8.5 (M*) *Check out this phase difference across the cell as you display wave functions from different energies in the same band. (To check quickly a number of functions in the same band, ZOOM the energy axis of the band structure plots so the single band fills the graph. Then DOUBLE-CLICK on the energies of interest to see the wave functions.) How does the phase difference vary as you move from the bottom to the top of the band?*

Exercise 8.6 (C) *What are the relative amplitudes at $z = -a/2$ and $z = a/2$? Picture what happens to the function as you move from cell to cell through the crystal? Is that consistent with what you see in the $|\psi|^2$ plot?*

Exercise 8.7 (M*) *What happens to the phase shifts across the unit cell $\Delta\phi(b, k)$ when you compare functions at the same k value but from different bands? Make a concise summary of the dependence of the phase shift as a function of both wave vector k and band index b.*

Exercise 8.8 (C*) *For PRESET 1 compare the wave functions at the top and bottom of the third band with that for a state near the middle of the band. Watch particularly carefully the change in character of the wave function as the band edge is approached. Here PSI-SQUARED and the PHASE OF PSI may be most useful. What evidence can you present that near the band edge the function has primarily standing wave character while in the middle of the band it is more like a simple traveling wave? (Hint: make $|\psi|^2$ plots and phase plots for both a traveling wave e^{ikz}, and a standing wave $\cos kz$.)*

\# 2   **Exercise 8.9 (M**)** *Using PRESET 2, COPY GRAPH the PSI-SQUARED graph and STEAL DATA the PHASE OF PSI plot to compare the two. You will find that wherever $|\psi|^2$ is small, the phase varies rapidly with z, and when $|\psi|^2$ is large the phase varies slowly. Explain why. (Hint: write $\psi(z) = |\psi(z)|e^{i\phi(z)}$ and think about conservation of particle current or about the invariance of the Wronskian.)*

\# 1 ♀ **Exercise 8.10 (C)** *Return to* PRESET 1 *and compare the z dependence of the electron density* $|\psi|^2$ *just above and just below the first energy gap. Which takes best advantage of the attractive potential of the square well?*

Exercise 8.11 (M**) *Use* PRESET 1 *with* E MIN *and* POT.DEPTH *equal* −50 eV *and* E MAX *equal* 100 eV. *The band energies for the high energy states are consistently higher than the free electron energies, while the lowest lying bands typically have energies lower than the free electron values. This is reflected in the* PSI-SQUARED *plots. Particularly for the lowest band, the wave functions are clearly "designed" to take advantage of the attractive potential. For the high lying bands the electron density is quite consistently higher in the barriers, not the wells. Find a classical argument for this (possibly) unexpected behavior. (Hint: classically, where does the electron spend most of its time?)*

8.3.3 Ramsauer effect**

Exercise 8.12 (M**) *For* PRESET 1 *change the* POTENTIAL WIDTH *to* 2 Å. *Then look at* PSI-SQUARED *and the* PHASE OF PSI *graphs for the two energies* $\mathcal{E} = 37.6$ *and* 27.6 eV. *You might want to* COPY GRAPH *one of them to have both available at once. In one, the solution in the well is a pure traveling wave; in the other it is a pure traveling wave in the barrier. What is the evidence for the claim of the pure traveling wave character of portions of each solution? The implication is that a wave incident on a single barrier (well) would be transmitted without scattering. Find the condition on the energy of the incident traveling wave to give zero reflection probability for a barrier (well) of width 2 Å and height (depth) 10 eV. Do the energies chosen for this exercise satisfy these conditions? (Remember that "bloch" measures state energies relative to the top of the barriers between the wells.)*

You are seeing the phenomenon of resonant transmission through a well (or across a barrier). The phenomenon is analogous to the resonant transmission of light through an ideal Fabry–Perot interferometer or of microwaves through a dielectric obstacle in a waveguide. A three-dimensional quantum analog is the Ramsauer effect in electron scattering by atoms. The scattering cross section for electrons by rare-gas atoms has a null for a critical value of the energy of the incident electron. At the critical energy the *s*-state wave function of the electron has one extra radial node relative to the function for a null potential. See Bohm [34, pp. 246, 568] for more on the Ramsauer effect.

Exercise 8.13 (M**) *Predict the energy for some other Ramsauer resonances and let "bloch" check your result.*

8.4   Gap states*

\# 3   When, as in PRESET 3, "bloch" chooses an energy $\mathcal{E}$ that happens to
yield a k value which is imaginary it calls that energy a forbidden energy
or an energy in a gap. This is because an imaginary part for k implies
that the corresponding Bloch function will increase exponentially moving
in one direction or the other along the one-dimensional crystal. Such
a function is inadmissible for an infinite medium on physical grounds,
though it may be a perfectly satisfactory mathematical solution to the
Schrödinger equation.

Are there situations in which there is physical significance to these
solutions? Yes, if we're concerned with surface, not bulk, properties of
the crystal! And "bloch" can show you the solutions if you double-click on
an energy in a gap; any k value will work, "bloch" reads only the energy
coordinate of the cursor.

Exercise 8.14 (C*) *Explore the wave functions corresponding to ener-*
gies in the gap. (In the PSI *display you need to rotate the axes to see*
anything.) How do these functions for the forbidden *states differ from*
the allowed *solutions? How do the gap states near the edge of the band*
gaps differ from the ones in the center of the gap?

Exercise 8.15 (M*) *Show formally that the gap solutions can carry no*
current.

Exercise 8.16 (M*) *In what sense do the gap functions still satisfy*
Bloch's theorem?

Exercise 8.17 (C**) *The eigenfunction for a state at the upper edge*
of a band may be considered either as the limit of a band function as the
band edge is approached from below, or as the limit of a gap function
as the band edge is approached from above. Explain how the band and
gap functions, though qualitatively different in their variation from cell
to cell, can nonetheless both approach the same limit at the band edge.

Do you see the connection between these solutions in the gaps for the
electron problem and the solutions in Chapter 4 for the lattice vibration
problem in which the chain was driven at a frequency above the maximum phonon frequency? Also, recall the discussion in physical optics of
the evanescent wave outside a dielectric when light is incident on the interface at an angle greater than the angle for total internal reflection. The
field extends outside the dielectric into the vacuum but is exponentially
attenuated. This is similar physics in a different context.

8.5 Nearly free electrons

In one dimension, numerical solutions are easily generated. In two and three dimensions we can do the same; the process is laborious but is the way to get useful numerical results. Such procedures give very little physical insight, however. On the other hand, there are two analytic approaches which, though rarely useful for serious calculations, are able to deepen our understanding of the physics of band structures in more dimensions. The first is the *nearly free electron* (NFE) approximation. Actually, for subtle reasons, this works better than we might expect for the valence electrons of many metals. We can try it out in one dimension by comparing the NFE picture with our numerical results. Later we will look at the second approach, the *tight binding* model.

The NFE approximation is a perturbation approach in which the small parameter is, roughly, the ratio of the strength of the periodic potential to the kinetic energy of the electrons. It is expected to be appropriate for ranges of energies in which the forbidden band gaps are narrow compared with the allowed band widths. The principal results are that:

1. far from the band edges, the dispersion relation is very nearly that of free electrons;
2. gaps develop at the Brillouin zone boundaries, the bisecting planes of the reciprocal lattice vectors; and
3. except near the edges of the zone face, the size of the gap at the boundary bisecting the reciprocal lattice vector $\mathbf{G}$ is twice the magnitude of the corresponding Fourier component of the periodic potential, $\Delta E_{\mathbf{G}} = 2V_{\mathbf{G}}$, with

$$V_{\mathbf{G}} \equiv \frac{1}{V_{\text{cell}}} \int_{\text{cell}} V(\mathbf{r}) e^{i\mathbf{G}\cdot\mathbf{r}} d^3 r \rightarrow (1/a) \int_{-a/2}^{a/2} V(z) e^{2\pi i n z/a} dz. \quad (8.8)$$

The second form is appropriate to calculate the gaps, $\Delta E_n = 2V_n$, for the one-dimensional case of "bloch" , where the reciprocal lattice vectors are $G_n = 2\pi n/a$, and the gaps can be labeled by the band index in ascending order in energy. "bloch" uses the $2\pi/a$ ambiguity in k to represent the dispersion relation in terms of k values in the first Brillouin zone, the *reduced zone scheme*, as in Figure 8.1(b). An alternative representation is the *extended zone scheme* of Figure 8.1(a), in which the $\mathcal{E}(k)$ is plotted as a single valued function of k. The two representations are equivalent and are related by translation of pieces of the dispersion relation parallel to the k-axis by appropriate reciprocal lattice vectors, $G = 2\pi n/a$.

Exercise 8.18 (M) *Use* PRESET 1 *and change the* POTENTIAL WIDTH #1 *to 0.1 Å and the* POTENTIAL DEPTH *to* −20 *eV.* CALCULATE BANDS *and obtain the widths of the first four band gaps from the data in the* BAND

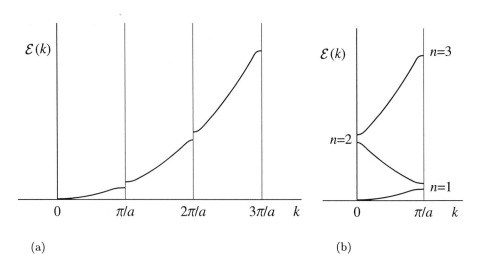

(a)   (b)

Figure 8.1: Different zone schemes: (a) extended zone scheme; (b) reduced zone scheme. Only positive k are illustrated.

LIST. *Why are they all nearly the same? Repeat for the pairs of (POT. WIDTH, POT. DEPTH) equal to (0.2 Å, −10 eV) and (0.4 Å, −5 eV). Why do the gaps decrease with increasing energy for the larger widths of the wells?*

Exercise 8.19 (M*) *Show for the potentials of Exercise 8.18 that the Fourier coefficients for $G \ll 2\pi/w$, where w is the width of the attractive well, are equal to 0.5 eV. How do the gaps calculated by "bloch" in Exercise 8.18 compare with the NFE prediction? At what value of G will the Fourier coefficient be reduced to roughly 0.25 eV? Is this reduction in the magnitude of the gap properly reflected in the band structure given by "bloch"?*

4 **Exercise 8.20 (M*)** PRESET 4 *defines a potential with a lattice constant of 4 Å, a square well width of 2 Å, and a well depth of −4 eV. Table 8.1 gives values from "bloch" of some of the energy gaps for a number of well depths. Take the necessary measurements to complete the table. Now use the NFE approximation (to first order only) to predict the magnitudes of the first three band gaps and compare your results with the numerical calculations. Comment on any discrepancies that you see.*

Exercise 8.21 (M**) *Work out the value of the magnitude of the second energy gap in Table 8.1 using the NFE picture to second order. (Hint: you need to include coupling of the states at the second gap both to the lowest state at $k = 0$ and to the states at the fourth gap.)*

Table 8.1: Gap widths in eV for 2 Å wells of various depths with a 4 Å lattice constant.

	Well depth			
	−1 eV	−2 eV	−4 eV	−8 eV
First gap	0.637	1.271		
Second gap	0.027	0.105		
Third gap			0.421	0.827

Exercise 8.22 (C**) *Have "bloch"* CALCULATE BANDS *for the* COSINE *choice from the* POTENTIAL *menu with the* POTENTIAL DEPTH *set to* −4 *eV. Why is the first gap so much larger than any of the others? Why aren't all the others precisely zero?*

For subtle reasons (see the discussion in reference [1] of the pseudopotential method) the band structure for the valence electrons in many metals is remarkably free-electron-like. A successful semiempirical approach to a description of the band structure is to take the few Fourier coefficients defining the gaps near the energies of interest as adjustable parameters. These gaps, determined by fits to a small set of experimental data, then suffice to define the relevant energy bands, allowing prediction of additional properties of the metals.

8.6   Tight binding (TB)

8.6.1   Background

For many metals the NFE picture gives an appropriate picture of the higher lying bands and of the critically important band(s) containing the Fermi surface. In other cases, however, the periodic potential is strong enough that the NFE picture is useless. An alternative approach starts from the opposite limit, a very strong potential which localizes the electron in atomic-like states near the potential minima. Band states are constructed by starting with atomic states and treating the overlap of the electron from one atom to the next as a perturbation. This is the *tight binding* (TB) approximation. If the interaction between atoms is weak, then the electron wave function near any atom should be well approximated by an atomic wave function on that atom. The extended wave functions may then be represented approximately as a linear combination of the atomic wave functions. Bloch's theorem tells us what linear combinations to take. For a band with index b, with wave functions formed as linear combinations of atomic functions $\varphi_b(z)$ for the bth atomic level, the TB Bloch function for wave vector k is

$$\psi_{b,k}(z) = \frac{1}{\sqrt{N}} \sum_{n=1}^{N} e^{ikna} \varphi_b(z - na). \qquad (8.9)$$

The number of atoms in the crystal is N and $\varphi_b(z - na)$ is the atomic function for the bth atomic level centered on the nth atom. If $V(z)$ is the atomic potential, the dispersion relation for the bth band obtained using this TB function is

$$\mathcal{E}_b(k) = \mathcal{E}_b^0 - \beta_b - 2\gamma_b \cos ka, \qquad (8.10)$$

with

$$\beta_b = -\int_{-a/2}^{a/2} dz |\varphi_b(z)|^2 V(z - a) \qquad (8.11)$$

and

$$\gamma_b = -\int_{-a/2}^{a/2} dz \varphi_b^*(z) V(z) \varphi_b(z - a). \qquad (8.12)$$

The picture which is derived from the TB model is of bands of allowed energies, each band corresponding to an atomic level from which it evolved. The bands are centered at energies which are lowered from the free atom values by β, the energy of interaction of the overlap of the electron density on one atom with the potential of a neighboring atom. The bands have a width 4γ, where γ is the energy of interaction of the *overlap electron density* $\varphi^*(z)\varphi(z-a)$ with the atomic potential on either site.

The validity of the TB picture relies on weak overlap: the band widths 4γ should be small compared with the band gaps. We see the two methods as representing opposite limits: the NFE approximation gives a convenient representation for systems with wide bands and narrow gaps, while the TB method is useful for narrow band systems with large energy gaps.

8.6.2   Square well potential

There are some simple predictions we can make using the TB approach without getting into the nitty-gritty of calculations. For a simple square well atomic potential, in one dimension, can we deduce how the widths of the low lying energy bands depend upon the interatomic separation? Let's explore the band structure for *square well atoms* with a well depth of -100 eV, a width $w = 2$ Å, and a variety of lattice constants $a = 2.5$, 3.0, 3.5, and 4.0 Å.

Table 8.2: Band width $\Delta\mathcal{E}$ and band energy $\mathcal{E}$ in eV for 2 Å wells of depth 100 eV with lattice constant a.

		Lattice constant a			
		2.5 Å	3.0 Å	3.5 Å	4.0 Å
Third band	$\Delta\mathcal{E}$				0.054
	$\mathcal{E}$				-43.25
Second band	$\Delta\mathcal{E}$			0.038	0.0041
	$\mathcal{E}$			-74.0937	-74.0936
First band	$\Delta\mathcal{E}$		0.058	0.0050	0.00030
	$\mathcal{E}$		-93.4432	-94.4431	-93.4431

Exercise 8.23 (M*) *First calculate the isolated atom energies for the three lowest states of a single square well of depth -100 eV. For the energies of interest, -100 eV $< \mathcal{E} < 0$ eV, the solutions within the well will be sinusoidal or cosinusoidal, and those outside the well will be decaying exponentials. At the boundary of the well, the wave function and its derivative must be continuous. You should be able to get an implicit equation for the energy which can be solved rather quickly with a calculator by trial and error. Be sure to take advantage of the symmetry of the problem and look explicitly for even and odd symmetry solutions. To start with, try energies near -40, -75, and -90 eV for the lowest three states.*

Armed with this information, change the LATTICE CONSTANT *of* PRE-SET 5 *to 2.5 Å and find, for the lowest three bands, their band centers and their band widths. To save you time, Table 8.2 gives some corresponding results for other* LATTICE CONSTANTS. *Accurate results for narrow bands require giving "bloch" a very narrow range of energies* E MIN *to* E MAX *in which to work. How well do the band center positions agree with the "free atom" energies? Plot the data in a fashion which tests the prediction of an exponential dependence of the band width on the barrier thickness,* (LATTICE CONSTANT$-$ POTENTIAL WIDTH).

5

Exercise 8.24 (M**) *Using the TB idea, predict quantitatively how the band widths should depend upon the lattice constant and compare this with your experimental data. (Hint: the first part of Exercise 8.23 should allow a prediction of the rate at which the wave functions fall off in the barrier between adjacent atoms in terms of the energies of these* states. Use this information to argue quantitatively, without doing the integrals, for the dependence of band width on lattice constant.)

Exercise 8.25 (C*) ZOOM in on the second band in PRESET 5, or RE-CALCULATE the band structure for a narrow range of energy which includes it, with the LATTICE CONSTANT = 3 Å. DOUBLE-CLICK on states at various points in the band. What changes do you see in the PSI plots as you move through the band? And in the PHASE OF PSI plots? Is $\Delta\phi(b, k)$, the phase shift across the cell required by the Bloch theorem, occurring primarily in the well or in the barrier? Explain why. Why does the PSI-SQUARED plot show so little variation?

8.7 'Atomic' and 'molecular' potentials

8.7.1 Atoms

The SQUARE WELL and COSINE potentials are useful for illustration, but they don't look much like atomic potentials. PRESET 6 uses the ATOM choice of POTENTIAL in "bloch", which is a cut-off attractive $1/z$ potential: $V(z) = A/z$ with A negative. It is cut off both at the cell boundary and at $z = \pm 0.1$ Å. (The cut-off at the cell boundary does *not* strictly correspond to the TB model, but is chosen to simplify the "bloch" algorithm.)

\# 6

For the narrow TB bands, the wave functions are essentially atomic functions. They are turned into band functions simply by writing down a TB linear combination of them, with a phase factor in each cell determined by the requirements of the Bloch theorem. Only in the third band and above is the effect of the overlap from one cell to the next apparent in the wave function.

Exercise 8.26 (M) Note down from the BAND LIST the positions of the edges of the three bands with negative energies, (or the single energy of the band for the flat one). Then decrease the LATTICE CONSTANT in steps of 0.5 Å, taking data for each lattice constant. Construct a graph showing the allowed and forbidden ranges of energy as a function of lattice parameter.

Exercise 8.27 (C**) If you increase the lattice constant to 5 Å, even the third band becomes fairly narrow. Check out the wave function for a state in this band. Explain carefully why it takes the shape it does, with the very small peak near the origin, a pair of nodes, and a larger amplitude relatively far out from the attracting center. (Hint: recall the orthogonality conditions on sets of eigenfunctions.)

8.7.2   Molecules*

"bloch" allows us to arrange pairs of 'atoms' to illustrate something of the electronic properties of a molecular crystal. In many molecular crystals the intermolecular interactions are weak compared with the intramolecular interactions. This is illustrated in PRESET 7 in which diatomic   # 7
molecules with interatomic separation (MOLECULE SEPARATION) $d = 1$ Å are arranged on a lattice of cell constant $a = 2.3$ Å. The lowest bands come in pairs, the two members of the pair corresponding to the bonding and antibonding molecular states made up from appropriate linear combinations of orbitals on the individual atoms.

Exercise 8.28 (C) *How does the band structure change if the molecular size,* MOLECULE SEPARATION, *is fixed and the distance between molecules is changed by varying the* LATTICE CONSTANT? *How does it change if the molecular size, (*MOLECULE SEPARATION*), is varied but the intermolecular spacing is fixed by changing both the* LATTICE CONSTANT *and the* MOLECULE SEPARATION *by the same amount?*

Exercise 8.29 (C*) *Investigate the nature of* PSI *in the first four bands. (*CALCULATE BANDS *for a narrow range of energies containing a pair of bands of interest. If these are too narrow to select the desired states by* DOUBLE-CLICKING, *then reduce the* LATTICE CONSTANT *to broaden the bands.) In what sense can these molecular functions be thought of as linear combinations of atomic orbitals? Which atomic states are used in which bands? What are the relative signs of the coefficients of these atomic states? Why are the energies of the lower one of each pair (the bonding levels) lower than those of the upper one of each pair (the antibonding states)?*

Exercise 8.30 (C**) *Look at the* WAVE FUNCTION *for a state near the middle of the lowest band, for several different* LATTICE CONSTANTS. *Explain why the phases of the* ATOMIC *functions on the two atoms in the cell are not quite the same. How does this phase difference depend upon the lattice constant and why?*

8.8   Summary

The periodicity of crystalline structures has a number of important implications about the nature of the quantum states of electrons in crystals. The wave functions themselves are not periodic, but rather have the Bloch property that they differ only by a phase factor on going from one cell to the next. Thus in many respects they behave like plane wave functions, despite their greater complexity. The states are characterized, in part, by the wave vector which defines the magnitude of the phase accumulated on moving from cell to cell.

The associated energy states fall into bands of allowed energies (the energy bands) and bands of forbidden energies (the energy gaps). Low lying band states closely resemble linear combinations of atomic states. The TB picture gives a representation of these bands which are characterized by being narrow with wide gaps. Higher lying bands are typically broad with narrow gaps, a picture suggested by the NFE model, though justified as a picture of real materials only by introduction of the ideas of the pseudopotential.

Well within the energy bands, the wave functions resemble traveling waves and correspond to an electron propagating without scattering through the lattice. At the band edges, the functions become standing wave in character, much as did the displacement functions for phonons at the energy of the maximum of the dispersion relation.

8.A   Deeper exploration

Barrier transmission model   Ashcroft and Mermin [1, p. 146] develop an alternative model for the one-dimensional band problem. The periodic potential is presented as an array of barriers between the minimum of the potential in one well and the minimum in the next. The conditions for allowed and forbidden states are formulated in terms of the transmission and reflection coefficients for electron waves incident on the barrier. Use "bloch" to test quantitative predictions based on this development.

Kronig–Penney model   The potential labeled SQUARE WELL in "bloch" is conventionally known as the Kronig–Penney model. Since the forms of the solutions of the Schrödinger equation are known both in the wells and the barriers, much of the analysis of the band structure can be done by matching of boundary conditions at the well/barrier boundaries. This model is discussed in Kittel [6, p. 180] and in many other texts. Use "bloch" as an experimental tool to verify predictions developed from one of these treatments of the Kronig–Penney model.

TB for the square well   The integrals needed to implement the TB treatment of the square well potential are straightforward. Develop TB predictions for the lowest few bands for a square well potential of fixed depth and varying lattice constant. Determine empirically, by comparison with "bloch", the range of validity of the TB method as the lattice constant is decreased.

Ramsauer effect Explore the Ramsauer effect [34, pp. 246, 568] in more detail, with particular attention to the analogs in physical optics, resonant transmission of microwave structures, and resonant tunneling of electrons through multiple barriers.

8.B "bloch" – the program

"bloch" computes the band structure and wave function of an electron in a one-dimensional periodic potential.

8.B.1 Electron states in a periodic potential

To determine the electronic states in a one-dimensional crystal, "bloch" assumes that the interaction of each electron with both the atomic nuclei and the other electrons is described by an effective, one-electron potential $V(z)$. In this *independent electron approximation* the state of an electron with energy $\mathcal{E}$ and mass m can be described by the Schrödinger equation

$$\left[-\frac{\hbar^2}{2m} \frac{d^2}{dz^2} + V(z) \right] \psi(z) = \mathcal{E}\psi(z). \qquad (8.13)$$

If the potential is periodic with periodicity a, $V(z) = V(z + a)$, then Bloch's theorem shows that solutions $\psi(z)$ of the Schrödinger equation (8.13) exist which take on the form

$$\psi_{b,k}(z + a) = e^{ika} \psi_{b,k}(z), \qquad (8.14)$$

where k is a wave vector to be determined. For each k there exist multiple solutions $\psi_{b,k}$ which are labeled by the band index b. Energies for which k is real give solutions for which $|\psi(z)|^2$ is periodic and are said to lie in a *band*. Energies for which k is imaginary give solutions with an exponentially decaying (or increasing) $|\psi(z)|^2$ and are said to lie in a *band gap*.

Four forms of the periodic potential $V(z)$ can be chosen (see Table 8.3): three modeling a periodic arrangement of single atoms, and one modeling the periodic arrangement of molecules.

8.B.2 Setting up the "bloch" algorithm

Given a periodic potential $V(z)$ and an energy eigenvalue $\mathcal{E}$, "bloch" uses a fifth order Runge–Kutta algorithm[1] to solve the Schrödinger equation (8.13) , for the wave function $\psi_{b,k}(z)$, with wave vector k and band index b. The band structure $\mathcal{E}_b(k)$ follows from determining k for a whole range of energies $\mathcal{E}$.

---

[1]For a short description of our `rkf` algorithm see Section 6.B.2.

Table 8.3: Potentials in "bloch". The form of the potentials is specified in the unit cell $-a/2 \leq z \leq a/2$, and the potentials are periodic in z with the periodicity of the LATTICE CONSTANT a. The negative POTENTIAL DEPTH A, the POTENTIAL WIDTH of the SQUARE WELL w, and the MOLECULE SEPARATION d of the two atoms in the MOLECULE can be specified on the panel.

	Potential $V(z)$
Square Well	$\begin{cases} A & \text{if } |z| < \frac{1}{2}w \\ 0 & \text{otherwise} \end{cases}$
Cosine	$\frac{1}{2} A \left[\left[\cos\left(\frac{2\pi}{a} z\right) + 1 \right] \right]$
Atom	$\begin{cases} A & \text{if } |z| < 0.1 \text{ Å} \\ 0.1 A/|z| & \text{otherwise} \end{cases}$
Molecule	$\begin{cases} A + 0.1 A/|z + \frac{1}{2} d| & \text{if } |z - \frac{1}{2} d| < 0.1 \text{ Å} \\ A + 0.1 A/|z - \frac{1}{2} d| & \text{if } |z + \frac{1}{2} d| < 0.1 \text{ Å} \\ 0.1 A \left(1/|z + \frac{1}{2} d| + 1/|z - \frac{1}{2} d| \right) & \text{otherwise} \end{cases}$

We start by noting that any solution to a second order linear differential equation such as the Schrödinger equation (8.13) can be written as a linear combination of any two linearly independent solutions ψ_1 and ψ_2:

$$\psi_{b,k}(z) = A\psi_1(z) + B\psi_2(z). \tag{8.15}$$

A solution ψ_i is uniquely determined by its initial values $\psi_i(-a/2)$ and $\psi_i'(-a/2)$, where the prime denotes the derivative with respect to z. "bloch" determines two independent solutions $\psi_1(z)$ and $\psi_2(z)$ by numerical integration of the Schrödinger equation (8.13) for two different (and conveniently chosen) sets of initial values at $z = -a/2$.

Next, we know from Bloch's theorem (8.14) that

$$\psi_{b,k}\left(-\tfrac{a}{2}\right) = e^{-ika}\,\psi_{b,k}\left(\tfrac{a}{2}\right) \quad \text{and} \quad \psi_{b,k}'\left(-\tfrac{a}{2}\right) = e^{-ika}\,\psi_{b,k}'\left(\tfrac{a}{2}\right). \tag{8.16}$$

These boundary conditions require the coefficients A and B in Eq. (8.15) to satisfy

$$\left[\psi_1\left(-\tfrac{a}{2}\right) - e^{-ika}\psi_1\left(\tfrac{a}{2}\right)\right] A + \left[\psi_2\left(-\tfrac{a}{2}\right) - e^{-ika}\psi_2\left(\tfrac{a}{2}\right)\right] B = 0, \tag{8.17}$$

$$\left[\psi_1'\left(-\tfrac{a}{2}\right) - e^{-ika}\psi_1'\left(\tfrac{a}{2}\right)\right] A + \left[\psi_2'\left(-\tfrac{a}{2}\right) - e^{-ika}\psi_2'\left(\tfrac{a}{2}\right)\right] B = 0, \tag{8.18}$$

where the values of $\psi_1(a/2)$, $\psi_1'(a/2)$, $\psi_2(a/2)$, and $\psi_2'(a/2)$ can be obtained through numerical integration of the Schrödinger equation. The pair of equations (8.17) and (8.18) has a non-trivial solution for A and B if and only if

$$\left[\psi_1\left(-\tfrac{a}{2}\right) - e^{-ika}\psi_1\left(\tfrac{a}{2}\right)\right] \times \left[\psi_2'\left(-\tfrac{a}{2}\right) - e^{-ika}\psi_2'\left(\tfrac{a}{2}\right)\right]$$

$$- \left[\psi_2\left(-\tfrac{a}{2}\right) - e^{-ika}\psi_2\left(\tfrac{a}{2}\right)\right] \times \left[\psi_1'\left(-\tfrac{a}{2}\right) - e^{-ika}\psi_1'\left(\tfrac{a}{2}\right)\right] = 0. \quad (8.19)$$

We are now in a position to determine the Bloch vector k. For the initial values chosen by "bloch", $\psi_1(-a/2) = 1$, $\psi_1'(-a/2) = 0$, $\psi_2(-a/2) = 0$, and $\psi_2'(-a/2) = 1/a$, it follows from Eq. (8.19) that

$$e^{ika} = \tfrac{1}{2}\left[\psi_1\left(\tfrac{a}{2}\right) + a\,\psi_2'\left(\tfrac{a}{2}\right)\right] \pm i\sqrt{1 - \tfrac{1}{4}\left[\psi_1\left(\tfrac{a}{2}\right) + a\psi_2'\left(\tfrac{a}{2}\right)\right]^2}, \quad (8.20)$$

where we used from Liouville's formula[2] that the Wronskian is independent of z and thus equal to $1/a$. We can find k from either the real or imaginary part of (8.20). If $-1 \le \tfrac{1}{2}[\psi_1(a/2) + a\psi_2'(a/2)] \le 1$ in (8.20) then k is real,

$$k = \cos^{-1}\left\{\tfrac{1}{2}[\psi_1\left(\tfrac{a}{2}\right) + a\psi_2'\left(\tfrac{a}{2}\right)]\right\}, \quad (8.21)$$

and the energy $\mathcal{E}$ lies in a band. In all other cases k is purely imaginary and the energy lies in a band gap.

8.B.3 "bloch" algorithm

Wave function With e^{ika} given by Eq. (8.20), it is straightforward to determine the coefficients A and B in Eq. (8.15) and thus the wave function $\psi_{b,k}(z)$. Disregarding the normalization of $\psi_{b,k}(z)$ we can choose $A = 1$ and solve (8.17) for B using the initial conditions specified above. The wave function $\psi_{b,k}$ is then given by

$$\psi_{b,k}(z) = \psi_1(z) + \left[e^{ika} - \psi_1(a/2)\right]\psi_2(z)/\psi_2(a/2). \quad (8.22)$$

Note that $\psi_{b,k}(z)$ in Eq. (8.22) indeed satisfies Bloch's theorem (8.14) since $\psi_{b,k}(-a/2) = 1$ and $\psi_{b,k}(a/2) = e^{ika}$. It also follows from Eq. (8.22) that $\psi_{b,k}$ is real for energies $\mathcal{E}$ in a gap (k imaginary) and complex for energies $\mathcal{E}$ in a band (k real). (Recall that $\psi_1(z)$, $\psi_2(z)$ and their derivatives are

---

[2]Liouville's formula tells us how the Wronskian $W(z) = \psi_1(z)\psi_2'(z) - \psi_2(z)\psi_1'(z)$ transforms under a translation, $W(z) \to W(z + z_0)$. More explicitly, we find

$$W(z) = W(z_0)\, e^{-\int_{z_0}^{z} f_1(t)\, dt},$$

where f_1 is the coefficient of the $d\psi/dz$ term in the Schrödinger equation (8.13) and hence zero. Therefore the Wronskian is independent of z, and we use its known value $W(z = -a/2)$.

real for any z since the Schrödinger equation (8.13) is real and so is the choice of initial ψs by "bloch".)

Band index   "bloch" also computes the index b of bands (and gaps). The index of a band is determined from $\Delta\phi(b, k)$, the difference in phase of $\psi_{b,k}(z)$ at the two ends of the unit cell $z = -a/2$ and $z = a/2$, as defined in Eq. (8.7). It follows from Bloch's theorem (8.14) that the difference in phase $\Delta\phi(b, k)$ increases by just π as one moves through any one band from the lowest to the highest energy in the band. The difference in phase increases in this manner by π for each successive band, and the number of half revolutions (per unit cell) of the wave function in the complex plane thus provides a means to enumerate the bands. In particular, the index of a band is the number of half-revolutions $\Delta\phi/\pi$ at the top of the band.

The choice of initial values in "bloch" allows us to count the half-revolutions of the phase of the wave function in a particularly simple way. Because the values of $\psi_1(z)$, $\psi_2(z)$, and their derivatives are real, it follows from Eq. (8.22) that the imaginary part of $\psi_{b,k}$ for a given energy $\mathcal{E}$ (in a band) is given by a constant times $\psi_2(z)$. Since $\psi_{b,k}(-a/2) = 1$, the number of zero crossings of $\psi_2(z)$ in the unit cell determines the number of half-revolutions and thus the index b of the band. (For a gap-state, "bloch" also computes the number of zero crossings of $\psi_2(z)$ per unit cell as a means to enumerate the gaps. This scheme is not foolproof, however, and leads to some awkwardness in the algorithm described below.)

Band structure   To determine the band structure $\mathcal{E}_b(k)$ "bloch" uses the following algorithm which computes the band edges with the user-defined accuracy $\Delta\mathcal{E}$ (see the CONFIGURE menu), finds all bands in the specified energy range, and minimizes the number of (time-consuming) numerical integrations of the Schrödinger equation (8.13).

"bloch" keeps a list of states indexed by i that have associated with them an energy $\mathcal{E}$, the corresponding Bloch vector k, the band index b, and the information whether the energy lies in a band or gap. This list is sorted by increasing energy. Initially, there are only two states in this list, corresponding to the specified E MIN and E MAX.

Given two states with indices i and $i+1$, "bloch" checks the conditions specified below to see whether it is necessary for a "good" resolution of the band structure to insert into that list an additional state with an energy $\mathcal{E} = \frac{1}{2}[\mathcal{E}(i) + \mathcal{E}(i+1)]$. When "bloch" inserts a new state into the list, the new state is indexed $i+1$ and all states with a higher index have their index increased by one. "bloch" then again checks the same set of conditions for the state i and the new state (which now has index $i+1$), and possibly inserts yet another state at the midpoint energy and so on. If the resolution of the band structure is satisfactory (see below), "bloch" repeats the process with the next two states $i+1$ and $i+2$ until the maximum energy (E MAX) is reached. To limit the computation time for an accuracy $\Delta\mathcal{E}$ which is too high or energy range E MAX−E MIN which is too large, the band structure calculation is interrupted and an error message displayed if the list contains more than 1000 points.

A new state with an energy half-way between its neighbors (state i and $i+1$) is inserted if any of the following conditions is true:

1. Both states lie in a band and any of the following conditions is true:

    (a) The band indices b of the two states differ by more than 1. This means that there is at least one band and two gaps between the two states.

    (b) The two states have band indices b that differ by 1, and their energy difference is more than the user-defined energy threshold $\Delta\mathcal{E}$. In this case there is one gap between the two states. The energy threshold is specified in order to limit the search if the gap is narrower than $\Delta\mathcal{E}$.

    (c) The two states have the same band index b and their energies differ by more than 50 $\Delta\mathcal{E}$. The value 50 is chosen so that the resolution in energy within a band is much coarser (and thus the computation much faster) than the resolution at a band edge.

2. One state lies in a band and the other in a gap and their energy difference is more than $\Delta\mathcal{E}$ which guarantees that the band edges are determined to within $\pm\Delta\mathcal{E}$.

3. Both states lie in a gap and either of the following conditions is true:

    (a) The gap indices b of the two states differ and their energy difference is more than $\Delta\mathcal{E}$ which occurs when there is at least one band, whose width is less than $\Delta\mathcal{E}$, between the two gaps.

    (b) The gap indices of the two states are the same and their energy difference is more than 100$\Delta\mathcal{E}$. This is a safety condition which will find a band even when the index of the upper gap changes at the top of the gap and the index of the lower gap changes at the bottom of the gap.

4. There is only one state computed in a band, i.e., state i is in the band but the states $i-1$ and $i+1$ are not. To draw a band in the graph as a line, however, we need at least two states in the same band.

The energy threshold $\Delta\mathcal{E}$ can be specified in the CONFIGURE menu via the RESOLUTION r as $\Delta\mathcal{E} = r(\mathcal{E}_{max} - \mathcal{E}_{min})$.

Because the above algorithm will fail to find flat bands with band widths less than $\Delta\mathcal{E}$ (typically found for the ATOM and MOLECULE POTENTIAL), "bloch" inserts a flat band into the graph whenever the (final) list of states indexed by i contains two neighboring gap states whose index b differs, provided the states do not lie within 0.5 eV of a band state.

8.B.4 Displayed quantities

The left hand display in "bloch" can show the complex wave function (PSI), its magnitude squared (PSI SQUARED), or its phase (PHASE OF PSI) for the ENERGY specified. The BAND or GAP INDEX b is given in the upper left hand corner of the display. The right hand display shows the band structure for both the chosen POTENTIAL and free electrons. A BAND LIST can also be displayed.

The energy range E MIN to E MAX in which the band structure is computed can be chosen on the panel. The ENERGY at which the wave function is computed can be chosen either from the panel or by DOUBLE-CLICKING with the left hand mouse button on the desired energy in the band structure plot.

Wave function   "bloch" draws the real (blue axis) and imaginary parts (green axis) of the wave function versus z (black axis) in a *three-dimensional* plot. To generate the three-dimensional effect, "bloch" computes the real and imaginary parts of the wave function $\psi_{b,k}$ at 100 z values in the unit cell $z = -a/2$ to $z = a/2$ and generates for each z a polygon in a three-dimensional space with the four corners

$$\left.\begin{array}{l} (0,0,z), \\ (\text{Re}[\psi_{b,k}(z)], \text{Im}[\psi_{b,k}(z)], z), \\ (\text{Re}[\psi_{b,k}(z+\Delta z)], \text{Im}[\psi_{b,k}(z+\Delta z)], z+\Delta z), \\ (0,0,z+\Delta z), \end{array}\right\} \qquad (8.23)$$

where $\Delta z = a/100$. The polygons are then projected into the xy-plane of the screen, i.e., only the x and y screen coordinates of each polygon are drawn as the "red fan". The three-dimensional picture can be rotated by grabbing the picture and dragging it with the left hand mouse button pressed. The rotation takes every corner (x,y,z) of each polygon into (x',y',z'), and only the x'- and y'-coordinates are drawn. Clicking on the picture with the right hand mouse button resets the rotation to the original view along the z-axis.

Magnitude squared of the wave function   The square of the wave function $|\psi_{b,k}(z)|^2$ is drawn at 400 z values over four unit cells $z = -a/2$ to

$z = 7a/2$. The normalization is chosen such that $\int_{-a/2}^{a/2} dz |\psi_{b,k}(z)|^2 = a$: "bloch" requires the average, not the integral, of $|\psi_{b,k}(z)|^2$ to be unity. For comparison, the potential is displayed over the same unit cells in arbitrary units.

Phase of the wave function The phase of the wave function (8.6) is drawn at 400 z values over four unit cells $z = -a/2$ to $z = 7a/2$. The phase is displayed in units of π in the range -1 to 1, together with the potential (in arbitrary units). The discontinuities of 2 (corresponding to 2π) in this plot have no physical significance.

Band structure The band structure is displayed as a red solid line in the reduced zone scheme. The energy range of each band is listed in the BAND LIST. For comparison with the computed band structure, the band structure for the free electron approximation, $E(k) = \langle V \rangle + \hbar^2 k^2 / 2m$, is displayed in blue. The free electron band structure is shifted by the average potential $\langle V \rangle$ so that the two band structures agree in the high energy limit.

8.B.5 *Bugs, problems, and solutions*

If you get the error message that "bloch" ran out of trial points in the band structure calculation (i.e., the maximum number of 1000 computed energies was exceeded) then either choose a smaller energy range E MAX− E MIN or increase the RESOLUTION parameter (implying a reduced energy resolution) in the CONFIGURE menu.

To find the low lying, flat bands typically encountered for ATOM or MOLECULE POTENTIALS, "bloch" inserts a flat band into the band structure whenever two of the computed neighboring gap states differ in their index by 1 and the states do not lie within 0.5 eV of another band state. Because of limited precision in the numerical integration, it can happen that this condition is true even when no band is missing in the band structure, and an additional (false) band is shown in the band structure plot. These spurious flat bands can easily be recognized as *false* since they appear mixed in with broad bands.

The BANDLIST will occasionally incorrectly split a band into two pieces: two successive band listings with the same band index are a single band.

9

"ziman" – Dynamics of a single band-electron

Contents

9.1   Introduction

Sommerfeld's picture of the electron gas gave a useful model for discussing the properties of metals. How can that be? The valence electrons in a real metal move not in a "flat-bottomed box" but in a box full of negative ions with which they interact strongly. What are the consequences of the scattering from the ion core potentials? "bloch" has given a hint of the solution. The eigenstates of the system with the periodic potential are extended throughout the crystal, and still have a plane wave character. A wave packet formed by superposition of these eigenstates corresponds to an electron propagating through the perfect crystal without scattering. The electrical current carried by such electrons, accelerated by an applied field, is limited only by the scattering from lattice imperfections (impurities, etc.) and the lattice vibrations but not by scattering from the periodically arranged ion cores. Thus, in one sense, the electrons are behaving just as they did in the Sommerfeld model.

However, there are two important features of the electron states, as modified by the periodic potential, which we must explore. First, these quantum mechanical electrons respond to the forces of applied electric and magnetic fields in a bizarre manner, often not free-electron-like at all. They may respond to an applied force with an acceleration opposite to the force. In a magnetic field they may move in orbits which never close. This strange dynamic behavior is the subject of "ziman". Second, the interaction with the ion cores breaks the energy continuum of translational states of the Sommerfeld model into bands of allowed and of forbidden energy states. Some of the implications of this band structure for electronic transport are illustrated by "peierls".

9.2 $\mathcal{E}(\mathbf{k})$ and the contour plot

"ziman" simulates the dynamics of a single band-electron in static electric and magnetic fields. The electron moves in two dimensions in an energy band described, for $\mathbf{k}$ within the first Brillouin zone ($|k_x|, |k_y| \leq \pi/a$), by

$$\mathcal{E}(\mathbf{k}) = \frac{\hbar^2}{2m} \left[\left(k_x^2 - \frac{a^2}{2\pi^2} k_x^4 \right) + A \left(k_y^2 - \frac{a^2}{2\pi^2} k_y^4 \right) \right]. \tag{9.1}$$

The quartic terms in $\mathbf{k}$ are added to give an $\mathcal{E}(\mathbf{k})$ with zero slope normal to the zone boundaries. For a $\mathbf{k}'$ outside the first zone, the energy is generated by using this same formula with the argument $\mathbf{k} = \mathbf{k}' + \mathbf{G}$ with $\mathbf{G}$ chosen such that $\mathbf{k}$ is within the first zone. This is *not* the energy obtained by inserting $\mathbf{k}'$ into Eq. (9.1). A is an ANISOTROPY parameter, which provides for different curvatures of $\mathcal{E}(\mathbf{k})$ in the x- and y-directions.

"ziman" uses a square structure with lattice constant $a = 2.00$ Å, which gives a bandwidth for the isotropic case, $A = 1$, of 9.4 eV $= 1.49 \times 10^{-18}$ J. The Brillouin zone boundaries, indicated in blue, are at $k_x, k_y = \pi/a = \pm 1.57 \times 10^{10}$ m^{-1} and the real-space display on the left is bounded by the lines $x, y = \pm 40$ μm, which may be adjusted in the CONFIGURE dialog box.

1   The dependence of energy on **k** is shown in PRESET 1 by the contour plot on the right. The nine green lines are the loci in **k** space of those **k** values such that $\mathcal{E}(k) = $ constant with the constant chosen at roughly 1 eV intervals starting at the bottom of the band. If you are unfamiliar with contour maps, imagine looking down along the $\mathcal{E}$-axis of a three-dimensional plot of $\mathcal{E}(\mathbf{k})$ versus k_x and k_y. That plot is a two-dimensional surface with a minimum at the origin, maxima at the corners of the Brillouin zone, and saddle points at the middle of the edges of the zone. Now draw on that surface a line that corresponds to all of the points on the surface that are the same distance above the k_x, k_y coordinate plane. This is a contour line. The dispersion relation $\mathcal{E}(\mathbf{k})$ is shown only in the first Brillouin zone. Recall that the dispersion relations for electrons, just as for phonons, may be considered to be periodic functions in reciprocal space with the periodicity of the reciprocal lattice: the ambiguity of $\mathbf{k} \rightarrow \mathbf{k} + \mathbf{G}$.

Is it sensible to focus, as "ziman" does, on just a single band? What about the states in higher and lower energy bands? In response to applied fields, an electron wave packet with energy initially in a specific band changes its **k**-state but remains within the same band for electric and magnetic fields of "reasonable" magnitude. The physics of fields of "unreasonable" magnitude, i.e., Zener and magnetic breakdowns, are not addressed by "ziman". For "reasonable" fields the electron motion is confined to a single band.

9.3   Group velocities

The relation between the wave vector **k** of an electron and its velocity is more complicated for a band-electron than for the free electron. The relation $v_g(q) = d\omega(q)/dq$ noted in the "born" simulation of classical lattice dynamics becomes for the three-dimensional quantum mechanical electron

$$dr/dt = \mathbf{v}_g(\mathbf{k}) = \nabla_{\mathbf{k}} \mathcal{E}(\mathbf{k})/\hbar. \qquad (9.2)$$

Exercise 9.1 (M) *Verify that Eq. (9.2) implies the familiar* $\mathbf{v} = \hbar\mathbf{k}/m$ *for the case of the free electron dispersion relation.*

The relation (9.2) can be checked using "ziman". There are no fields applied in PRESET 1 so the electron remains stationary in **k**-space and moves with constant velocity, its group velocity $\mathbf{v}_g$, in real space. By clicking at different points in k-space you can reinitiate the program with the electron in any desired **k**-state.

Exercise 9.2 (C) RUN *"ziman" and test the relation between the velocity and the gradient of the dispersion relation $\mathcal{E}(\mathbf{k})$. In particular verify that:*

1. *For **k** values near the origin, the direction of the velocity is parallel to the direction of **k** and increases with increasing $|\mathbf{k}|$.*

2. *For **k** values at the origin, at the corners of the zone, and at the centers of the edges of the zone, the velocity is zero. Why is that?*

3. *For **k** values near the corner of the zone, the velocity is in the direction opposite to the vector connecting the zone corner to the **k**-point. Why would you expect that?*

4. *For **k** values on the zone edges, but not at the center or ends of the edges, the velocity is parallel to the edge of the zone. Why is that?*

Exercise 9.3 (M) *Use the cursor readout in real space and the clock to determine quantitatively the velocity vector for an electron in the states $(k_x\ k_y)$ = (0.1 0.1) π/a, (0.2 0.2) π/a, (0.9 0.9) π/a, and (1.1 1.1) π/a. Calculate these velocities from the dispersion relation for comparison.*

Exercise 9.4 (M) *Show by calculation from the dispersion relation that "ziman" gives free electron behavior for **k** values suitably close to the origin.*

9.4 Acceleration in electric fields

9.4.1 Acceleration theorem

How do these electrons respond to applied fields? The essential result is simple: the time derivative of the wave vector is $(1/\hbar)$ times the applied force, or

$$d(\hbar\mathbf{k})/dt = \mathbf{F} = -e(\mathbf{E} + \mathbf{v} \times \mathbf{B}).\qquad(9.3)$$

This may seem like black magic, but is suggestive of Newton's second law in the form $\mathbf{F} = d\mathbf{p}/dt$. Do remember, however, that $\hbar\mathbf{k}$ is *not* a real momentum, so the result is *not* a trivial one. The derivation for the one-dimensional case given in many texts shows the power of arguments based on energy! For the harder problem of more dimensions, Appendix H of Ashcroft and Mermin [1] gives a careful explanation which follows from the following observation.

Exercise 9.5 (M**) *Show that the two first order differential equations (9.2) and (9.3) for dr/dt and d($\hbar$k)/dt, take the form of Hamilton's equations of motion if* **p** *is replaced by $\hbar$k and the kinetic energy $|\mathbf{p}|^2/2m$ is replaced by $\mathcal{E}(\mathbf{k})$. (Hint: try it first for* **B** $= 0$.)*

 # 2   The **k**-space trajectory illustrated in PRESET 2 shows the response of the electron to an electric field in the positive x-direction. It is a very simple uniform motion. The one potential source of confusion is the apparently discontinuous jump from the left hand to the right hand side of the Brillouin zone. However, remember the periodicity in **k** of the function $\mathcal{E}(\mathbf{k})$. You may think of the wave vector **k** of the electron as moving forever to the left, moving over the periodic landscape defined by $\mathcal{E}(\mathbf{k})$. This representation is the *repeated zone scheme*. Choose REPEATED from ZONE SCHEME to see this representation. The alternative usually adopted by "ziman" recognizes the equivalence of the points **k** and **k** + **G** and requires **k** to remain within the first Brillouin zone. This is the *reduced zone scheme*.

The real-space trajectory is more interesting. Initially the trajectory is parabolic, as would be expected for an electron moving in a constant field in the x-direction with initial velocity components in both the x- and the y-directions. But what happens as the motion continues? For the particular dispersion relation used by "ziman", the x- and y-components of the velocity depend only upon the x- and y-components of the wave vector respectively. Hence, for the electric field in the x-direction the y-component of the velocity remains constant, and the y-component of the position vector increases uniformly in time.

Exercise 9.6 (M*) *Prove from the form of the dispersion relation (9.1) that v_x depends only upon k_x and that v_y depends only upon k_y.*

The periodic motion in x, illustrated by PRESET 2 is referred to as a *Bloch oscillation*. Bloch oscillations have little relevance to natural crystals: experimentally achievable electric fields are nowhere near adequate to accelerate an electron through the Brillouin zone in the time between successive scatterings. The issues *may* become important in semiconductors with artificially constructed periodic potentials. So why, with no scattering, does the x-component of the position vary periodically with time?

Exercise 9.7 (C) *Check the velocity at the various points along the k-space orbit. Remember that widely spaced contour lines imply a small value for the gradient, hence velocity. What is the direction of the velocity just before the electron reaches the left hand edge of the zone, and just after it leaves the right hand side of the zone?*

Exercise 9.8 (M) *Calculate the Bloch period from the dimensions of the Brillouin zone and the magnitude of the applied field. Does it agree with "ziman"?*

Exercise 9.9 (M**) *A careful look will show that the curvatures at the two x-extrema of the real-space trajectory are different. (Increase the initial k_y if you're not convinced.) Calculate the ratio of the two curvatures from the dispersion relation $\mathcal{E}(\mathbf{k})$.*

Exercise 9.10 (C) *Click on the right hand display to initialize the electron with a variety of $\mathbf{k}$ values and interpret the trajectories in real space.*

The trajectories become even more interesting if the electric field is not parallel to one of the axes.

Exercise 9.11 (C) *In* PRESET *3 the field is at 60° to the x-axis.* RUN   # 3
"ziman" for several Bloch periods. Now the orbit in $\mathbf{k}$-space no longer closes on itself after one round trip through the zone. In real space the electron is confined to a rectangular region and does not wander off the screen. Does the $\mathbf{k}$-trajectory ever close on itself? Why doesn't the electron wander off in real space as it did in PRESET *2? All that has changed is the direction of the applied field.*

Exercise 9.12 (M**) *What is the condition on the orientation of the electric field for which the real space orbits, for the $\mathcal{E}(\mathbf{k})$ used by "ziman", will (eventually) close? (Hint: use the fact that there are no cross terms in k_x and k_y in the dispersion relation (9.1): see Exercise 9.6.)*

9.4.2 Effective mass

Exercise 9.13 (C) *For the two initial $\mathbf{k}$ values $(k_x \, k_y) = (0 \, 0) \, \pi/a$ and $(1 \, 1) \, \pi/a$, compare the directions of the initial accelerations in real space in the electric field of* PRESET *2. If you were to interpret the trajectory*   # 2
in terms of the familiar $\mathbf{F} = m d^2\mathbf{r}/dt^2 = m\mathbf{a}$, what would you have to conclude about the mass of the electron at the point $(k_x \, k_y) = (1 \, 1) \, \pi/a$?

The familiar $\mathbf{F} = m\mathbf{a}$ is *not* useful in discussing the general problem of dynamics of electrons in crystals. One can define an effective mass m^* at any point in the Brillouin zone (see the next section), but it's hardly a convenient tool when its value depends upon where the electron is in $\mathbf{k}$-space. If, however, the electrons of interest remain near the top or the bottom of a band, where the effective mass is nearly constant, it does become a useful concept. The idea of an effective mass, commonly used in discussions of semiconductors, is less useful for metals. Near a band maximum or minimum, for spherical (circular in two dimensions) energy contours, the effective mass is defined by

$$1/m^* = (1/\hbar^2)d^2\mathcal{E}(k)/dk^2. \tag{9.4}$$

Exercise 9.14 (C) *Show from the expression (9.4) that:*

1. *The effective mass is positive near the bottom of a band and negative near the top.*
2. *A small curvature, characteristic of narrow bands, implies a large mass.*

Exercise 9.15 (M*) *For initial* **k** *values near* $(0\ 0)\ \pi/a$ *and* $(1\ 1)\ \pi/a$, *use the clock and cursor readout to measure from "ziman" the acceleration of the electron in real space. What values do you deduce for the effective mass* m^* *at the two points? How do these values compare with those calculated from the dispersion relation (9.1)? (Beware of the rapid variation of* m^* *with* **k**. *It helps to keep the* **E***-field small which limits the displacements in* **k***-space while still allowing time for easily measured displacements in real space.)*

9.4.3   Band anisotropy*

The band structure we have seen so far in "ziman" has square symmetry, and the contour lines near the origin, full circular symmetry. What happens if the symmetry of the crystal is lowered. By choosing the ANISOTROPY parameter A to be 0.2, PRESET 4 gives a potential with five-fold lower curvature in the k_y-direction than in the k_x-direction. The constant energy contours near the origin are now ellipses. What are the consequences for the dynamics?

\# 4

Exercise 9.16 (C) RUN *this* PRESET *which applies a field in the* $[1\ 1]$ *direction. What is the direction of the displacement of the electron in* **k***-space? In what direction is its initial acceleration in real space?*

Exercise 9.17 (M*) *Construct an argument to predict from the dispersion relation (9.1) the direction of the initial acceleration of the electron in real space, in response to an applied electric field* **E**. *(Hint: recall from Exercise 9.6 that the* x*- and* y*-components of the motion for "ziman" are independent.)*

Accelerations which are not parallel to the applied force are unsettling but may be incorporated succinctly by introducing the inverse effective mass tensor,

$$\frac{1}{\mathbf{m}^*(\mathbf{k})} \equiv \frac{1}{\hbar^2}\nabla_\mathbf{k}\nabla_\mathbf{k}\mathcal{E}(\mathbf{k}). \tag{9.5}$$

Newton's $F = ma$, in two dimensions, then takes the form

$$
\mathbf{a} = \frac{1}{\mathbf{m}^*(\mathbf{k})} \cdot \mathbf{F} = \frac{1}{\hbar^2}
\begin{pmatrix}
\dfrac{\partial^2 \mathcal{E}}{\partial k_x^2} & \dfrac{\partial^2 \mathcal{E}}{\partial k_x \partial k_y} \\[2mm]
\dfrac{\partial^2 \mathcal{E}}{\partial k_y \partial k_x} & \dfrac{\partial^2 \mathcal{E}}{\partial k_y^2}
\end{pmatrix}
\cdot
\begin{pmatrix}
F_x \\
F_y
\end{pmatrix}
\tag{9.6}
$$

Note that it is more natural to work in terms of the *inverse mass tensor* since we typically are interested in the acceleration a produced by a force F and therefore write Newton's law as $a = (1/m^*)F$.

Exercise 9.18 (M**) *Calculate the inverse effective mass tensor for the dispersion relation $\mathcal{E}(\mathbf{k})$ for $\mathbf{k}$ at the three points: the center of the zone, the corner of the zone, and the centers of the edges. Show that it becomes a multiple of the unit tensor at the zone center and corner if the* ANISOTROPY *constant is taken to be $A = 1$.*

9.5 Magnetic field

It's fun to imagine the effects of an electric field strong enough to drive an electron through $\mathbf{k}$-space from one boundary of the Brillouin zone to another, but it's not practicable to achieve such an effect in real systems. The electric field required to complete a period of the Bloch oscillation before the electron is scattered would be prohibitively large.

Exercise 9.19 (M**) *For an energy band in a typical solid, calculate the electric field required for an electron to complete a Bloch period within a scattering time of 10^{-13} s. Few materials can support fields as large as 10^7 V/m, and scattering times for energetic electrons will be much shorter than the value of 10^{-13} s.*

On the other hand, by applying large magnetic fields it is often possible in metals to force electrons to travel large distances in $\mathbf{k}$-space between scattering events. First, the Lorentz force on an electron moving with a typical Fermi velocity is high. Second, the magnetic field does not carry the electron to energies far from the Fermi energy: as a consequence, at low temperatures scattering times may be kept much longer than 10^{-13} s.

Exercise 9.20 (M) *Consider an electron at the Fermi surface of a free electron metal with a Fermi energy of 10 eV. What electric field would be required to give a force on the electron equal to the Lorentz force in a magnetic field of 10 T (10^5 G).*

9.5.1 Orbits of positive and negative mass

5   PRESET 5 illustrates an electron moving in a field of 1 T, directed out of
the screen. The electron energy is chosen to be small and the behavior
is that of the electrons in "drude". More interesting is the behavior for
larger values of the energy. Click near the fourth contour line to start an
electron at a higher energy. The orbit is no longer circular: the electron
dynamics is being influenced by the complexities of the band structure.

Exercise 9.21 (M) *No matter where you start the electron, the trajectory in* **k**-*space runs parallel to the contour lines. Explain both physically,
and mathematically from the form of the Lorentz force (9.3), why this is
so. It is a very convenient observation, since once having constructed
the contour lines we know immediately what the magnetic field orbits in*
k-*space will be.*

Exercise 9.22 (C) *The magnetic field is perpendicular to and points
out of the screen. Argue from knowledge of the group velocity from the
dispersion relation, that a counterclockwise travel around the orbit, for a
particle near a minimum of* $\mathcal{E}(\mathbf{k})$, *implies that the particle is carrying a
negative charge.*

The picture looks quite different if you click on the fifth contour line
or above. Now the electron keeps jumping from one quadrant of the zone
into another. A glance at the real-space orbits reminds us that these apparent discontinuities represent no physical discontinuity: the real-space
orbits are continuous. Switching the ZONE SCHEME from REDUCED to
REPEATED allows the electron to move in **k**-space without restriction to
the first Brillouin zone and the discontinuities are removed. A noteworthy
feature is that the orbits in both real and **k**-space are now traversed in
the *clockwise* direction, though we've *not* changed the sign of the charge.

Exercise 9.23 (M) *Show that for classical particles a clockwise sense of
traversal is appropriate to a particle of either negative charge and negative
mass, or positive charge and positive mass. It is the sign of the charge to
mass ratio that determines the sense of circulation.*

In Section 9.4.2, looking at the **E**-field response of an electron near a
maximum of $\mathcal{E}(\mathbf{k})$, we noted the strange behavior that could be described
in terms of a negative effective mass. Here we are seeing the same thing
in response to a **B**-field, a *negative* mass describing the electron dynamics
near the top of an energy band. Unfortunately, in a real system with
a single, or a few, electrons in states near the top of an empty band,
the electrons would scatter too quickly to states near the bottom of the
band. The behavior so easily illustrated in "ziman" is not accessible in

the laboratory. However, "peierls" will illustrate the way such orbits are evidenced by having a band which is nearly filled with electrons.

Exercise 9.24 (C*) RUN PRESET 6. *This special choice of the initial* k   # 6
value arises naturally while exploring the dividing line between the two types of orbits (positive and negative mass). Explain how the electron manages to get stuck at the edge of the zone. (Ultimately, because of finite computational precision, the electron will break loose again.)

Exercise 9.25 (M**) *Calculate an alternative initial* k *value which will lead to such an orbit, type it in, and see if it works.*

9.5.2 *Anisotropy and real space orbits**

To complicate things a little, PRESET 7 adds ANISOTROPY to the system.   # 7
As noted on page 188, the energy contours near the band maximum and minimum are ellipses instead of circles. The effective mass has become a tensor quantity with the principal values of the inverse effective mass tensor being

$$\frac{1}{m_{xx}} = \frac{1}{\hbar^2}\frac{\partial^2 \mathcal{E}(\mathbf{k})}{\partial k_x^2} \quad \text{and} \quad \frac{1}{m_{yy}} = \frac{1}{\hbar^2}\frac{\partial^2 \mathcal{E}(\mathbf{k})}{\partial k_y^2}. \tag{9.7}$$

How can these masses be determined experimentally in real systems? A *cyclotron mass* can be determined from the cyclotron resonance experiment as $m_c^* \equiv eB/\omega_c$. Measurements of the cyclotron mass for a number of different orientations of a three-dimensional crystal with respect to the magnetic field allow determination of the principal components of the effective mass tensor. How the components combine to give the cyclotron mass is not obvious, however.

Exercise 9.26 (M*) *An "experimentalist's approach" is to measure the cyclotron mass for a known system and see which works best of the three most obvious options for the two-dimensional system:*

$$m_c^* = (m_{xx} + m_{yy})/2, \tag{9.8}$$
$$m_c^* = 2m_{xx}m_{yy}/(m_{xx} + m_{yy}), \tag{9.9}$$
$$m_c^* = \sqrt{m_{xx}m_{yy}}. \tag{9.10}$$

Try it out on "ziman" using an ANISOTROPY $= 0.2$. *Which form best fits the data? Choose an initial* k *value close to the origin to get the true limiting value of* m_c^*, *and a small value of the* **B***-field to get a large enough real-space orbit to work with.*

Exercise 9.27 (M**) *Show analytically that the cyclotron frequency for an orbit near a band extremum is given by* $\omega_c^2 = e^2B^2/m_{xx}m_{yy}$,

(Eq. (9.10) above). (Hint: use the effective masses (9.7) in the classical equations of motion for the two components of dv/dt and assume a sinusoidal time dependence of the velocities.)

Stranger yet are the *open orbits* that appear when the initial energy is chosen in the vicinity of the fifth contour line. The **k**-space orbit repeats itself, but the real-space orbit never closes. Again we can get rid of the apparent discontinuities in **k** by using the REPEATED choice of the ZONE SCHEME. Now, however, in the REPEATED zone picture, we see the electron wandering off indefinitely in k_y while oscillating about a nonzero mean in k_x, somewhat reminiscent of the response to the electric field on page 186. By using the REDUCED zone scheme, "ziman" forces the electron's wave vector to remain in the first Brillouin zone where we can keep track of it.

Exercise 9.28 (M*) *From the qualitative behavior in **k**-space, deduce the shape of the real space trajectory. Be sure to explain why, with the **k**-space orbit extended in the y-direction, the real-space orbit is extended in the x-direction.*

Exercise 9.29 (C*) *Generate a variety of trajectories by clicking on a number of initial **k** values. What general relationship can you see between the shapes and orientations of the **k**-space trajectories and those of the real-space trajectories?*

The similarity, with rotation by 90°, of the orbits in the two spaces follows directly from the equation of motion (9.3) for $d\mathbf{k}/dt$ since it expresses a proportionality between the components of the velocities, $d\mathbf{k}/dt$ and $d\mathbf{r}/dt$, in the two spaces.

Exercise 9.30 (M**) *Show explicitly from the equation of motion (9.3) for $d\mathbf{k}/dt$ that the orbit in real space may be obtained from that in the repeated zone scheme in **k**-space by rotating by 90° and multiplying by a scale factor of $\hbar/eB$. Is that 90° rotation clockwise or counterclockwise for a negatively charged particle? Does the direction of this rotation between the two spaces depend on whether the **k**-orbit is near the top or near the bottom of the band?*

There are experimental techniques, referred to as *caliper experiments*, which are sensitive to the real-space dimensions of the cyclotron orbits. They may be described in terms of a *dimensional resonance* between the wavelength of a sound wave or the thickness of a sample and an appropriate diameter of the cyclotron orbit, see Ashcroft and Mermin [1, pp. 275–281]. The mapping relationship between **k**- and real-space orbits allows a conversion of the experimentally determined spatial distance to a corresponding dimension of the Fermi surface of the material. The shape of the Fermi surface of a metal, a seemingly purely abstract concept, can be determined experimentally!

9.6 Crossed fields**

If you found the open orbits for the anisotropic band bizarre, try the orbit in PRESET 8 with crossed electric and magnetic fields. A number of interesting "border designs" can be generated in real space by choosing different initial **k** values or by changing the field parameters. Ashcroft and Mermin [1, p. 233] give a geometrical scheme for interpreting such orbits.

\# 8

Exercise 9.31 (M**) *For the parameter values of* PRESET 8, *calculate the critical value of the electric field for which the orbits just touch the zone boundary. Compare the prediction with results from "ziman". (Hint: assume the* **E**-*field to be small enough that the saddle point of the "tilted* $\mathcal{E}(\mathbf{k})$ *surface" is not significantly displaced from the midpoint of the zone boundary.)*

9.7 Summary

Quantum mechanical electrons moving in a periodic potential respond in unfamiliar ways to the application of electric and magnetic fields. The most striking of these is the acceleration of the electron in a direction opposite to the force, for an electron near the top of an allowed energy band. The negative effective mass suggested by this behavior is confirmed by the clockwise sense of travel in cyclotron orbits for an electron near the top of the band. The key to the analysis is knowledge of the dispersion relation $\mathcal{E}(\mathbf{k})$.

These bizarre predictions follow from two basic equations. The first is the familiar relation (9.2) connecting the group velocity **v** to the gradient with respect to **k** of the dispersion relation $\mathcal{E}(\mathbf{k})$. The second, Eq. (9.3), expresses the acceleration of the electron by relating the time derivative of the wave vector **k** to the usual Lorentz force.

For motion near a band maximum or minimum, the electron dynamics is conveniently described in terms of an effective mass related to the curvature of $\mathcal{E}(\mathbf{k})$. In principle, for large electric fields, the electron position and velocity oscillate in time, the Bloch oscillation, in response to a time-independent electric field. This follows from the equations of motion as a consequence of the periodicity in **k** of the energy $\mathcal{E}(\mathbf{k})$. Electron trajectories in a magnetic field can take on quite bizarre shapes, and in some situations will not be closed. These strange orbit shapes appear both in position space and reciprocal space, the two orbits being related by a 90° rotation and a scaling factor.

9.A   Deeper exploration

Bloch oscillators   Although we do not expect to see Bloch oscillations in conventional materials, periodic arrays have been fabricated in semi-conductor materials in an effort to exploit the phenomenon for device applications [36]. Explore with "ziman" the features of the band structure which, along with the **E**-field, determine the Bloch frequency and also the amplitude of the oscillatory motion in real space.

One-dimensional conductors   Some materials [39] show extreme anisotropy in their electrical conductance, with ratios of conductivities in different directions being as large as 1000 or more. Learn about some of these materials and use "ziman" to illustrate features of the electron dynamics in the presence of extreme anisotropy.

Dimensional resonance   Learn about some of the dimensional resonance experiments [35, 37, 38]. Choose parameters for an anisotropic band structure in "ziman" to simulate an experiment and work out the required sample size or acoustic frequency, and geometry to determine the shape of the **k**-space orbit for electrons of different energies in the band.

9.B   "ziman" – the program

"ziman" simulates, in two dimensions, the semi-classical motion of a single band-electron in static electric and magnetic fields.

9.B.1   Electron dynamics in "ziman"

"drude" and "sommerfeld" simulate the motion of free electrons: the energy of each electron is given by Eq. (7.13) and the real-space velocity **v** of the electron is proportional to its crystal momentum $\hbar\mathbf{k}$. "ziman" simulates the dynamics of an electron in a periodic potential. The dispersion relation $\mathcal{E}(\mathbf{k})$ consists of a single band,

$$\mathcal{E}(\mathbf{k}) = \frac{\hbar^2}{2m}\left[\left(k_x^2 - \frac{a^2}{2\pi^2}k_x^4\right) + A\left(k_y^2 - \frac{a^2}{2\pi^2}k_y^4\right)\right], \qquad (9.11)$$

for wave vectors **k** in the first Brillouin zone ($|k_x|, |k_y| \leq \pi/a$). The energy $\mathcal{E}(\mathbf{k})$ is taken to be periodic in **k** with the periodicity $2\pi/a$, and for wave vectors $\mathbf{k}'$ outside the first Brillouin zone the energy is given by $\mathcal{E}(\mathbf{k}' + \mathbf{G})$ where the reciprocal lattice vector $\mathbf{G}$ is chosen such that $\mathbf{k}' + \mathbf{G}$ lies in the first Brillouin zone. A is the ANISOTROPY, specified on the panel.

The dynamics of an electron in "ziman" is purely deterministic (no probabilistic scattering), and is described by two semiclassical equations of motion with the force given by the Lorentz force. In terms of the wave vector $\mathbf{k}$, these equations take on the forms

$$d(\hbar\mathbf{k})/dt = -e\,(\mathbf{E} + \mathbf{v} \times \mathbf{B}), \qquad (9.12)$$
$$\mathbf{v} = \nabla_{\mathbf{k}}\mathcal{E}(\mathbf{k})/\hbar, \qquad (9.13)$$

where $\hbar\mathbf{k}$ and $\mathbf{v}$ are the electron's crystal momentum and velocity respectively. Both the electric and magnetic fields, $\mathbf{E}$ and $\mathbf{B}$, are static in "ziman". The electric field lies within the plane of motion (the xy-plane) and the magnetic field is perpendicular to that plane.

With the energy $\mathcal{E}(\mathbf{k})$ given in Eq. (9.11), the equation of motion (9.12) is a non-linear, first order, differential equation in $\mathbf{k}$. Once $\mathbf{k}(t)$ is known, we can use Eq. (9.13) to determine the position $\mathbf{r}$ of the electron by integration of $\dot{\mathbf{r}} = \mathbf{v}(t)$. "ziman" uses a fifth order, Runge–Kutta routine to integrate numerically both the equations of motion (9.12) and (9.13).[1]

9.B.2 Displayed quantities

The current position and past trajectory of the electron are shown in the left hand display, the corresponding wave vectors in the right hand display. A new dot is drawn every time step Δt. For the SPEED chosen to be 1, that time step is $\Delta t_0 = 0.3$ ps. Changing the SPEED changes the time step, $\Delta t = \text{SPEED} \times \Delta t_0$.

Press the letter 'e' to erase the past trajectory in real and reciprocal space *without* initializing the electron's position and wave vector. (As in most of the SSS, press the letter 'i' to initialize the program and SPACE BAR to RUN/STOP.)

Real-space display The xy-position of the electron, in μm, is shown in the left hand display. The bounds of the visible region are chosen to be $x, y = \pm 40$ μm. These bounds can be changed by changing the REAL SPACE RANGE in the CONFIGURE menu.

Reciprocal-space display The right hand display shows the wave vector of the electron. If the ZONE SCHEME is chosen to be REDUCED, the wave vector remains within the first Brillouin zone. The Brillouin zone boundaries are shown in blue. Wave vectors outside the first Brillouin zone are

---

[1] For a short description of the fifth order, Runge–Kutta routine `rkf` see page 133.

translated back by a reciprocal lattice vector **G** into the first zone. If
the ZONE SCHEME is chosen to be REPEATED, the electron's wave vector
can take on any value. The initial wave vector **k** can be chosen either
on the panel, or by CLICKING with the left hand mouse button in the
reciprocal-space display.

The energy contours are drawn in green at nine equidistant energies
between 0 and $\Delta\mathcal{E}$. Here $\Delta\mathcal{E} = 4.7(1 + A)$ eV is the width of the band
and A the ANISOTROPY. Note that the energy contours are just the
trajectories of an electron in a magnetic field perpendicular to the plane
of motion. "ziman" therefore draws each contour line simply by letting an
electron evolve in a magnetic field for three free electron cyclotron periods
$T = 6\pi m/eB$.

9.B.3   Bugs, problems, and solutions

For some choices of the ANISOTROPY A, an energy contour line may not be
drawn completely. This can happen if the energy of that line is such that
the electron (whose time evolution in a magnetic field gives the contour
lines) becomes stuck at the edge of the Brillouin zone, see Exercise 9.24,
and therefore cannot complete a whole orbit within three cyclotron periods.

10

"peierls" – Transport in a partially full electron band

Contents

10.1   Introduction

"ziman" has shown that for electrons moving in a periodic potential the response to applied fields is qualitatively different from that of free electrons, except near the bottom of a band. Even there the electrons may react as if they had an effective mass different from that of the free electron. What can we say about electronic transport in metals and semiconductors in which the energy bands are filled with an arbitrary concentration of electrons n? When, if ever, might we expect the expression $\sigma = ne^2\tau/m$ for the electrical conductivity to have any relevance?

"peierls" addresses this question by working with the same energy band used in "ziman", with the dispersion relation,

$$\mathcal{E}(\mathbf{k}) = \frac{\hbar^2}{2m}\left[\left(k_x^2 - \frac{a^2}{2\pi^2}k_x^4\right) + A\left(k_y^2 - \frac{a^2}{2\pi^2}k_y^4\right)\right]. \tag{10.1}$$

The bandwidth, with the "peierls" choice of lattice constant $a = 2$ Å and with the ANISOTROPY parameter $A = 1$, is 9.4 eV. Application of periodic boundary conditions, as in "sommerfeld", restricts the allowed $\mathbf{k}$ values to a discrete grid in $\mathbf{k}$-space with spacing $2\pi/Na$, for a crystal of $N \times N$ cells. These states are filled with electrons up to some specified FERMI

\# 1   ENERGY, about 5 eV for the example of PRESET 1. The prescription is the same as for "sommerfeld", but the picture is different both because the energy contours, as in "ziman", are no longer a series of concentric circles, and because there is an upper bound to the allowed energies. Change the value of the FERMI ENERGY to see how the shape of the Fermi surface changes with filling level. For now, ignore the display panel on the left.

The contrast between the Sommerfeld and band models becomes more evident when fields are applied. As in "ziman", in "peierls" each of the electrons responds according to the semiclassical equation,

$$d(\hbar\mathbf{k})/dt = \mathbf{F} = -e(\mathbf{E} + \mathbf{v} \times \mathbf{B}). \tag{10.2}$$

RUN PRESET 1, in which both electric and magnetic fields are applied, with a FERMI ENERGY of 5 or 6 eV. The collection of orbits for the states near $\mathbf{k} = 0$ is familiar from "sommerfeld", but for the states further out the individual electron trajectories are no longer circular arcs. Indeed, the addition of the band structure has complicated things dramatically. Is there any hope of making sense of it?

10.2   Nearly empty band

\# 2   PRESET 2, on the other hand, in which the Fermi energy is chosen near the bottom of the band, presents no great problem. RUNNING this PRESET,

with just the ELECTRIC FIELD as programmed or with the addition of a
MAGNETIC FIELD, the display is indistinguishable from "sommerfeld" as
long as the electrons remain in the region in which the contour lines are
circular.

In "sommerfeld" the green dot in the display gives the average wave
vector of the electrons, and, when scaled by $\hbar/m$, is graphed as the average
or drift velocity of the electrons. "peierls" also shows the average wave
vector as a green dot. However, for band-electrons, the electron velocity
is *not* proportional to wave vector. Therefore, to generate the average-
velocity graph in "peierls", a velocity must be calculated from the gradient
of $\mathcal{E}(\mathbf{k})$ for each electron, and these are then averaged and plotted.

Exercise 10.1 (M) *For* PRESET 2 *in steady state, what is the average
wave vector of the electrons? You can estimate this by adjusting the
cursor position to the average position of the green dot as it fluctuates
during the running of the program. Is it roughly equal to $eE\tau/\hbar$? Is it
consistent with the average velocity displayed in the graph?*

Exercise 10.2 (C) *Roughly, for what range of values of the* FERMI EN-
ERGY *would you expect the average velocity in "peierls" to be approximately proportional to the wave vector, $\mathbf{v} = \hbar\mathbf{k}/m$?*

Exercise 10.3 (C**) *Why is the graph so much noisier than the one in
"drude" even though the number of electrons is considerably greater?*

We begin to see why the Sommerfeld model might work as a rough
description of metals, even with the additional complication of the band
structure. If over the range of occupied states, e.g., near the bottom
of a band, the curvature of $\mathcal{E}(\mathbf{k})$ is roughly constant, then the effective
mass introduced in Chapter 9 is constant: $m^*(\mathbf{k}) \approx m^* = $ constant. In
that case the Sommerfeld (and Drude!) expression for the conductivity
$\sigma = ne^2\tau/m$ should remain valid with the replacement $m \to m^*$.

10.3   Half-full band

10.3.1   Qualitative

What happens, however, if the FERMI ENERGY is in the vicinity of the
middle of the band, say about 5 eV as in PRESET 1? In a qualitative   # 1
sense, it is still the "sommerfeld" picture. Change the MAGNETIC FIELD
to $B = 0$ and in steady state you will again see that the effect of the
electric field is to displace the Fermi sea to the left. Associated with the
displacement is a drift velocity of the electrons and hence a net electric
current. Now, however, we cannot simply compute the average $\mathbf{k}$ and
multiply by $\hbar/m$ to get the drift velocity.

Exercise 10.4 (M*) *To demonstrate the difficulty,* RUN *for a while with* PRESET 2, *and estimate the average velocity in steady state by an "eyeball time average" from the velocity graph. Now increase the* FERMI ENERGY *to 4.09 eV and repeat. The displacement $eE\tau/\hbar$ from zero of the average wave vector is roughly the same for both cases. How do the average electron velocities compare? (Warning! In addition to problems with noise, the average velocities in "peierls" are artificially sensitive to changes in Fermi energy small enough not to change the number of states enclosed by the Fermi surface. If you attempt more careful measurements, check Section 10.B.)*

\# 2

Exercise 10.5 (C) *The average electron velocity, for a given average wave vector, decreases with increasing Fermi energy. Rationalize this result with what you know about electron band structure. (Hint: how does $\mathcal{E}(\mathbf{k})$ for a typical energy band differ from the parabolic energy dependence $\mathcal{E}(\mathbf{k}) \propto |\mathbf{k}|^2$? What, then, are the implications for the variation of $\mathbf{v}$ with $\mathbf{k}$?)*

As the FERMI ENERGY is increased further, the decrease in curvature of $\mathcal{E}(\mathbf{k})$ with increasing $|\mathbf{k}|$ is supplemented by another source of decreasing average velocity, namely the contact of the Fermi surface with the zone boundary.

\# 3

Exercise 10.6 (C) PRESET 3 *increases the* FERMI ENERGY *even further, to 6.11 eV. Now with contact of the Fermi surface with the Brillouin zone boundary the average velocity is reduced by an even larger factor. Explain why none of the states in the range -0.6 $\text{Å}^{-1} < k_y < 0.6$ Å^{-1} contribute to the average velocity, and hence to the electric current.*

The same idea can be expressed differently. Again, focus on the Fermi surface. The current contributions come only from electrons near the Fermi surface, where substantial changes in the electron occupation probability $f(\mathbf{k})$ occur, as discussed on page 146. For $\mathcal{E}_F > 4.7$ eV, the area of the Fermi surface decreases with increasing Fermi energy: adding more electrons lowers, rather than increases, the conductivity because the area of the Fermi surface is decreasing. Both the decreasing curvature of the bands with increasing energy and the decreasing area of the Fermi surface when the surface is interrupted by the zone boundaries contribute to a decreasing average electron velocity with increasing energy.

10.3.2   Quantitative**

These ideas are formalized by extending the argument for the electrical conductivity started on page 146. The probability of finding a state occupied by an electron is given by a steady state distribution function

$f(\mathbf{k})$, determined by balancing the effects of acceleration by the electric field against the restoration of equilibrium by scattering, i.e., by solving the Boltzmann transport equation (7.7). In three-dimensions, the conductivity tensor is obtained from the distribution function $f(\mathbf{k})$ (see, for example, Ashcroft and Mermin [1, Eq. (13.25)]) as an integral over the Brillouin zone,

$$\sigma = \frac{e^2\tau}{4\pi^3}\int_{\text{BZ}} d^3k\, \mathbf{v}(\mathbf{k})\,\mathbf{v}(\mathbf{k})\left[-\frac{\partial f_0(\mathcal{E})}{\partial\mathcal{E}}\right]_{\mathcal{E}=\mathcal{E}(\mathbf{k})}. \tag{10.3}$$

Since the Fermi function $f_0(\mathbf{k})$ varies rapidly with energy only near the Fermi energy, this integral receives contributions only from states near the Fermi surface. From this follows the statement that it is the electrons near the Fermi surface that determine the electrical properties of a metal. The integral may be converted to an integral over the Fermi surface to make the point even more explicit. Whatever the form, knowledge of the dispersion relation $\mathcal{E}(\mathbf{k})$ is required for its numerical evaluation. In the formal result (10.3) we see expression of the two effects noted on page 200: the dependence of the conductivity on both the Fermi velocity and the area of the Fermi surface. In particular, since the area of the Fermi surface approaches zero as the Fermi energy approaches either the top or the bottom of a band, almost full and almost empty bands must have low conductivity. Is there a way to see any connection of this general formalism to the Drude or Sommerfeld results?

Exercise 10.7 (M**) *Show that an integration by parts converts the integral in Eq. (10.3) to the alternative form*

$$\sigma = \frac{e^2\tau}{4\pi^3}\int_{\text{BZ}}\frac{d^3k}{\hbar}\nabla_{\mathbf{k}}\mathbf{v}(\mathbf{k})\,f_0[\mathcal{E}(\mathbf{k})] = \frac{e^2\tau}{4\pi^3}\int_{\substack{\text{occupied}\\\text{states}}} d^3k\,\frac{1}{m^*(\mathbf{k})}. \tag{10.4}$$

(Hint: start by finding a relation between $\nabla_{\mathbf{k}}f_0(\mathcal{E}(\mathbf{k}))$ and a portion of the integrand in (10.3).)

If $m^*(\mathbf{k})$ is independent of $\mathbf{k}$ for the occupied states (e.g., for a band with electrons only near the bottom of the band) this formula gives the Drude result, $\sigma = ne^2\tau/m^*$. The free electron mass is replaced by the effective mass characterizing the curvature of the dispersion relation at the bottom of the energy band. This is a reasonable approximation in "peierls" for a Fermi energy of 2 eV. For 5 eV, however, we would have to keep track of the variation of effective mass with position in the band and result (10.4) is no longer helpful. Can we say anything useful for a nearly full band, say for a Fermi energy of 7 or 8 eV in the band used by "peierls"?

10.4   Nearly full band: holes

10.4.1   Electric field

2   **Exercise 10.8 (M)** *Change the* FERMI ENERGY *in* PRESET 2, *to 6.9 eV,* RUN *for a while, and determine the average velocity of the electrons from the velocity graph. It is now much smaller than for the almost empty band. Explain why the average velocity is so small. What will its value be when the band becomes full?*

In Figure 10.1 the solid curve, labeled "current", gives the sum of the electron velocities in arbitrary units. Multiplied by the charge this sum would be proportional to the electric current carried by the sample. The "current" is divided by the number of electrons to give, in the dotted curve, the plot of average electron velocity $\langle v_e \rangle$. Ignore the dashed curve for the present. The data were taken with a 14×14 array of states (K-SPACE GRID $= 14$), $E_x = 1 \times 10^6$ V/m, and $\tau = 1$ ps.

Figure 10.1 illustrates in more detail, using data from "peierls", the observations you've already made: for a fixed electric field and scattering time, the average velocity of the electrons gradually decreases as the Fermi energy moves further up in the band. What is strange at first sight is that the average velocity of the electrons goes to zero as the band becomes full. However, this has to be since the current carried by a full band is zero. For a full band, for every electron moving with some velocity parallel to the field, there is another electron with just the opposite velocity and the sum of all the velocities is zero. In your consideration of these curves remember that we're talking of velocity in real space. Even though the electrons march through the Brillouin zone in **k**-space at the rate $eE/\hbar$, the average real space velocity is zero and the full band of electrons conducts no electricity.

The formal expressions (10.3) and (10.4) are, of course, still appropriate for the nearly full band. However, a formal integration of the inverse effective mass over the occupied states (10.4) does not offer much insight. Soon we will see that Eq. (10.4) can be modified to give a much simpler interpretation for a nearly full band.

Before working with the equations, however, let's take a look at PRESET # 4   4 in "peierls", and focus on the empty rather than the full states. We will call these empty states HOLES. In the left hand display, which is activated by clicking on the SHOW HOLES toggle, are plotted the positions in **k**-space of the unoccupied states, including one critical change in sign: if, in the display on the right, the state **k** is not occupied by an electron then the state $\mathbf{k}' = -\mathbf{k}$ is shown as occupied by a hole in the left display. Kittel [6, pp. 206–209] describes carefully and in detail the rationale for the hole representation of the electron transport for an almost full band.

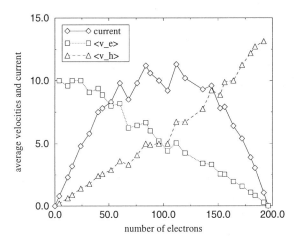

Figure 10.1: Average velocities and current (arbitrary units) as a function of the number of electrons in the band.

You will find there the rationale for the change in sign of the vector **k**.

Exercise 10.9 (C*) RUN PRESET 4 *briefly until you clearly see some empty electron states below the Fermi surface in the display on the right. Pick one of these empty states and suppose its wave vector is **k**. Look next in the hole display at the point **k'** = −**k**. There should a white dot here indicating occupation of this state by a hole. Now find an electron above the Fermi surface in the electron display. If it is in the state **k** what do you expect to see at −**k** in the hole display?*

It's somewhat confusing to see these holes in four distinct groups near the four corners of the Brillouin zone. "peierls" takes advantage of the periodicity of the dispersion relation to give a different representation. Click on the SHIFT ORIGIN toggle and three quadrants (A, B, C) of the Brillouin zone in Figure 10.2 are translated up and/or to the right to form a new cell (A', B', C', D) centered on the corner $(0')$ of the original cell. This array is then shifted back to the center of the panel display. The blue lines in the display (the heavy weight lines in Figure 10.2) remain portions of the boundaries of the first Brillouin zone after these translations. Now the occupied hole states behave like occupied states of a simple Sommerfeld model of positively charged carriers with positive mass!

Exercise 10.10 (C) RUN *the program briefly with the shifted origin. In which direction in **k**-space are the holes displaced by the* ELECTRIC FIELD? *Is that appropriate to positively or negatively charged particles?*

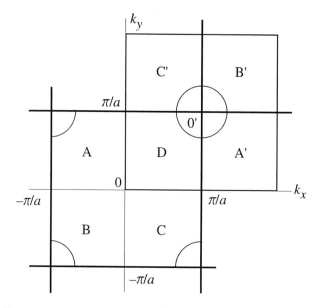

Figure 10.2: Shift of the origin of the hole space display.

We can get some feeling for the source of this alternative point of view from Eq. (10.4).

Exercise 10.11 (M)** *Verify the following variation on Eq. (10.4):*

$$
\sigma = \frac{e^2 \tau}{4\pi^3} \int_{\substack{\text{occupied} \\ \text{states}}} d^3k \, \frac{1}{m^*(\mathbf{k})}
$$

$$
= \frac{e^2 \tau}{4\pi^3} \left\{ \int_{\substack{\text{all} \\ \text{states}}} d^3k \, \frac{1}{m^*(\mathbf{k})} - \int_{\substack{\text{unoccupied} \\ \text{states}}} d^3k \, \frac{1}{m^*(\mathbf{k})} \right\}
$$

$$
= \frac{e^2 \tau}{4\pi^3} \int_{\substack{\text{unoccupied} \\ \text{states}}} d^3k \, \frac{-1}{m^*(\mathbf{k})}. \tag{10.5}
$$

(Hint: you need the formal definition of the inverse effective mass in order to show that the integral over all states in the band is zero.)

Exercise 10.12 (C) *Use "peierls" to show experimentally that the full band gives zero current, by setting the* FERMI ENERGY *to 10 eV and* RUNNING *the program with an applied* ELECTRIC FIELD. *What is the average velocity now? (The small oscillation in average velocity is an artifact of the simulation. How does the oscillation amplitude compare with the average velocities being measured earlier?)*

Exercise 10.13 (M)** *What is the source of the small oscillation seen in the preceding exercise? What would you predict for its amplitude?*

Eq. (10.5) gives a formal picture of what we've seen in "peierls". For an almost full band, the unoccupied states are all near the top of the band, a region in which the effective mass is roughly constant and *negative*. The expression above then reduces to the Drude or Sommerfeld form, $\sigma = pe^2\tau/m_h^*$ if we define p, the number density of holes, to be the density of *unoccupied* states, and m_h^*, the effective mass of the holes, to be $m_h^* \equiv -m_{\text{top of band}}^* > 0$.

When the SHOW HOLES toggle is clicked to activate the hole display on the left, the velocity graph is changed to show the average hole velocity instead of the average electron velocity. It is this average hole velocity which is shown in the dashed curve in Figure 10.1.

Exercise 10.14 (M*) *Check the average hole velocity for several values of the* FERMI ENERGY *and plot it to show that the average velocity decreases as the Fermi energy decreases. Compare this with what was found for the electron representation (see Figure 10.1).*

Exercise 10.15 (M*) *For the same value of the* ELECTRIC FIELD *and scattering time* TAU, *find the ratio of the hole drift velocity for the band nearly full to the electron drift velocity for the band nearly empty. Use Fermi energies of 8.32 eV and 0.65 eV, corresponding to 12 holes or electrons, respectively. How do you expect this ratio to be related to the ratio of the electron and hole effective masses?*

The picture of the almost full band in terms of holes fits well with the Sommerfeld picture except that the charge carriers, the missing electron states, have positive, not negative, charge. The electrical conductivity experiment by itself, of course, does not allow us to determine, for a given material, whether a nearly full or nearly empty band model is the more appropriate. The crucial test comes with the application of a magnetic field.

10.4.2 Magnetic field

RUN "peierls" using PRESET 5 to see the response of band electrons to   # 5
a magnetic field when the band is more than half full. For low electron energies, the orbits are counterclockwise as expected for negative charge/mass. At the Fermi surface, the general sense of circulation is also counterclockwise about the origin of reciprocal space. However, these trajectories have the strange discontinuities associated with forcing the electrons to remain in the first Brillouin zone. Use the SHOW HOLES representation with the SHIFT ORIGIN option to give continuous orbits. The sense of circulation about the interior of the orbit is now clockwise. These

are the orbits which, in "ziman", are associated with a negative charge and negative effective mass, and hence a positive charge/mass ratio.

Exercise 10.16 (C) *Argue that in the hole representation in the left hand display, the observed trajectories are those expected for particles of positive charge and positive mass.*

Exercise 10.17 (M) *Use* PRESET 5 *of "peierls" to measure the hole and electron effective masses from the cyclotron period of the holes and electrons. The effective mass changes rapidly with distance from the edge of the band, so use the four holes at the top of the band and the four electrons near the bottom of the band to determine the effective mass at the band edges. What does the dispersion relation (10.1) predict for these masses? (Hint: expand $\mathcal{E}(\mathbf{k})$ about the $\mathbf{k}$ at the band maximum or minimum and keep only the quadratic term.)*

If an electric field $E_x = 2 \times 10^6$ V/m is now added, we have the Hall effect experiment. RUNNING and watching the electron display should convince you that to develop a full description in the electron language would not be a pleasant task. In the hole language, there's no problem!

Exercise 10.18 (C) *Show that the centers of the circular arcs of the* $\mathbf{k}$*-space trajectories of the holes (left hand display) should, as observed, be displaced downwards for this crossed field configuration. Explain why this displacement downwards is the* same *for both the electron and the hole orbits!*

Exercise 10.19 (C) *Use the velocity graph with the* SHOW HOLES *selected to determine the x- and y-components of the average hole velocity. What will be the direction of the electrical current with respect to the applied* ELECTRIC FIELD? *How do these directions compare with the corresponding quantities for the nearly empty band? (Be sure to turn off the* SHOW HOLES *toggle to check out the graph for the electron picture.)*

The puzzle of the sign of the Hall voltage is no longer a mystery. The effect of the periodic potential in breaking the one-electron energies into bands of allowed and forbidden states complicates the discussion of transport considerably. However, with the Fermi energy near the top of a nearly full band, the introduction of the hole language removes the complexity and gives back the Sommerfeld picture, though with the mysterious but important change in sign of the effective charge of the carriers. Miraculously, the awkward question of particles of negative mass has been finessed: the holes have both positive charge and positive mass.

10.5   Anisotropy*

Many materials, including the high temperature superconductors, show a large anisotropy of the electrical conductivity. As with "ziman", "peierls" can introduce ANISOTROPY into the band structure. First, open PRESET 6 to see a nearly empty anisotropic band.

6

Exercise 10.20 (C) RUN *the* PRESET. *The field is applied at 45° to the coordinate axes. What angle do the individual electron* **k***-space trajectories make with respect to the coordinate axes? Check the graph of the average velocities. Are the x- and y-components the same? How can it be that the displacement of the electron distribution in* **k***-space is in one direction (45°) but the average velocity is in another direction?*

Exercise 10.21 (M*) *From the graph, measure the average velocity for the* **E***-field applied in each of the x- and the y-directions. The current is proportional to the average velocities, so what is the ratio of the electrical conductivities for the two directions of applied field? What would you predict for that ratio, knowing the dispersion relation* $\mathcal{E}(\mathbf{k})$, *Eq. (10.1)? Does the prediction work, at least within the limits of our very noisy data? (Hint: recall the Drude formula for the conductivity and get the components of the effective mass tensor from the dispersion relation.)*

Exercise 10.22 (M)** *Change the* FERMI ENERGY *to 3.5 eV. Now measure again the anisotropy of the conductivity (of the drift velocity). Explain why the anisotropy is substantially greater than the effective mass ratio. (Hint: focus your attention on Eq. (10.3), remembering that* $\partial f_0 / d\mathcal{E}$ *is large only at the Fermi surface.)*

In PRESET 6 we've reduced the symmetry of the dispersion relation from that of a square to that of a rectangle. But even for the square symmetry, the dispersion relation is not isotropic. The energy of a state **k** depends not only on |**k**| but also on its direction. Wouldn't we therefore expect the current to be different for the electric field applied along the coordinate axes and along the line at 45° to the axes?

Exercise 10.23 (M*) *If you return the* ANISOTROPY *slider to 1, you can verify that, to within the considerable noise, the magnitude of the average velocity for given magnitude of* ELECTRIC FIELD, *is independent of the orientation of the field. Compare the ratio* |**v**|/|**E**| *for the* ELECTRIC FIELD **E** = $[E_x \; E_y]$ *in the* [1 0] *and the* [1 1] *directions. Can you detect a significant difference?*

For the system with square symmetry, or cubic symmetry in three dimensions, the magnitude of the electric current is the same whatever the orientation of the electric field.

Exercise 10.24 (M**) *Show formally that the electrical conductivity experiment cannot distinguish between a system with cubic (square in two dimensions) anisotropy and one with full isotropy as in the Sommerfeld model. (Hint: recall the use of a conductivity tensor to describe an anisotropic medium.)*

10.6   Summary

The complexities of the dynamics of band-electrons, as illustrated in "ziman", raise serious questions about the range of validity of the Sommerfeld model. "peierls" has shown that the model is more useful than one might first suspect. The incorporation of two new ideas, the effective mass and the hole, provides a rationale for accepting a slightly generalized Sommerfeld language to describe both nearly empty and nearly full bands. In particular, that major difficulty of the free electron theory, the anomaly of the sign of the Hall coefficient, is beautifully disposed of. For roughly half-filled bands, the full formal apparatus is required in principle, but in practice numbers for electrical conductivities based on the Sommerfeld picture will not be far out of line.

10.A   Deeper exploration

Density of states   Figure 10.3 shows the number of states enclosed by an energy contour of energy $\mathcal{E}$. The derivative of this curve is the electron density of states for the "peierls" band. It shows a hint of a van Hove singularity in the center of the band. Increase substantially the number of points in the K-SPACE GRID, in the CONFIGURE menu, enough to allow such a plot to show the singularity more convincingly. Explore the analytic nature of the singularity and compare with the experiment [23].

One-dimensional conductors   Some materials have very large anisotropy of the electrical conductivity, with the ratio of conductivities in different directions being as high as thousands. Learn about the behavior of some of these [39] and simulate some of their behavior with "peierls". The influence of magnetic fields probably offers the most interesting possibilities. Be cautious, however, of artifacts related to the coarseness of the k-space grid.

10.B   Preferred Fermi energies

The coarse graining required for the numerical simulations gives artifacts in some of the proposed exercises. A minor example is seen in Exer-

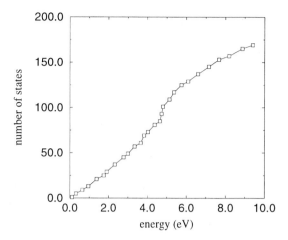

Figure 10.3: Number of enclosed states as a function of the Fermi energy for a 14 × 14 **k**-space grid. Note the kink at the half-filled band which gives rise to a van Hove singularity in the density of states.

cise 10.12 in which a small current is observed for a full band. The coarse graining also makes the scattering rates sensitive to the available area within the Fermi surface. The rates are significantly changed by variations in the Fermi energy too small to alter the total number of enclosed states. This problem may be ameliorated by selecting FERMI ENERGIES such that the area enclosed by the Fermi surface is strictly proportional to the number of enclosed states on the **k**-grid. An appropriate set of energies and the corresponding numbers of enclosed states are given in Tables 10.1 and 10.2 for two grid sizes; the larger is that used by default by "peierls". These energies are the energies that give the Fermi surface contour lying well separated from the **k**-values of the allowed states.

A by-product of the tables is figure 10.3, showing the number of enclosed states as a function of energy for a 13 × 13 array of states. The derivative of this plot is the density of states. The kink at 4.7 eV is the source of a logarithmic Van Hove singularity in the density of states, characteristic of a saddle point in the dispersion relation $\mathcal{E}(\mathbf{k})$ of a two-dimensional system.

10.C   "peierls" – the program

"peierls" simulates, in two dimensions, the motion of band electrons and holes in a periodic potential and static electric and magnetic fields. Periodic boundary conditions give rise to quantized **k**-states, and the Pauli exclusion principle allows each state to be occupied by at most one (spinless) electron.

Table 10.1: Pairs of electron numbers and Fermi energies which give areas enclosed by the Fermi surface proportional to the number of electrons. This table is for a 14×14 array of states (K-SPACE GRID=14).

$\mathcal{E}_F$	0.28	0.65	1.01	1.36	1.82	2.17	2.36	2.80	3.15
N	4	12	16	24	32	40	44	52	60
$\mathcal{E}_F$	3.34	3.78	4.09	4.16	4.32	4.54	4.84	5.08	5.43
N	68	76	84	88	96	104	112	120	136
$\mathcal{E}_F$	5.90	6.11	6.41	6.88	7.39	7.86	8.32	8.92	9.40
N	144	152	156	164	172	180	184	192	196

Table 10.2: A table similar to Table 10.1 for an 11×11 array of states (K-SPACE GRID=11).

$\mathcal{E}_F$	0.15	0.45	0.89	1.32	1.90	2.35	2.53
N	1	5	9	13	21	25	29
$\mathcal{E}_F$	3.12	3.61	3.81	4.26	4.67	4.80	4.85
N	37	45	49	57	61	65	73
$\mathcal{E}_F$	5.30	5.89	6.48	7.13	7.78	8.68	9.4
N	81	89	97	105	109	117	121

10.C.1   Electron dynamics in "peierls"

The "peierls" simulation is, crudely speaking, a mixture of "ziman" and "sommerfeld": the band-electrons move in the same potential and according to the same equations of motion as in "ziman"; and, as in "sommerfeld", the initial quantized **k**-states are filled with one electron each up to the Fermi energy, and the motion of the electrons is interrupted by scattering, subject to restriction by the Pauli exclusion principle.

In particular, the states in "peierls" move independently of each other according to the two semiclassical equations of motion

$$d(\hbar \mathbf{k})/dt = -e\left(\mathbf{E} + \mathbf{v} \times \mathbf{B}\right), \tag{10.6}$$
$$d\mathbf{r}/dt = \mathbf{v} = (1/\hbar)\nabla_{\mathbf{k}}\mathcal{E}(\mathbf{k}), \tag{10.7}$$

with the dispersion relation $\mathcal{E}(\mathbf{k})$ given in Eq. (10.1). Both the electric and magnetic fields, **E** and **B**, are static with **E** restricted to lie within the plane of motion (the xy-plane) and **B** perpendicular to that plane. Inelastic scattering is crudely modeled by allowing only electrons with energies above the Fermi energy $\mathcal{E}_F$ to scatter, and then only into holes with energies below the Fermi energy.

10.C.2   "peierls" algorithm

Initial configuration   The periodic boundary conditions restrict the allowed **k**-states to lie on a discrete grid. For example, the x-components of the initial wave vectors **k** can take on any of the N values

$$k_x(t = 0) = i\Delta k, \quad i = -\frac{(N-1)}{2}, -\frac{(N-1)}{2} + 1, \ldots, \frac{(N-1)}{2}, \quad (10.8)$$

and similarly for k_y. The grid spacing Δk is given by $\Delta k = 2\pi/Na$ with the lattice constant a set to 2 Å, and the K-SPACE GRID N an adjustable parameter in the CONFIGURE menu. The value N of the K-SPACE GRID implies an $N \times N$ array of states. In the initial ground state, **k**-states with energies (10.1) below the FERMI ENERGY $\mathcal{E}_F$ are occupied by electrons. All other states are initially empty.

Time evolution of states   At time t "peierls" uses a fifth order Runge–Kutta routine[1] to integrate the equation of motion (10.6) numerically and determine the wave vectors **k** at time $t + \Delta t$. Because we are only interested in the wave vectors and velocities of the states, there is no need to integrate $\dot{\mathbf{r}} = \mathbf{v}(t)$ in Eq. (10.7).

Scattering   "peierls" uses a particularly simple algorithm to implement inelastic scattering of the electrons. After the states have evolved by one time step Δt, "peierls" first checks for every electron state *above* the Fermi energy whether that electron scatters into an empty state. Second, it fills as many randomly chosen empty states *below* the Fermi energy with electrons as electrons have scattered in the first step:

1. Each electron above the Fermi energy scatters with the probability $p = \Delta t/\tau$ per time step Δt; i.e., a **k**-state occupied by an electron becomes empty. The scattering time τ can be specified on the panel, and the time step Δt is chosen to be always smaller than $\frac{1}{5}$ of the scattering time. (For details on the computation of the time step see page 135.) "peierls" keeps track of how many electrons are annihilated in this way during each time step.

2. Next, "peierls" tries to fill as many empty states below the Fermi energy with electrons as it annihilated electrons in the previous step. To do so, it picks a random state on the grid, and fills this state with an electron provided that state is empty and below the Fermi energy. This procedure is repeated until either enough empty states are filled to restore the proper number of electrons, or "peierls" has made more than $N \times N$ attempts to find empty states below

---

[1] For a short description of the fifth order Runge–Kutta routine `rkf` see page 133.

the Fermi surface. (Recall that $N \times N$ is the number of states in the system.) In the latter case, the number of missing electrons is carried over into the next time step and added to the number of additional states emptied in that next time step.

Note that this algorithms guarantees the cooling of the electron gas since the electrons lose energy in every scattering event.

10.C.3 Displayed quantities

Displays   The wave vectors **k** of the electrons are shown in the repeated zone scheme in the right hand display. Each **k**-state is shown as a *worm* whose WORM LENGTH dots represent the electron's wave vector at the last WORM LENGTH time steps. The WORM LENGTH can be specified in the CONFIGURE menu. If SHOW HOLES is toggled on, the wave vectors **k** of the hole states are shown in the left hand display and the hole, instead of the electron, velocities are shown in the graph. The origin of the hole representation can be shifted from the center of the Brillouin zone to its corner by clicking the SHIFT ORIGIN toggle. See Figure 10.2 on page 204 for an illustration.

The Brillouin zone boundaries are shown in blue in both displays, and the Fermi surface in red. The average **k**-vector of all electrons (holes) is indicated by a larger green dot, the origin of **k**-space by a larger red dot.

Graph   Click on the SHOW GRAPH button to show the x- and y-components of the average velocity of the electrons (if SHOW HOLES is *not* selected) or of the holes (if SHOW HOLES is selected). For example, the x-component of the average electron velocity is given by

$$\langle v_x \rangle = \frac{1}{N_e} \sum_{\text{electrons}} \frac{1}{\hbar} \frac{\partial \mathcal{E}}{\partial k_x} = \frac{1}{N_e} \sum_{\text{electrons}} \frac{\hbar}{m} \left(k_x - \frac{a^2}{\pi^2} k_x^3 \right), \qquad (10.9)$$

where N_e is the number of states occupied by electrons.

10.C.4 Bugs, problems, and solutions

Occasionally in implementing the scattering algorithm, "peierls" fails to find an empty state into which to place a scattered electron. Thus electron number is not quite conserved. The NUMBER OF ELECTRONS, shown over the right hand display, allows you to monitor this problem. Its severity is reduced by use of Fermi energies from the lists in Tables 10.1 and 10.2.

As with most of the animations in SSS, the speed varies with the platform. You will be forced to compromise speed against accuracy in your choice of SPEED and K-SPACE GRID.

11

"fermi" – Carrier densities in semiconductors

Contents

11.1   Introduction

Modern electronics is based on semiconductor technology developed over the past 50 years. The potential for device applications arises from our ability to control and to modify the concentrations and spatial distributions of the charge carriers. Many devices require working simultaneously with two quasi-independent kinds of charge carriers, electrons and holes. "fermi" illustrates some of the physics underlying the required control of carrier concentrations.

There is no sharp line to be drawn between the two classes of solids: insulators and semiconductors. A semiconductor is simply an insulator in which the temperature is high enough to excite electrons into the bottom of an empty band, or from the top of a full band, to give a medium with mobile charge carriers. Impurity doping of the material provides the control of concentrations essential for device fabrication. "fermi" considers a homogeneous sample of semiconducting material and explores the effect on the carrier concentrations of changing the material parameters. The most important of these are the impurity concentrations and the energy gap of the material. Effective masses and impurity ionization energies play a relatively minor role. Temperature dependence of concentrations is a major issue if devices are to work effectively over reasonable ranges of ambient temperatures.

11.2   Pure (intrinsic) material

Thermal excitation of electrons from the valence band to the conduction band provides a semiconductor with the two types of charge carriers, holes and electrons. It is these carriers which provide the electrical conduction in *pure* semiconductors. Given the temperature and the band structure, the carrier concentrations are calculated by straightforward application of the Fermi–Dirac distribution,

$$f(\mathcal{E}) = 1 \Big/ \left(e^{(\mathcal{E}-\mu)/k_B T} + 1 \right). \tag{11.1}$$

This function gives the probability, in terms of the temperature T and chemical potential μ, that a one-electron state of energy $\mathcal{E}$ will is filled. Note that there is *not* universal consistency in the use of the terms *Fermi energy*, *Fermi level*, and *chemical potential*. The Fermi energy usually means the limit of the chemical potential as $T \to 0$, or the energy of the highest filled or lowest unfilled state in the case of metals. Fermi level and chemical potential tend to be used synonymously. Don't be surprised, however, to see Fermi energy used when what is meant is chemical potential or Fermi level.

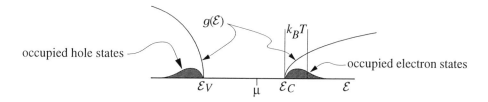

Figure 11.1: DOS $g(\mathcal{E})$, plotted vertically, near the top of the valence band and the bottom of the conduction band. The dark areas represent states occupied by thermally excited holes and electrons.

Figure 11.1 shows the density of states (DOS) near the top of the valence band and the bottom of the conduction band, $g(\mathcal{E})$. The shaded areas represent those states in the two bands which are occupied (by electrons in the conduction band and by holes in the valence band) as a consequence of thermal excitation across the energy gap. As long as the chemical potential μ lies in the gap between the bands, only the states within a few $k_B T$ of the band edge have appreciable probability of being occupied, and the number of such states is defined as $N_C(T)$ for the conduction band and $N_V(T)$ for the valence band. Section 11.B gives expressions (11.12) and (11.13) for these, which depend on both the effective masses of the bands and the degeneracies of the band edges. "fermi" assumes both bands to be non-degenerate and determines $N_C(T)$ and $N_V(T)$ based on the effective masses that can be specified on the panel. We will refer to this fictitious semiconductor as *ferminium*. The numerical magnitude of $N_C(T)$ and $N_V(T)$ at $T = 300$ K is 2.5×10^{19} cm^{-3} for a band degeneracy $g = 1$ and effective mass $m^* = m$.

If the chemical potential μ lies at least a few $k_B T$ below the bottom of the conduction band, or above the top of the valence band, then the electron concentration n, or the hole concentration p, may be written to a good approximation as

$$n = N_C(T)e^{-(\mathcal{E}_C - \mu)/k_B T} \quad \text{or} \quad p = N_V(T)e^{-(\mu - \mathcal{E}_V)/k_B T}. \qquad (11.2)$$

These expressions allow "fermi" to calculate n and p once μ is known; but how is μ determined? The condition that defines the chemical potential, both in this calculation of intrinsic concentrations and later in determining the extrinsic, impurity-dependent, concentrations, is the *charge neutrality condition*. For the intrinsic semiconductor it is simply the requirement that $n = p$: the number of electrons must equal the number of holes if the crystal is to remain neutral.

How stringent is this condition? With semiconductors, we may be interested in carrier concentrations as small as 10^{13} cm^{-3}. That's only one carrier for 10^{10} atoms. Can we really speak of charge neutrality to that level of precision?

Exercise 11.1 (M*) *Consider a piece of conductor about 1 mm^3 in volume. Suppose it is connected to a circuit and that its potential with respect to its surroundings is less than a millivolt. What bound can you place on its total charge and on the excess density of carriers of one sign over that of the other? (Hint: you will need to know the capacitance of the piece to its surroundings. Just make a very rough estimate of that capacitance: e.g., don't worry about the difference between a cube and a sphere.)*

"fermi" uses the charge neutrality condition in the following way. At a given temperature, it assumes a value for μ, calculates n and p from Eqs. (11.2), and sees which is larger. It then adjusts μ to try to make them equal and repeats the cycle until charge neutrality is achieved. It then moves to a different temperature and repeats until it has developed a picture of the full temperature dependence of the chemical potential and the carrier concentrations. PRESET 1 gives the result of such a calculation for pure material.

1

In the upper graph, "fermi" plots the electron concentration in red, and the hole concentration, sometimes obscured by the electron plot, in blue. The vertical scale can be chosen to be either LINEAR in concentration or LOGARITHMIC. The lower graph gives the band edges in red and blue, the donor and acceptor levels in black, and the chemical potential in green. The horizontal scale in both plots can be chosen to be either TEMPERATURE T or INVERSE TEMPERATURE $1/T$.

Exercise 11.2 (C) *With PRESET 1 explore the various ways in "fermi" that you can display the data, plotting both versus T and versus $1/T$ and, for the carrier densities, both LINEAR and LOG. Which is most useful to display the exponential dependence on the inverse temperature?*

Exercise 11.3 (M*) *Use PRESET 1 to determine for pure ferminium the change in temperature ΔT required to increase the carrier concentration by a factor of 100, starting at $T = 300$ K. From this ΔT let's try to deduce the energy gap for ferminium. First, as many people do, note that the carrier concentration n is proportional to $e^{-\mathcal{E}_g/2k_BT}$. Then show that $\mathcal{E}_g = 2\ln(n_2/n_1)k_BT_1T_2/(T_2 - T_1)$ and deduce the gap from the measured temperatures.*

However, they've forgotten that in the expression for n there's a pre-exponential factor proportional to $T^{3/2}$; or else, as we frequently do, comment that it's unimportant. So second, correct the expression above to *include this additional temperature dependence. By how much is the deduced energy gap changed?*

Now that you have an answer for the gap, what does it mean? In PRESET 1, "fermi" uses for the gap a temperature-dependent relation which approximates that for germanium,

$$\mathcal{E}_g(T) = (0.74 - 2.74 \times 10^{-4}\, T)\ eV \Rightarrow \mathcal{E}_g(300\ K) = 0.66\ eV \qquad (11.3)$$

Note that the gap you have deduced from the data is closer to the "fermi" value for the gap at 0 K than at 300 K.

What's going on? Rationalize this apparent discrepancy. (Hint: using the temperature-dependent form of the gap and temporarily neglecting the temperature dependence of $N_C(T)$, calculate $(k_B T^2/n)(dn/dT)$. What is the effect of including the linear temperature dependence of $\mathcal{E}_g(T)$?) Can you show the same thing more easily without taking derivatives?

Determining activation energies from semilog versus $1/T$ plots, i.e., *Arrhenius plots*, can be dangerous. Remember these lessons: the linear temperature dependence of activation energies cleverly hides itself as a constant pre–exponential factor; and power law pre–exponentials do not always have a negligible effect in the analysis. (A third danger is, of course, too little data. Over a sufficiently narrow range of temperature, nearly any set of data will fit a proportionality to $e^{-A/T}$.)

Exercise 11.4 (C) *To see the effect of the linear temperature dependence of the gap energy, compare* PRESET 1 *with* PRESET 2, *in which the temperature dependence of the gap is deleted.* COPY DENSITIES *for one* PRESET *and then* STEAL DATA *for the other to make comparison easy. Remember that the gap energies are different in the temperature range for which the concentrations are being plotted! How do the slopes of the two compare? What about their intercepts at $T = \infty$?*   # 2

A numerical method is hardly necessary to obtain an expression for the carrier concentrations, n and p, for pure materials. Taking the product of the two equations (11.2) gives the result

$$np = N_C(T)N_V(T)\, e^{-\mathcal{E}_g/k_B T} \equiv n_i^2. \qquad (11.4)$$

Results of this nature are referred to by chemists as the law of mass action. The result (11.4) is valid, *independent of impurity doping*, depending only upon the assumptions required for Eqs. (11.2). The product np is fixed at a value n_i^2, where n_i is called the *intrinsic concentration*, and any change in impurity concentration which implies a change in concentration of electrons automatically implies a compensating change in the concentration of holes.

Exercise 11.5 (M) *Combine Eq. (11.4) with the charge neutrality condition to show that the intrinsic electron and hole concentrations are given by*

$$n = p = \sqrt{N_C(T)N_V(T)}\, e^{-\mathcal{E}_g/2k_BT} = n_i. \qquad (11.5)$$

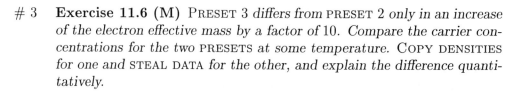

First observe that the concentrations are thermally activated with an energy that is one half of the energy gap, not the full gap energy: a factor of 2 that is easy to forget. Also note that the result (11.4) follows without needing to evaluate the value of the chemical potential! It is quite generally valid, even in doped materials, depending only on the assumption that the chemical potential lies well within the energy gap.

 # 3   **Exercise 11.6 (M)** PRESET 3 *differs from* PRESET 2 *only in an increase of the electron effective mass by a factor of* 10. *Compare the carrier concentrations for the two* PRESETS *at some temperature.* COPY DENSITIES *for one and* STEAL DATA *for the other, and explain the difference quantitatively.*

 # 2,3   **Exercise 11.7 (C*)** *Compare the temperature dependence of the chemical potential at high temperatures for the two* PRESETS 2 *and* 3. *(*ZOOM *the μ-axis of the chemical potential plot to the range 0.3–0.45 eV and switch between the two* PRESETS *to see more clearly the variation of μ with T.) Explain why there is a temperature dependence of the chemical potential for* PRESET 3 *but not for* PRESET 2. *(Hint: Figure 11.1 shows the chemical potential displaced from the center of the gap, but equal concentrations of electrons and holes. What feature of the DOS g(ε) forces this shift of μ?)*

11.3   Doped (extrinsic) material

Most semiconductor devices rely for their operation on the incorporation of impurities, called *donors* and *acceptors*, in a carefully controlled fashion. The impurities contribute to the electrical properties in two ways. Most importantly they provide the carriers that make the material conducting at operating temperatures. Secondly, since they are ionized at operating temperatures, their charge is an important source of scattering of the carriers. Remarkably, the binding energy of a carrier to one of these impurities (or the impurity ionization energy) can be estimated using a very simple hydrogenic model, modified both to include the effect of the dielectric constant of the crystal and to use the effective mass, typically less than the free electron mass, of the carriers. "fermi" accepts the binding energy of the carriers to the impurities, EC−ED for donors and EA−EV for acceptors, as input parameters.

Be aware of the inconsistencies of notation concerning energies. Some texts (e.g., Kittel [6]) choose the edge of the valence band as the origin of energy and reference all other energies to that. Others (e.g., Ashcroft & Mermin [1], as well as "fermi") assign no reference energy. Thus energies always appear in differences so that expressions are independent of the choice of origin. The latter is to be preferred because it maintains the symmetry between electrons and holes. An exception is required for the energy graph for which an arbitrary choice has been made for the zero as the low temperature value of the valence band edge. An additional minor problem concerns the distinction between the *donor energy*, $\mathcal{E}_D$, and the *donor binding energy*, $(\mathcal{E}_C - \mathcal{E}_D)$. Context should remove any ambiguity!

11.3.1   Charge neutrality with impurities

At room temperature, typical binding energies of the carriers to the impurities are small enough to permit thermal excitation of the carriers into the conduction or valence bands. For moderate doping levels and at operating temperatures, the impurities contribute a much higher concentration of carriers than would be present in the pure crystal. There is a range of temperature in which this carrier concentration, the *extrinsic* concentration, is nearly independent of temperature. At higher temperatures carriers excited across the gap begin to dominate and we speak of a transition into the *intrinsic* regime. Although impurities are present, the carrier concentrations are essentially the same as for the pure material.

The trapping of carriers at low temperatures gives the phenomenon of carrier freeze-out. "fermi" keeps track of all this. Besides the expressions above for the concentration of electrons and holes, "fermi" needs the expressions (11.6) for the concentrations of ionized donors and acceptors. In these N_D and N_A are the donor and acceptor concentrations, and superscripts indicate whether the impurity is ionized (high temperature) or neutral (low temperature). The concentrations of ionized donors and acceptors are given by

$$N_D^+ = \frac{N_D}{2e^{(\mu-\mathcal{E}_D)/k_BT}+1} \quad \text{and} \quad N_A^- = \frac{N_A}{2e^{(\mathcal{E}_A-\mu)/k_BT}+1}. \tag{11.6}$$

These appear to be the usual Fermi–Dirac expressions except for a strange factor of 2. This factor properly accounts at high temperature for the statistical weight associated with the two-fold spin degeneracy of the neutral impurity states. The form of Eqs. (11.6) also insures that at low temperature only a single carrier can be trapped by the impurity.

As for case of the intrinsic semiconductor, evaluation of the carrier concentrations requires knowing the chemical potential μ. "fermi" assumes a value for μ, calculates the concentrations of carriers and of ionized impurities based on that μ, and then uses the charge neutrality condition,

$$p + N_D^+ - n - N_A^- = 0, \tag{11.7}$$

to test the validity of the initial assumption for the value of μ. If charge neutrality is not obeyed, the guess for μ is adjusted and the trial repeated until charge neutrality is achieved.

11.3.2   Extrinsic regime

4   **Exercise 11.8 (M)** PRESET 4 dopes the ferminium with donors of binding energy $0.05 \, eV$ at a concentration of 10^{15} donors/cm^3. Determine from "fermi" the range of temperature for which the electron concentration in this material is within 1% of its room temperature value. By how much does the chemical potential vary with temperature in this range? Explain how there can be so little variation in electron concentration when there's so much variation in the chemical potential.

The qualitative picture of donation of carriers by shallow impurities describes well what are called the *majority carriers*: electrons if donor impurities dominate, holes if acceptors dominate. This picture is consistent with the temperature independence of the majority carrier concentration at intermediate temperatures, the carrier freeze-out at low temperatures, and the transition to intrinsic behavior at high temperature. Not so clear is what is going on with the *minority carriers*, i.e., holes for the example of PRESET 4. Instead of following the intrinsic behavior, the concentration plummets in the temperature range in which the majority (electron) concentration is constant.

Exercise 11.9 (C) For the ferminium of PRESET 4, what is the hole concentration at room temperature? Why is it orders of magnitude smaller than the intrinsic concentration? After all, the energy cost required to create an electron–hole pair remains just $\mathcal{E}_g$ and is the same as for the intrinsic case. If the energy gap is unaltered by the presence of the donors, why isn't the hole concentration also independent of the donor concentration?

The law of mass action, Eq. (11.4), gives a formal statement relating the electron and hole concentrations. It is usually more reliable than your intuition, Exercise 11.9 being perhaps a case in point, and is quite generally valid requiring only that the chemical potential remain well within the gap.

Exercise 11.10 (M) Find the approximate activation energy for the concentration of holes in the extrinsic range of PRESET 4 from the DEN-

SITY *plot. Is this equal to the full energy gap as it should be, instead of half the gap?*

11.3.3 Limits of the extrinsic range

At high enough temperatures the carriers thermally excited across the gap completely swamp those provided by the impurities. We can define this transition to the intrinsic regime as when the intrinsic carrier concentration becomes comparable to the impurity concentration.

Exercise 11.11 (M) *Develop an expression for the temperature of the transition from extrinsic to intrinsic behavior by setting the impurity concentration equal to the intrinsic concentration. For the system of* PRESET 4, *if you were to increase the donor concentration by a factor of 10, what would be the change in the temperature of the transition from extrinsic to intrinsic? Does "fermi" agree with you?*

Exercise 11.12 (M) *For* PRESET 4 *compare the extrinsic behavior for the* PRESET *value of the energy gap of 0.74 eV with that for a gap of 1.1 eV, appropriate to silicon. What is the effect on the value of the extrinsic concentration? What is the effect on the value of the cross-over temperature from extrinsic to intrinsic conduction? Does that agree with your expression from Exercise 11.11? Explain the physics of this change in cross-over temperature. The operation of typical semiconductor devices relies on the material being in the extrinsic regime.*

Do you see why silicon won out over germanium as the material of choice in semiconductor electronics? And why some people are seriously pursuing the possibility of using diamond based electronics for specialized high temperature applications [51]?

The extrinsic regime is also limited as the temperature is lowered, now by the trapping of the carriers by the charged impurities. This loss of carriers by trapping at low temperature is called *carrier freeze-out*.

Exercise 11.13 (M*) *For n-type ferminium with 10^{15} donors/cm^3, use* PRESET 4 *of "fermi" to find the freeze-out temperature T_{fo}, the temperature at which roughly half of the donors have trapped electrons. A naive guess for the freeze-out temperature would be $k_B T_{\text{fo}} \approx (\mathcal{E}_C - \mathcal{E}_D)$. Does the prediction of "fermi" agree with this? Explain why this guess is so wrong and show that the correct freeze-out temperature for n-type material is given by*

$$k_B T_{\text{fo}} \approx (\mathcal{E}_C - \mathcal{E}_D)/\ln[N_C(T_{\text{fo}})/N_D]. \qquad (11.8)$$

Exercise 11.14 (M) *Change the donor binding energy to 0.25 eV and determine the effect on the extrinsic temperature range. Make a quantitative comparison with Eq. (11.8).*

11.3.4   Compensation

Up to now we've supposed there have been either donor impurities or acceptors, but not both. When both types are present we speak of *compensated* material.

Exercise 11.15 (C) *A naive hand-waving argument would suggest that a crystal with $N_D = N_A = 10^{16}$ cm^{-3} would have, at room temperature, $n = p = 10^{16}/\text{cm}^3$. Argue from the law of mass action (11.4) that this would be a wrong guess.*

\# 5   **Exercise 11.16 (M*)** PRESET 5 *is a material, with $N_D = 10^{15}$ cm^{-3} and $N_A = 3 \times 10^{14}$ cm^{-3}. Predict with a sound argument, and check with "fermi", the value of the electron and hole concentrations in the extrinsic regime. Explain why the hole concentration is not equal to the acceptor concentration in the extrinsic temperature range.*

Exercise 11.17 (C**) ZOOM *the μ graph to low temperature to compare the temperature dependences of the chemical potential at the lowest temperatures for* PRESET 5, *with and without the compensation (i.e., compare the two cases $N_A = 3 \times 10^{14}$ cm^{-3} and $N_A = 10^0$ cm^{-3}). Explain the differences.*

Exercise 11.18 (M*) *Predict the carrier concentration as a function of temperature for fully compensated material, i.e., material with $N_D = N_A$. Use "fermi" to fabricate some material of this sort in order to check your prediction.*

For many purposes, compensated material with donor and acceptor concentrations N_D and N_A behaves identically to material doped to a concentration $|N_D - N_A|$ of whichever impurity dominates. Thus there is nothing much here of interest for physics. However, for technical applications the result is of critical importance. Nearly all semiconducting devices are based on controlled spatial variations in local donor and acceptor concentrations. To see the utility of compensation, consider the problem of fabricating a *pn*-junction. In making the junction, it is not necessary to have material with only donors on one side of the junction and only acceptors on the other. One can start with a single crystal of lightly doped *p*-type material and diffuse in a high concentration of donor impurities from one surface to convert a thin layer of material near the surface to n-type to form the junction. The n-layer near the surface is partially compensated, but behaves as if it were n-material with an effective donor concentration of $(N_D - N_A)$.

11.4   Summary

To provide the charge carriers for electrical conduction, semiconductor devices depend on thermal excitation either across the energy gap or, more often, from shallow impurities. Given the chemical potential μ, the carrier concentrations and impurity ionization states can be deduced from the Fermi distribution, the band DOS, and impurity concentrations and binding energies. The appropriate value of μ is determined numerically by "fermi" from the charge neutrality condition.

Intrinsic behavior, of pure materials and of doped materials at high temperatures, is characterized by exponentially activated conductivity with an activation energy of one half of the gap energy. Extrinsic behavior in suitably doped materials, by contrast, is characterized by a majority carrier concentration sensibly independent of temperature over the useful operating range of temperature. The high temperature end of the extrinsic range is higher for semiconductors with larger band gaps. The extrinsic majority carrier concentration is equal to the net doping concentration $|N_D - N_A|$. In the extrinsic temperature range, equilibrium minority carrier concentrations are thermally activated with an activation energy equal to the full energy gap, and over most of the extrinsic range are negligibly small.

11.A   Deeper exploration

Diamonds   Let "fermi" help you consider diamond as a possible material for high temperature devices. Take for the material parameters: dielectric constant $= 5.7$, energy gap $= 5.5$ eV, and $m^*/m = 0.2$. Use the hydrogenic model to estimate the shallow impurity binding energy. (You might consider the issues concerning the validity of this model for diamond.) Consider material doped to an impurity concentration of 10^{16} cm^{-3}. Why would this not be useful in a room temperature device? What would be the lowest temperature at which most of the donors would be ionized? At what temperature would the material become intrinsic? See reference [51] for arguments supporting a research effort on the use of diamond for high temperature devices.

Low temperature   Explore in more detail, both analytically and using "fermi", the behavior of the chemical potential in the freeze-out regime and the influence of compensation on the low temperature behavior as T approaches 0. Why is the low temperature variation of the chemical potential so qualitatively different depending upon the presence or absence of compensation?

11.B   "fermi" – the program

"fermi" computes the chemical potential and carrier (electron and hole) densities in a homogeneous semiconductor as a function of temperature.

11.B.1   Homogeneous semiconductors

Quantities of central interest in a homogeneous semiconductor are the number of electrons in the conduction band, $n(T)$, and the number of holes in the valence band, $p(T)$. In the model used by "fermi", an implicit equation for the chemical potential μ is obtained from the application of Fermi–Dirac statistics. This equation for μ, however, has no simple analytical solution, and "fermi" uses a binary search algorithm to solve it. Once μ is known, the determination of the carrier densities is straightforward.

Charge neutrality condition   Let $N_A^-(T)$ be the number of acceptor impurities per unit volume that have contributed a hole to the valence band and let $N_D^+(T)$ be the number of donor impurities per unit volume that have contributed an electron to the conduction band. Charge neutrality in the semiconductor requires

$$n(T) + N_A^-(T) = p(T) + N_D^+(T). \tag{11.9}$$

Given the temperature, the gap energy, the band densities of states, and the donor and acceptor concentrations and binding energies as input parameters, "fermi" searches for the chemical potential μ that balances the charge neutrality condition (11.9).

Carrier densities in thermal equilibrium   As shown in standard texts, the densities of electrons in the conduction band, $n(T)$, and holes in the valence band, $p(T)$, are given (with the assumptions explained in the next section) by

$$n(T) = N_C(T)\, e^{-(\mathcal{E}_C - \mu)/k_B T}, \tag{11.10}$$
$$p(T) = N_V(T)\, e^{-(\mu - \mathcal{E}_V)/k_B T}, \tag{11.11}$$

with the two μ-independent quantities

$$N_C(T) \;=\; \frac{g_C}{4}\left(\frac{2m_e k_B T}{\pi\hbar^2}\right)^{3/2}, \tag{11.12}$$

$$N_V(T) \;=\; \frac{g_V}{4}\left(\frac{2m_h k_B T}{\pi\hbar^2}\right)^{3/2}. \tag{11.13}$$

Here $\mathcal{E}_C$ and $\mathcal{E}_V$ are the energies of the conduction and valence band edges respectively, and m_e and m_h are the effective masses of the electrons and holes in these bands. The parameters g_C and g_V are the degeneracies of the conduction and valence bands at the band edges and are set to 1 in "fermi".

Donors and acceptors Much of the interesting physics occurs when the semiconductor is doped with donor and acceptor impurities. Given the donor and acceptor densities N_D and N_A, and the donor and acceptor binding energies $(\mathcal{E}_C - \mathcal{E}_D)$ and $(\mathcal{E}_A - \mathcal{E}_V)$, it follows from Fermi–Dirac statistics that the mean number of ionized donors N_D^+ and acceptors N_A^- per unit volume is

$$N_D^+ \;=\; \frac{N_D}{2e^{-(\mathcal{E}_D - \mu)/k_B T} + 1}, \tag{11.14}$$

$$N_A^- \;=\; \frac{N_A}{2e^{-(\mu - \mathcal{E}_A)/k_B T} + 1}. \tag{11.15}$$

The neutral donors and acceptors have a two-fold spin degeneracy, but, because of the strong Coulomb repulsion between the carriers, each impurity is only able to accommodate a single electron or hole. This effect is accounted for by the factor 2 in Eqs. (11.14) and (11.15).

11.B.2 Assumptions and limitations of the model

Chemical potential The simple form and the validity of Eqs. (11.10) and (11.11) rely on the important assumption that the chemical potential μ must be above the maximum of the valence band $\mathcal{E}_V$ and below the minimum of the conduction band $\mathcal{E}_C$ by more than several $k_B T$,

$$\mathcal{E}_C - \mu \gg k_B T \quad \text{and} \quad \mu - \mathcal{E}_V \gg k_B T. \tag{11.16}$$

This is a reasonable assumption for most semiconductors at room temperature: the gap energy in common semiconductors is of the order of 1 eV. With the chemical potential typically in the middle of the energy gap (except at heavy doping), this gives rise to an energy difference that is many times the room temperature value of $k_B T = 0.025$ eV. The chemical potential in "fermi" is always computed with the assumption (11.16), and it is left to the user to check its validity from the displayed quantities μ, $\mathcal{E}_C$, and $\mathcal{E}_V$.

Band structure Eqs. (11.12) and (11.13) are the result of assuming a particularly simple band structure, consisting of one valence and one conduction band.[1] In a typical semiconductor the electrons are found only near the minimum of the conduction band and the holes only near the maximum of the valence band. It therefore suffices to approximate the band structure $\mathcal{E}(\mathbf{k})$ by a quadratic function. For the one-dimensional case we take the conduction band to be of the form

$$\mathcal{E}(k) = \mathcal{E}_C + \frac{\hbar^2}{2m_e^*} k^2 \qquad (11.17)$$

and the valence band of the form

$$\mathcal{E}(k) = \mathcal{E}_V - \frac{\hbar^2}{2m_h^*} k^2. \qquad (11.18)$$

The effective masses m_e^* and m_h^* of electrons and holes, respectively, are input parameters determining N_C and N_V.

Gap energy The valence and conduction band energies in "fermi" are given in terms of the gap energy, $\mathcal{E}_g = \mathcal{E}_C - \mathcal{E}_V$. In real semiconductors the gap energy has a weak temperature dependence which is typically quadratic in T for low temperatures and linear in T for higher (room) temperatures. "fermi" assumes a purely linear temperature dependence for $\mathcal{E}_g$,

$$\mathcal{E}_g(T) = \mathcal{E}_g^0 + (d\mathcal{E}_g/dT)\, T, \qquad (11.19)$$

where $\mathcal{E}_g^0$ and $d\mathcal{E}_g/dT$ can be specified on the panel.

Because the absolute energy scale is arbitrary, in "fermi" we have chosen the valence band energy at $T = 0$ to be zero, $\mathcal{E}_V^0 = 0$. We arbitrarily share the temperature dependence equally between the two bands and use for the energy of the conduction band

$$\mathcal{E}_C(T) = \mathcal{E}_g^0 + \tfrac{1}{2}(d\mathcal{E}_g/dT)\, T \qquad (11.20)$$

and for the energy of the valence band

$$\mathcal{E}_V(T) = -\tfrac{1}{2}(d\mathcal{E}_g/dT)\, T. \qquad (11.21)$$

---

[1] With g_C and g_V set to 1 in Eqs. (11.12) and (11.13), one can adjust the effective masses to take into account a more complicated band structure with multiple minima in the conduction bands and degenerate valence bands.

11.B.3   Computation of the chemical potential μ

Given the electron and hole effective masses, the energy gap, and the donor and acceptor densities and binding energies as input parameters, the chemical potential μ can be determined from the charge neutrality condition (11.9). It follows from the expressions for the carrier densities (11.10) and (11.11) and the ionized donor and acceptor densities (11.14) and (11.15) that Eq. (11.9) is a fourth order polynomial in $e^{\mu/k_B T}$, and thus has no simple analytical solution. "fermi" uses the following binary search algorithm to determine the chemical potential at some temperature T:

1. The valence and conduction band energies (11.21) and (11.20) are determined from the input parameters $\mathcal{E}_g^0$ and $d\mathcal{E}_g^0/dT$.
2. The lower and upper bounds for the chemical potential are set to $\mu_{\text{low}} = \mathcal{E}_V$ and $\mu_{\text{high}} = \mathcal{E}_C$.
3. The midpoint, $\mu = \frac{1}{2}(\mu_{\text{low}} + \mu_{\text{high}})$, between the lower and upper bound is used as a guess for the chemical potential. The "charge density" $\rho = p + N_D^+ - n - N_A^-$ is computed for this μ.
4. If $\rho = 0$ (up to machine precision) then the charge neutrality condition holds and the guess for μ was correct and the search algorithm returns that μ. If ρ is positive then the guessed μ was too low. The new lower bound for the chemical potential μ_{low} is then set to that μ ($\mu_{\text{low}} = \mu$), and the algorithm continues with step 3. Similarly, if ρ is negative the guessed μ was too high and the upper bound for the chemical potential is set to that μ ($\mu_{\text{high}} = \mu$) for the next computation of ρ in step 3. The search is stopped if either the exact value for μ is found or 16 search steps have been carried out. After 16 steps the value of μ is determined to within $\pm 2^{-16}(\mathcal{E}_C - \mathcal{E}_V)$.

11.B.4   Displayed quantities

"fermi" displays the chemical potential μ (lower graph) and the carrier densities $p(T)$ and $n(T)$ (upper graph) as functions of temperature or inverse temperature. The carrier densities can be displayed in a linear or logarithmic scale.

Chemical potential   The chemical potential μ, shown as a solid green line in the lower graph, is computed according to the algorithm given above at POINTS IN GRAPH temperatures in the range MIN TEMPERATURE to MAX TEMPERATURE. (All three parameters are specified in the CONFIGURE menu.) The steps in temperature are equidistant in $1/T$. On the same graph the valence band energy $\mathcal{E}_V(T)$ (lower blue line), the acceptor energy $\mathcal{E}_A(T)$ (lower black line), the donor energy $\mathcal{E}_D(T)$ (upper black line), and the conduction band energy $\mathcal{E}_C(T)$ (upper red line) are shown. The valence and conduction band energies are given by Eqs. (11.21) and (11.20) in terms of the $\mathcal{E}_g^0$ and $d\mathcal{E}_g/dT$ specified on the panel. The acceptor and donor binding energies can be chosen on the panel.

Carrier densities   For a number of equidistant steps in $1/T$ (specified by POINTS IN GRAPH in the CONFIGURE menu) the hole and electron carrier densities p and n in the valence and conduction band are computed from the chemical potential μ determined above. Carrier densities, with a lower cutoff, of 10^8 cm^{-3} are shown in the upper graph in blue (holes) and red (electrons). Only the electron curve is visible in the intrinsic regime where the two concentrations are nearly identical.

12

"poisson" – Band bending in semiconductors

Contents

12.1   Introduction

Semiconductor devices rely on control of the spatial distribution of carrier concentrations through a combination of electrode geometry and inhomogeneous composition, achieved both by alloying (e.g., GaAs/Al$_x$Ga$_{1-x}$As structures) and doping (e.g., As doped Si). "fermi" considered the doping aspect of the problem but was restricted to homogeneous systems. Inhomogeneities allow local violation of charge neutrality and require finding a self-consistent solution to Poisson's equation, which gives the electrostatic potential in terms of the local charge density, and equations, e.g., Eqs. (11.2), giving the charge density in terms of the chemical potential and the local energies of the band edges.

In order to illustrate the essential ideas, "poisson" looks at two configurations. The first is an idealized *pn-junction*, the basic structure of such diverse devices as the light emitting diode (LED), the voltage tunable capacitor, the laser diode, the diode rectifier, the photovoltaic cell, and, when two diodes are placed back to back, the junction transistor. The second is a structure used for diagnostics or characterization, the *metal–insulator–semiconductor (MIS) diode*. With the addition of source and drain contacts, the MIS diode becomes the *metal–insulator–semiconductor field effect transistor* (MISFET), the device that comes by the millions on computer chips. (Commonly the insulator is an *oxide*, in which case the MISFET is the familiar MOSFET.) In both DEVICES, the PN-JUNCTION and the MIS DIODE, we deal with a planar structure so that we need consider the variations in electrostatic potential and concentration only along the single direction perpendicular to the planes.

With the introduction of inhomogeneous doping, or with electric fields applied, the condition of charge neutrality, which played so important a role in "fermi", need not be locally satisfied. It is replaced by the requirement that the local electrostatic potential $\phi(\mathbf{r})$ satisfy Poisson's equation,

$$\nabla^2 \phi(\mathbf{r}) = -\frac{e}{\kappa \epsilon_0}(N_D^+ + p - N_A^- - n), \qquad (12.1)$$

where κ is the dielectric constant of the material and ϵ_0 is the permittivity of vacuum. The concentrations of carriers are determined from Fermi–Dirac statistics as was done in "fermi", Eqs. (11.2), and the donors and acceptors are assumed to be fully ionized. The analysis becomes complicated because the variation with position of the electrostatic potential $\phi(\mathbf{r})$ implies corresponding variations of the energy of an electron found at the bottom of the conduction band at the position $\mathbf{r}$. The band edge energies now depend upon position as

$$\mathcal{E}_C(\mathbf{r}) = \mathcal{E}_C(0) - e[\phi(\mathbf{r}) - \phi(0)], \qquad (12.2)$$

with similar expressions for the edge of the valence band and the impurity levels. The values of the energy of the conduction band edge and the electrostatic potential at some convenient reference point are denoted by $\mathcal{E}_C(0)$ and $\phi(0)$. (An alternative approach is noted in Ashcroft and Mermin [1, p. 594].)

12.2   *pn*-junction

An idealized analysis of the *pn*-junction diode introduces many of the important concepts relevant to the operation of bipolar devices, i.e., devices in which both electrons and holes play an important role. So, what happens near a planar junction at which the doping changes abruptly from *n*-type to *p*-type?

12.2.1   *Equilibrium*

Imagine joining together two pieces of semiconductor, one *p*-type, the other *n*-type. There will be a redistribution of carriers which is governed by three requirements:

1. the potential $\phi(x)$ must be consistent, through Poisson's equation (12.1), with the charge density;
2. the carrier concentrations must be consistent with the thermal-equilibrium expressions

$$n(x) = N_C(T)\, e^{-(\mathcal{E}_C(x)-\mu)/k_B T}, \qquad (12.3)$$
$$p(x) = N_V(T)\, e^{-(\mu-\mathcal{E}_V(x))/k_B T}, \qquad (12.4)$$

   with the energies of the band edges given by Eq. (12.2);
3. in equilibrium, the chemical potential μ must be the same everywhere.

Figure 12.1 summarizes the results of such a self-consistent calculation of the redistribution of charge. The horizontal axis gives the distance from the plane of the junction, and the vertical axis gives the energies of electrons. The conduction and valence band edges are drawn as functions of position to reflect the contributions $-e\phi(x)$ of the electrostatic potential to the total energy of the electrons as in Eq. (12.2). In using these diagrams, it's helpful to think of "electrons falling downhill" and "holes falling uphill" along the band edges. Figure 12.1 is sketched for an *asymmetric junction*, one in which the doping concentration is different on each side of the junction. We assume there is no compensation in either region.

Open PRESET 1 of "poisson" and you see in the lower graph a plot   # 1
similar to Figure 12.1. The upper graph shows the spatial variation of

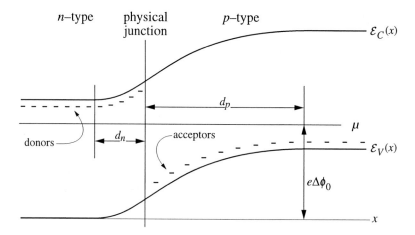

Figure 12.1: Band diagram for the *pn*-junction.

three quantities: the electron concentration in red, the hole concentration in blue, and, in black, the net charge density divided by the magnitude of the electronic charge e. We see that to the left of the junction there is a region, the *carrier depletion region*, in which the electron concentration is zero. The net charge density, $\rho(x) = N_D e$, determined by the ionized donor density, is positive. There is a similar depletion region in the *p*-material on the right, with a negative charge density given by the acceptor concentration, $\rho(x) = -N_A e$. Although the two homogeneous materials are good conductors, the redistribution of charge results in a planar region near the interface which is devoid of mobile charge carriers and is hence an insulator. The redistribution of charge creates a dipole layer, with negative charge in the *p*-material on the right and positive charge in the *n*-material on the left. The dipole layer gives a step $\Delta\phi_0$ in the potential as one moves from one side of the junction to the other. $e\Delta\phi_0$ is referred to as the *equilibrium barrier height*.

"poisson" calculates the charge and potential distributions across the junction in two different ways. PRESET 1 opens with an 'EXACT' solution, the fully self-consistent solution obtained as outlined on page 231 and in Section 12.B. It is not truly exact because it requires the chemical potential everywhere to be at least a few $k_B T$ from the band edges, in order to use Eqs. (12.3) and (12.4) for the carrier densities. The APPROXIMATE choice from the SOLUTION menu yields the solution which is given in many texts, e.g., Ashcroft and Mermin [1, p. 595] or Sze [46, p. 74]. Here the charge density is approximated by a step function, changing from zero in the bulk material to $N_D e$ or $-N_A e$ at the edges of the depletion zones, $x = -d_n$ and $x = +d_p$ in Figure 12.1. This is well justified in some instances because of the exponential dependence of carrier concentrations on the energy of the band edges, Eqs. (12.3) and (12.4). The resulting solution gives a potential in the depletion zone, quadratic in x, which is derived from the assumed charge distribution. However, the charge distribution is *not* that which would be derived from that potential and Eqs. (12.3) and (12.4): the solution is *not* self-consistent.

Exercise 12.1 (C) PRESET 1 *gives the 'EXACT' solution for the case in which the doping concentration is four times higher on the n-side than on the p-side. On which side of the junction does most of the potential jump $\Delta\phi_0$ occur? On which side is the depletion layer wider? Construct a convincing qualitative physical argument for each of these observations. (Hint: Gauss's law is a good starting point.)*

Exercise 12.2 (C) PRESET 2 *gives a symmetric junction with the same   # 2 doping concentration on each side, and the doping concentration is considerably less than in* PRESET 1. *Set the* AUTOSCALE *toggle to off and watch the absolute and relative widths of the p- and n-depletion layers as you click in the slot of the N_A or N_D* SLIDER *to change the impurity concentration. (If you leave the* AUTOSCALE *engaged and change both N_A and N_D by the same amount there is apparently very little change in the graph: until you notice that the x-axis scale has changed.)*

Exercise 12.3 (M) *To get a more quantitative view, turn on the* AUTOSCALE *and note down the thickness of the depletion region for* PRESET 2. *Then change the doping* CONCENTRATIONS, *N_A and N_D, on both sides of the junction by a factor of 10 and note the new depletion widths. Repeat a couple of more times and deduce the rough dependence of depletion widths on doping concentration.*

The depletion width varies roughly as the inverse of the square root of the doping CONCENTRATIONS, N_A and N_D: high doping levels imply small depletion lengths and large electric fields in the barrier. A more careful look shows a weak deviation from the square root law for the equilibrium junction.

Exercise 12.4 (M**) *Use the 'EXACT'* SOLUTION *to take data more carefully, again in steps of powers of 10, for equal doping concentrations on both sides in the range from 10^{14} cm^{-3} to 10^{19} cm^{-3}. (You will need to adopt some sensible criterion for defining d_n and d_p.) Use the approximate method to predict analytically how the depletion width should depend on doping and compare with the data.*

Let's pursue briefly a comparison of the APPROXIMATE and 'EXACT' approaches to the problem.

Exercise 12.5 (C) *Open* PRESET *2 again and switch between the* 'EX-ACT' *and the* APPROXIMATE SOLUTIONS *as you watch first the potential distribution and then the charge distribution. Are you content with the* APPROXIMATE *solution? Change the energy gap* $\mathcal{E}_{\mathrm{gap}}$ *to 0.4 eV and repeat.*

Exercise 12.6 (M*) *Starting from* PRESET *2, develop experimentally a condition on the energy gap and other relevant parameters for when the* APPROXIMATE SOLUTION *seems to be reasonably good. Don't concern yourself with values of numerical constants of order unity. Do find out which parameters matter and which do not. For example, how does the quality of the approximation scheme change if you:*

1. *vary the doping* CONCENTRATIONS, N_A *and* N_D, *by two orders of magnitude up or down?*
2. *vary the energy gap* EGAP *by a factor of 2 larger or smaller?*
3. *vary the* TEMPERATURE *(in the* CONFIGURE *menu) by* ± 100 *K?*

Although the APPROXIMATE SOLUTION for the charge densities is poor, the prediction for the potential is quite good and adequate for most purposes over a wide range of parameters.

12.2.2   Debye screening*

Do we have to use the computer to find a self-consistent solution to the *pn*-junction problem? Can't we do it analytically? We can get part way there, Eq. (12.8) in Section 12.B, but not all the way. Instead let's just look carefully at the tails of the distributions, near the outer boundaries of the depletion zone, where an analytic solution is possible. What is the detailed variation in carrier density at the edges of the depletion zone near $x = d_p$ and $x = -d_n$, and what are the implications of using the approximation of a sharp cut-off?

\# 3   **Exercise 12.7 (C)** *Open* PRESET *3 and* COPY ENERGIES *for the* AP-PROXIMATE SOLUTION. *Now switch to* 'EXACT' *and* STEAL DATA *in order to compare the two* SOLUTIONS. ZOOM *in on the tail in the n-region to be sure to see the difference in the two solutions. The quadratic dependence for the* APPROXIMATE *case is clearly distinguishable from the tailing off of the* 'EXACT' *solution. On the other hand, the general agreement in the energy plot is remarkably good when you see the substantial difference in the two charge densities.*

Exercise 12.8 (C) *We can follow the nature of the tailing more easily by looking at the densities plot instead of the energies. Now* COPY DENSITIES *for the* 'EXACT' SOLUTION, *open the* CONFIGURE *dialog box for the copied* *graph and click on the Y* LOG *toggle to get a semilog plot of the densities.* ZOOM *in to that portion of the plot of the net charge density, ρ/e, near the edge of the depletion zone on the n-side. The semilog plot seems to be a nice straight line over several decades variation in the charge density. What does that imply for the position dependence of the charge density?*

You have just shown that the tail in the charge density falls off exponentially with distance. What is the characteristic length for this exponential decay?

Exercise 12.9 (M*) *Measure the decade-length, which is 2.3 times the $(1/e)$-length, and also note the value of x at which the straight line fit becomes poor. Then repeat the measurement for a donor concentration N_D varying either way by a factor of 10 to determine how the characteristic length depends upon the carrier concentration.*

What could we predict theoretically? Although we can't solve the equations throughout the depletion region, we can look analytically at a limited range of x in which $\phi(x) \ll k_B T$, a range near the boundaries of the depletion zone, which we just looked at experimentally.

Exercise 12.10 (M*) *Consider the edge of the depletion zone in the n-region and define, for convenience, $\lim_{x \to -\infty} \phi(x) \equiv 0$. Then use the inequality $|e\phi(x)| \ll k_B T$ to expand the exponential in the expression (12.3) for $n(x)$ and use this approximate $n(x)$ in Poisson's equation (12.1). Show that, in this approximation, the potential varies exponentially with position. What is the length characterizing this decay? Does your predicted dependence of screening length on carrier concentration match the experiment of Exercise 12.9?*

The screening at the edge of the n-depletion zone is exponential with a characteristic length, called the *Debye screening length*, given by $\lambda_{\text{Debye}} = \sqrt{\kappa \epsilon_0 k_B T / n_0 e^2}$. Here, $n_0 = N_D$ is the electron concentration in the bulk n-material far from the inhomogeneity and κ is the dielectric constant.

Exercise 12.11 (M*) *Find the x value for which the conduction band edge has shifted by about $k_B T$ from its value deep in the bulk of the n-material. Compare this with the x value, determined in Exercise 12.9, for which the charge density has deviated significantly from the limiting exponential behavior in the semilog plot. Explain why these are about the same.*

There are numerous examples of exponential *screening* of electric fields, or local charge perturbations, by mobile charges. In a common idealization, an external electric field normal to a metal surface is thought of as terminated by a sheet charge density of zero thickness on the surface of the metal. More realistically, the field is screened in metals over a depth of a few angstroms by the conduction electrons. The degeneracy of the electron gas leads to the *Fermi–Thomas screening*, with a screening length similar in form to the Debye screening, but with the replacement of the $k_B T$ by a Fermi energy. Screening in ionic solutions, by ions not electrons and called *Debye-Hückel screening*, is closer to the semiconductor case.

In Exercise 12.6 we noted that the APPROXIMATE SOLUTION began to be unconvincing for small band gaps, but that its quality was insensitive to the doping level except at low doping concentrations. In particular, the approximate solution was most satisfactory for the regime defined by $e\Delta\phi_0 \gg k_B T$ where $\Delta\phi_0$ is the band offset between the bulk n- and p- regions. We can see why this should be so by comparing the depletion length with the Debye screening length.

Exercise 12.12 (M*) *Find an expression for the ratio of the depletion length to the Debye length. Knowing that typical barrier heights are fractions of an eV, show that the typical (depletion length/Debye length) ratio is of the order of 4. Thus the standard approximation is acceptable, though not highly accurate: the charge density varies from its maximum value to zero over a length of the order of $\frac{1}{4}$ of the full depletion length. How does the theoretical criterion for small (Debye length/depletion length) compare with the experimental criterion from Exercise 12.6 for validity of the approximate scheme? (See Sze [46, p. 77] for corrections to the approximate result.)*

12.2.3   Einstein relation*

At first sight, a puzzling feature of the equilibrium junction is that in the depletion region there are at least a few carriers, and these are in a very large electric field. Why isn't there a *drift current* $j_{\text{drift}} = ne\mu E$ as a consequence? (Here, μ is a mobility, not a chemical potential.) The answer is that there is also a large concentration gradient dn/dx. This concentration gradient, according to the diffusion equation, drives a *diffusion current* in the opposite direction $j_{\text{diff}} = -eD(dn/dx)$, where D is the diffusion constant. In equilibrium j_{drift} and j_{diff} must be in opposite directions and just cancel.

Exercise 12.13 (M**) *Show that equality of these two currents requires the diffusion constant D and the mobility $\mu = e\tau/m^*$ to be related by the condition,*

$$\mu = eD/k_B T. \tag{12.5}$$

(Hint: you will need to use Eqs. (12.3) and (12.2) to figure out the dependence of the carrier concentration on the local electrostatic potential.)

This relation between the diffusion constant and the mobility is referred to as the *Einstein relation*. As for the case of the Debye screening, the Einstein relation has an analog for the case of metals, in which the k_BT is replaced by the Fermi energy, and in ionic solutions.

12.2.4 Applied voltage

What happens if a battery is connected to the opposite ends of the sample? Not surprisingly, there is a current flow. What may be unexpected is that the magnitude of the current flow is very different for the two signs of the applied voltage. "poisson" is only able to predict the effect of the voltage on the charge distribution and potential variation within the material; it cannot address the question of the current characteristic. However, over a wide range of parameters, the potential and net charge distribution are accurately given by "poisson" without consideration of the current flow that results from the application of the voltage. To insure that the voltage remains within this range of validity, "poisson" restricts eVAPPLIED in the forward direction to be less than one half of the energy gap. However, *neither* the minority carrier concentrations near the barrier *nor* concentrations of either carrier deep in the depletion region will be given properly. Since it is these concentrations which determine the current flow, we must be content to have "poisson" give us only the potential distribution, not the current.

A battery of voltage V forces the chemical potentials at the two ends to differ by eV. Note that it is essential to recognize that the voltage we talk about with batteries is a difference of chemical potentials at the two terminals, *not* a difference in electrostatic potentials, as it may appear from freshman physics. The effect on the band diagram is to change the height of the barrier $\Delta\phi_0$ by $\pm eV$. To avoid a hopeless tangle of sign conventions, it is convenient to refer to the applied voltage as *forward*g if it combines with the "built-in barrier" $\Delta\phi_0$ to decrease the barrier height, and as *reverse* if it increases the barrier height. Thus the barrier potential becomes $\Delta\phi = (\Delta\phi_0 - V)$ and $V > 0$ is forward. (Convince yourself, from these definitions, that a forward biased diode has the n-material connected to the negative voltage supply.)

The calculation of the space charge distribution is basically the same as for the equilibrium case sketched on page 231. The required generalizations of the analysis are the following:

1. The combination barrier height $\Delta\phi = (\Delta\phi_0 - V)$ is used in place of the equilibrium barrier height $\Delta\phi_0$.
2. In evaluating the charge densities, a different chemical potential is used in the n- and the p-regions. The two differ by the amount of the applied voltage times the electronic charge.

Watch the variation in band edges and chemical potentials in any PRESET as the VAPPLIED slider is changed to check that the changes in display are consistent with these rules. Minority carrier concentrations and carrier concentrations in the depletion region are determined by steady state conditions on dynamic equations and cannot be deduced from equilibrium arguments. However, since the total charge density is determined essentially entirely by the ionized impurities and the majority carriers, the equilibrium analysis just outlined is still viable.

4   **Exercise 12.14 (C)** *Open* PRESET 4 *to display the* APPROXIMATE SOLUTION *for an applied reverse bias of 3 V, and be sure the* AUTOSCALING *is turned off. Now use the* SLIDER *and vary the applied voltage* VAPPLIED. *What is the effect on the thickness of the depletion region? And on the magnitude of the depletion charge which is given in the* READOUT *between the graphs?*

Exercise 12.15 (M) *Use the approximate theory to develop an expression for the depletion width as a function of applied voltage. Devise a graph of width and voltage for which the theory would predict a straight line plot. Use "poisson" in the* 'EXACT' *mode and plot a few values of depletion width versus voltage to see whether the theory works.*

Additional insight may be gained from the READOUT of the depletion charge Q_{depl}, the charge in the depletion layer on either side of the physical junction. The variation of charge with voltage defines an incremental capacitance of the junction $C_{inc}(V) \equiv dQ/dV$. One of many device applications of the pn-junction structure is the voltage tunable capacitor. A reverse biased junction may be incorporated in an AC circuit as a capacitor whose incremental capacitance may be varied by application of a variable DC voltage.

Exercise 12.16 (M) *Use the standard approximation to find an expression for $Q(V)$, the charge stored in the depletion regions on either side of the physical junction. From it determine the incremental junction capacitance dQ/dV as a function of applied voltage V. Apply a variety of voltages to the junction and use the Q_{depl}* READOUT *to compare with your expectation. Why are Q/V and dQ/dV not the same?*

12.2.5   *Diode characteristic*

The pn-junction discussed here is the basic structure used in a variety of applications: rectifier diodes, light emitting diodes (LEDs), voltage tunable capacitors, laser diodes, photodiodes, and photovoltaic cells. We have carried the discussion of the pn-junction far enough to get a glimpse of its operation as a rectifier, a device which, crudely speaking, carries current in one direction but not the other.

In the *n*-region the electrons have a high concentration as majority carriers. In the absence of the potential barrier, $\Delta\phi$, they would quickly flood the *p*-region, giving a gigantic electron diffusion current from left to right. The equilibrium barrier height, $\Delta\phi_0$ allows just enough electrons to pass over the barrier to balance an exceedingly small flow, from right to left, of minority electrons in the *p*-material which diffuse to the barrier region and are swept across by the large barrier field. (See the comments on the Einstein relation on page 236.)

If a forward voltage is applied, the barrier is lowered and more electrons are allowed to flow across the barrier from left to right. This current depends exponentially, through a Boltzmann factor, on the changes in barrier height. The compensating current from right to left is independent of barrier height and is completely swamped by the large forward current. In reverse bias, the majority flow across the barrier is completely blocked by the large barrier; the reverse current flow is independent of voltage and in many diodes too small to be detectable. With zero applied voltage, these two currents just cancel, as they must since there can be no current flow in equilibrium.

Exercise 12.17 (C*) *Construct the analogous argument for the hole currents. It is critical in your argument to show that the sign of applied voltage which is forward for electrons is also forward for holes.*

12.3   Metal–insulator–semiconductor (MIS) diode

The MIS diode is a planar structure consisting of a bulk semiconductor covered first by a thin insulating layer, and on top of that a metal film called a *gate*. With the gate limited in extent, and source and drain contacts fabricated at either end of the gate, the structure becomes a *field effect transistor* (FET). Understanding the physics of the MIS diode is the first step in understanding the operation of the MISFET. We will also see hints of methods of testing device fabrication techniques, methods based on capacitance measurements on MIS diodes. Remember that in MISFET devices the current flow of interest, from source to drain, is parallel to the surface in contrast to the *pn*-junction for which the flow is perpendicular to the planar structure.

Click on PRESET 5 to see in the lower plot the band energies and chemical potentials, or Fermi levels, as a function of depth under the gate of the diode. (This chapter uses Fermi level and chemical potential interchangeably.) The energy band diagram of the lower graph is reproduced   # 5

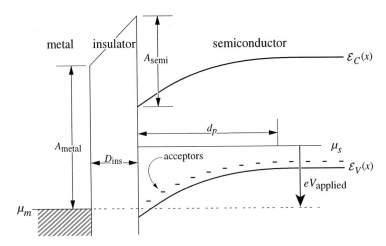

Figure 12.2: Band diagram for the MIS diode.

in Figure 12.2. The origin of the x-axis is at the interface between the insulator and the semiconductor. To the left of the insulator is the metal gate with the dashed green line indicating its Fermi level. The insulator is supposed by "poisson" to contain no free charge. The sloping line in the insulator denotes the edge of the conduction band for the insulator, with the valence band edge off-scale below the diagram. The step of 3 eV from the metal Fermi level to the insulator conduction band edge, the A_METAL in the CONFIGURE menu, denotes the metal–insulator work function: the energy required to carry an electron from the Fermi level in the metal to the lowest conduction band state in the insulator. In the semiconductor on the right, "poisson" keeps track of the edges of both the conduction (red) and valence (blue) bands. The step between the conduction bands in the insulator and the semiconductor is the semiconductor–insulator band offset, A_SEMI in CONFIGURE. Finally the dashed green line in the semiconductor is the semiconductor chemical potential which is constant throughout the semiconductor.

The upper graph in "poisson" gives the carrier and total charge densities as a function of position. The behavior to the right of the insulator–semiconductor interface should remind you of the depletion region on the p-side of the pn-junction. Moving from the semiconductor into the insulator, as illustrated in Figure 12.2, there is a change in slope of the band edge produced by the change in dielectric constant. Then, moving from the insulator into the metal the electric field drops to zero; it is shielded from the interior of the metal by a surface charge density on the metal surface. More properly, we need an analysis similar to that used for the Debye screening, but one appropriate to a metal. The result in the simplest treatment, the *Fermi–Thomas model*, is an exponential decay of the field with characteristic length of a few angstroms. The corresponding density would be way off-scale on our graph. To suggest the presence of this surface charge, the "poisson" plot gives the vertical line which is proportional to the surface density of the screening charge. The proportionality constant is adjusted by the SPIKE WIDTH in the CONFIGURE dialog, and is described on page 253.

12.3.1 Accumulation and depletion

There is a range of applied voltages, 0–1 V, for which the physics of the MIS diode is very similar to that for one side of the *pn*-junction. Note the variation in thickness of the depletion layer in the semiconductor as the VAPPLIED is varied. The Q_{depl} READOUT gives the charge per unit area (in units of e/cm^2) associated with the acceptors and holes, $Q_{\text{depl}} = \int_0^{d_p} [p(x) - N_A] \, dx$. This is essentially the same as the depletion charge readout for the *pn*-junction. Q_{gate} gives the charge per unit area on the gate.

In the *pn*-junction the field at the physical junction always has the same sign, independent of applied voltage, and the screening of the field is always by depletion. With the MIS diode, however, we can change the sign of the electric field at the insulator–semiconductor interface which then requires the shielding charge to be of opposite sign.

Exercise 12.18 (C) *What happens to the screening charge in* PRESET 5 *as the voltage* VAPPLIED *is changed, in small steps, to* −0.5 V? *Instead of depletion of carriers near the interface, extra carriers are "sucked" near to the interface by the negative potential on the gate. What is the relative concentration of holes and acceptors near the interface?*

Depending upon the value of the gate voltage, the screening of the gate field by the semiconductor can be either by carrier *depletion* near the interface, much as in the *pn*-junction, or by carrier *accumulation* near the interface. "poisson" continues to label the net charge associated with the holes and acceptors QDEPL, even though the name is hardly appropriate when the diode is shielded by an accumulation layer.

Exercise 12.19 (C) *Use the toggle to change the semiconductor from* P-TYPE *to* N-TYPE *and vary* VAPPLIED *in the range 0–1 V. What now is the source of the shielding charge? And how does the source depend upon* V_{applied}?

In "poisson" QDEPL will always refer to shielding by the combination of majority carriers and ionized impurity, regardless of whether the majority carriers are in depletion or accumulation.

Exercise 12.20 (C) *For* PRESET 5 *(be sure the material is returned to p-type), vary the thickness of the insulator* DINS *and note the effect on the magnitude of the screening charge,* Q_{depl}. *Explain why, for fixed applied voltage, making the insulator thinner increases the amount of screening charge.*

Exercise 12.21 *The gate voltage* $V_{applied}$ *for which the depletion charge* Q_{depl} *is zero is called the flat-band voltage,* V_{fb}. *With one eye watch the* Q_{depl} READOUT, *and with the other watch the band diagram, as you vary* VAPPLIED. *Why is the transition between depletion and accumulation called the flat-band voltage? The flat-band voltage is an important parameter in device design.*

For small deviations of the applied voltage from the flat-band value, the screening by the majority carriers follows the Debye law, Section 12.2.2.

Exercise 12.22 (M*) *With* PRESET 5 *and* VAPPLIED $= -0.38$ V, COPY DENSITIES *and display on a* SEMILOG *plot. Is the screening exponential? (Remember the* ZOOM.*) What is the value of the characteristic screening length? Does it agree with the predicted Debye screening length? Does the screening length depend upon whether the screening is by accumulation or by depletion?*

Exercise 12.23 (C*) *Repeat the preceding experiment, but with the voltage changed to* -0.9 V. *Why is the screening no longer exponential?*

Considerable information is available about the structure of the MIS diode from a plot of the gate charge Q_{gate} as a function of voltage. In the real laboratory, we're not so lucky: the charge on the gate is not something we can measure. The incremental device capacitance, dQ_{gate}/dV, can be measured, however. Apart from a constant of integration, it provides the same information.

Exercise 12.24 (M*) *Make a plot of gate charge versus voltage over the range* -0.4 *to* -1.0 V. *The slope, or the incremental capacitance, becomes very nearly constant at the most negative end of the range.*
Relate that limiting capacitance to the dimensions and physical properties of the materials used in the "poisson" diode. PRESET 5 *uses the* DIELECTRIC CONSTANTS 4 *and* 12 *for the* INSULATOR *and the* SEMICONDUCTOR *respectively (see the* CONFIGURE *menu).*

Exercise 12.25 (M)** *Extend the range of the plot of gate charge versus voltage to* $0 <$ VAPPLIED < 1.3 V. *Show how could you use the measurement of the incremental capacitance,* dQ/dV, *of this device as a function of applied DC voltage in this range to determine the doping concen-*

tration of the semiconductor. (Hint: watch the density plot as the voltage
is varied or recall Exercise 12.16.)

12.3.2 Inversion

What happens as the depletion layer is driven even deeper, as you take
the voltage to even larger positive values?

Exercise 12.26 (C) *Choose* PRESET 6, *and increase the* VAPPLIED *in*   # 6
small increments by CLICKING *in the triangles at the end of the* SLIDER.
*Note the new feature which appears in the density plot. Explain how it
comes about that electrons can populate the conduction band near the
interface, though this is supposed to be a p-type semiconductor.*

We refer to the layer of minority carriers at the insulator–semiconductor
interface as a *channel* or an *inversion layer* and the accumulation of
minority carriers at the interface as the phenomenon of *inversion*. The
p-type material has been converted by the applied field to a thin layer in
which conduction is by electrons. The charge density associated with the
inversion layer is given in the READOUT as the Q_{inv}.

Exercise 12.27 (M) *Plot the three charge densities,* Q_{gate}*,* Q_{depl}*, and*
Q_{inv} *as a function of gate voltage for* $1 <$ VAPPLIED < 2.4 V. *The sum
of the three is shown as* Q_{tot}*. Explain why* Q_{tot} *should be zero to within
the numerical precision of the algorithm. What is the limiting slope, or
incremental capacitance, at the higher voltages? How is it related to the
device parameters? Does your plot give a reasonable value?*

The inversion charge turns on abruptly as a function of voltage and
is linear with voltage at higher voltages. The *threshold voltage* V_t is the
voltage at which inversion begins: the extrapolation to the voltage axis
of a straight line fit to the inversion charge as a function of voltage. It is
convenient to use the approximate relation $Q_{\text{inv}} = C_{\text{inv}}(V - V_t)$.

Exercise 12.28 (M*) *Make a new plot of* $Q_{\text{gate}}(V)$*, but with the insulator thickness doubled. Don't take a lot of data, but take enough to
identify roughly the threshold and flat-band voltages. How does the dif-
ference* $|V_t - V_{\text{fb}}|$ *depend upon the thickness of the insulator? What about
the slope,* C_{inv}*? Explain the reasons for these dependences.*

The ability to use the gate voltage to control the inversion layer charge
is the key to the operation of the MISFET. The device fabricators are
able to make electrical contact, via the source and drain contacts of the
FET, to the inversion layer without electrically contacting the p-type
substrate. The source–drain conductance, assuming the electron mobility

is independent of channel charge density, is proportional to the channel charge density. Hence a plot of conductance versus gate voltage will be the same as the plot you've made of the inversion charge, Q_{inv}, except for a scale factor.

Exercise 12.29 (C) *Why do engineers push very hard to reduce the gate oxide thickness in the fabrication of MOSFETs?*

Although adequate for illustrating many problems of device physics, there is a shortcoming of the "poisson" analysis of the MIS diode. "poisson" has treated the electrons and holes as classical particles, except for the use of the Fermi statistics and the possible inclusion of an effective mass for the carriers. A comparison of the wavelength of electrons of thermal energy with the length scale over which the potential varies substantially should cause some alarm.

Exercise 12.30 (M*) *What is the wavelength of an electron with effective mass $m^* = 0.1m$ and an energy of $k_B T$ at room temperature? What is a typical thickness of the inversion layer for the examples we've been looking at?*

"poisson" has tried to make the potential variation and the local charge density consistent but has oversimplified the real problem by neglecting the wave like nature of the electrons. To explore the implications of the wave nature of the carriers in the inversion layers read the review article by Ando *et al.* [48].

12.3.3   Interface states*

As described in Section 12.3.2, the idea of the MISFET (or MOSFET) seems a quite simple one, and it is! So why did a number of years elapse between the inception of the idea and the fabrication of a useful MOS-FET device? In PRESET 7 we have modified the "poisson" model by introducing interface states at the boundary between the insulator and semiconductor. At the interface we picture localized states at a variety of different energies within the gap. If the chemical potential lies above the energy of the state it will be filled with an electron, if below, it will be empty.

7

"poisson" uses a picture in which the number of interface states per unit energy is constant at a value set by the N_INT slider. Further, the states above the center of the gap are assumed neutral if empty of electrons and negatively charged if they've trapped an electron, rather like an acceptor impurity. Below the center of the gap, they are neutral if filled with an electron and positive if empty, rather like a donor impurity. This particular model implies a neutral surface if the chemical potential at the surface is in the middle of the gap, negative if it lies above it and positive if it lies below. Obviously one could invent many other scenarios. In real materials the distribution in energy of the surface states and their charges depends on details of processing. The trap density is expressed in units of $(eV)^{-1}$ cm^{-2}, and is the number of available interface states in a 1 eV range of energy for an interface area of 1 cm^2. We will soon see why the most important single element of quality control in MOSFET technology involves limiting the density of these interface states.

In PRESET 7 the interface state density has been set to a "high" value. "High" means this device would not make a useful transistor, rather like the first FETs that were tried at Bell Laboratories. On opening this PRESET you see an added feature on the charge density plot, a negative spike at the position of the interface. This represents the areal density of charge in the interface states, which has been divided by an arbitrary SPIKE WIDTH of 1000 Å from the CONFIGURE dialog, in order to get a measure that will be "on scale" on the densities plot. The net charge of the interface, in units of the electronic charge, is given in the panel READOUT as Q_{int}.

Exercise 12.31 (M) *Make a plot of the inversion layer charge $Q_{inv}(V)$ versus voltage for* PRESET 7. *Choose a range such that both V_t and C_{inv} may be determined from the plot. Now remove the interface states by changing the* N_INT *(interface) slider to* 1e+01 *and repeat. How does the interface trap density* N_INT *influence the threshold voltage? How does it affect C_{inv}, the slope of the $Q_{inv}(V)$ plot?*

These examples show that too high an interface state density makes the construction of an effective MOSFET impossible. A high density of interface states shields the interior of the material from the fields applied by the gate: we speak of *Fermi level pinning* by the interface states. Even with the state density reduced to a value that allows construction of a practical device, variations in state density from one production run to the next can yield unacceptable variations in threshold voltages of the devices.

Exercise 12.32 (M*) *If the device of* PRESET 7 *is to be manufactured with a threshold voltage held at 1.6 V within a tolerance of ±0.2 V, what is the maximum density of interface states that can be tolerated?*

12.4   Contact potential and work functions**

Let's replace the insulator in the "poisson" model by a vacuum gap between two parallel surfaces, a metal and a semiconductor. We define the

work function of each to be the energy difference between the chemical
potential in the material and the energy of an electron at rest in vacuum
immediately outside the surface of the material. For the semiconductor
we distinguish between the three quantities:

1. the *work function*: the difference between the chemical potential
   and the vacuum level;
2. the *ionization energy*: the difference between the valence band edge
   and the vacuum level; and
3. the *electron affinity*: the difference between the conduction band
   edge and the vacuum level.

For the metal, the ionization energy, the work function, and the electron affinity are all equal, since the Fermi level lies at an energy where the
DOS is finite. (For the metal the work function is usually first met as the
threshold energy for the photoelectric effect.) In the "poisson" model with
vacuum as the insulator, the A_METAL and A_SEMI in the CONFIGURE dialog box are the work function for the metal and the electron affinity
for the semiconductor respectively. We develop here a picture of a technique used in surface physics for the determination of work functions of
semiconductors, or, more precisely, the work function of a semiconductor
relative to that of a reference metal.

*12.4.1   Clean surface***

8   In PRESET 8 we have arranged the metal–vacuum–semiconductor system
with zero applied voltage so that the Fermi levels in the two materials are
the same. In this situation there is an electric field in the vacuum and
associated screening charge on both the metal surface and the semiconductor. The *contact potential* between the metal and semiconductor is
the difference in electrostatic potential at the two boundaries of the vacuum slab. The field in the vacuum gap is the contact potential divided
by the vacuum thickness, D_{ins}. The band diagram of Figure 12.2 should
convince you that the contact potential is equal to the difference in work
functions of the two materials.

Exercise 12.33 (M) *What is the electric field in the vacuum gap? Double or halve the thickness* DINS *of the vacuum layer and determine how the
magnitude of the screening charge is related to the metal–semiconductor
spacing. By what factor is the electric field in the gap changed?*

How can the contact potential be measured? The electrostatic potential
is easy enough to talk about, but not to measure! In the *Kelvin method*
for measuring differences of work functions, an adjustable DC voltage is
maintained between the sample and a reference electrode, which form

the two plates of a capacitor. The reference electrode is mechanically driven sinusoidally to give a time-dependent plate spacing (the DINS of "poisson"), and hence a sinusoidally varying capacitance. Because the charge on the capacitor, for fixed voltage, depends upon the capacitance as $Q = CV$, as D_{ins} is varied charge must flow in the external circuit in order to maintain the constant applied voltage. The trick of the Kelvin method is to measure this AC current and slowly vary the applied DC voltage. When the applied DC voltage is adjusted to give zero AC current, the *Kelvin condition*, there is no charge on the capacitor and no field in the gap. With this condition satisfied, the applied DC voltage equals the difference in work functions.

Exercise 12.34 (M) *Vary* VAPPLIED *for the conditions of* PRESET 8 *until the screening charge is zero. (Where have we met this condition before?) What is the electric field in the vacuum gap? Confirm that the electric field and screening charge remain zero as the gap thickness is changed. Construct a band diagram which shows that this condition implies equality of the applied voltage and the work function difference.*

Exercise 12.35 (M*) *Derive an expression relating the work function of a semiconductor to its electron affinity A_{semi} and the doping level N_A or N_D. Does it agree with experimental data obtained with "poisson"? Try some other doping levels.*

Exercise 12.36 (C) *Show that, in absence of surface states, the Kelvin condition for no motion induced AC current is the flat-band condition: i.e., there is no band bending at the surface.*

You have just shown that the Kelvin method gives a way to map out the dependence of the Fermi level on doping for the semiconductor; at least for the case of no surface states.

*12.4.2 Dirty surface***

We saw the importance of the localized electronic interface states in determining the properties of the MIS diode and the MISFET by their ability to shield partially the interior of the semiconductor from the gate field. It is not surprising, then, that localized surface states can also influence the work function of the semiconductor.

Exercise 12.37 (M*) *Note the value of the work function difference for the semiconductor–metal combination of* PRESET 8, *determined by using the Kelvin method, i.e., from the voltage that gives no field in the vacuum. What is the effect on the work function difference of increasing the interface state density to 10^{14} $(\text{eV})^{-1}\text{cm}^{-2}$?*

We see that the work function is not a unique property of a bulk material but is rather a property of the material surface and of the way in which the surface is prepared.[50] In fact different crystal faces of a crystal, even if atomically clean, will typically have different work functions. This raises the interesting question of the electric field in the vacuum just outside a polycrystalline metal surface. How does the field vary in the region near a grain boundary on a planar surface, with different crystal faces exposed to the vacuum on the two sides of the boundary? You are forced to the conclusion that there are electric fields in the vicinity of a polycrystalline surface which go from grains of small work function to grains of large work function.

12.5   Summary

A key element in semiconductor device operation is the ability to control the spatial distribution of the mobile charge carriers, the electrons and holes. This is achieved by a combination of inhomogeneous doping and appropriate electrode geometries and applied fields. "poisson" illustrates the important physics for configurations amenable to a one-dimensional analysis. The basic problem is to solve self-consistently Poisson's equation (12.1), which gives the potential when the charge distribution is known, and the relations (12.3), (12.4), which give the densities of charge carriers once the potential is known. Understanding the potential and charge distributions in the pn-junction and the MIS diode is fundamental to a study of the two main classes of semiconductor amplifiers, the bipolar transistor and the field effect transistor.

12.A   Deeper exploration

Schottky barrier   Typical metal work functions are of the order of 4 eV, while the electron affinity of silicon is 0.9 eV. This can cause a large space charge barrier, the *Schottky barrier*, to the flow of charge between metal and semiconductor in the absence of any fabricated insulating barrier. Explore the relationship between the properties of the Schottky barrier and the work function of the metal, the electron affinity of the semiconductor, the nature (n- or p-type) of the doping, and the presence of surface states. See Sze [46] for a discussion of Schottky diodes.

Capacitance analysis of MIS diodes   Expand on the ideas introduced in this chapter concerning the use of capacitance measurements [46, 47] to characterize semiconductor structures, developing illustrative examples using "poisson".

Kelvin method The Kelvin method is used in high vacuum studies of surfaces [50] to study work function changes associated with changes in surface coverage. Design an experiment, using a series of samples with different but well-controlled doping levels, to determine from contact potential measurements the surface state density as a function of energy, for energies between $\mathcal{E}_V$ and $\mathcal{E}_C$. Apply the Kelvin method, using the $Q_{\text{gate}} = 0$ substitution for the usual Kelvin condition, to the "poisson" model to confirm the assertions made on page 244 concerning the distribution in energy of the interface states [49].

12.B "poisson" – the program

"poisson" computes the self-consistent solutions of Poisson's equation and the equations for carrier concentrations for two planar heterogeneous semiconductor devices, the *pn*-junction and the MIS diode.

12.B.1 Bending of energy bands

In the presence of a non-uniform electrostatic potential $\phi(x)$, the energy bands in a semiconductor are displaced locally up or down by an amount $-e\phi(x)$. Since the carrier densities depend upon the energies of the band edges relative to a fixed chemical potential, variations in electrostatic potential will induce local changes in carrier densities. In turn, the charge density determines, through Poisson's equation, the curvature of the potential as a function of position, and therefore the position dependence of the band edges. Writing Poisson's equation (12.1) in terms of the conduction band energy $\mathcal{E}_C(x)$ gives

$$\frac{d^2 \mathcal{E}_C(x)}{dx^2} = -e\frac{d^2\phi(x)}{dx^2} = \frac{e\rho(x)}{\kappa\epsilon_0}, \tag{12.6}$$

with κ the dielectric constant and ϵ_0 the permittivity of free space. The charge density, $\rho(x) = e(N_D^+ + p - N_A^- - n)$ in turn depends on the relative positions of the band edges and the chemical potential. "poisson" works in the approximation that the chemical potential is far from both the conduction and valence bands, and also the donor and acceptor levels. In this case the donors and acceptors are fully ionized, and the electron and hole densities are given by exponential functions of the band energies, Eqs. (12.3) and (12.4). These are the fundamental equations that must be solved, self-consistently, to find the structure of the energy bands in semiconductor devices.

The equations are combined into the form

$$\frac{d^2\mathcal{E}_C}{dx^2} = \frac{e^2}{\kappa\epsilon_0}\left(-N_C e^{-\frac{(\mathcal{E}_C - \mu)}{k_B T}} + N_V e^{-\frac{(\mu - \mathcal{E}_C + \mathcal{E}_{\text{gap}})}{k_B T}} + N_D^+ - N_A^-\right). \tag{12.7}$$

In a region in which the impurity concentration is constant, Eq. (12.7) may be integrated once, to give an equation for the slope $d\mathcal{E}_C/dx$ as a function of $\mathcal{E}_C$, Sze [46, Section 7.2]:

$$\left(\frac{d\mathcal{E}_C}{dx}\right)^2 = \frac{2e^2 k_B T}{\kappa \epsilon_0}\left[N_C e^{-\frac{(\mathcal{E}_C-\mu)}{k_B T}} + N_V e^{-\frac{(\mu-\mathcal{E}_C+\mathcal{E}_{\text{gap}})}{k_B T}}\right.$$

$$\left. + (N_D^+ - N_A^-)\frac{\mathcal{E}_C}{k_B T}\right] + C. \qquad (12.8)$$

The constant of integration C can be determined by the boundary condition that the slope be zero in the bulk semiconductor where $(\mathcal{E}_C - \mu)$ is known from the methods of "fermi", Section 11.B. This will be a different constant for the n- and p-regions. Eq. (12.8) then gives the slope of $\mathcal{E}_C(x)$ at any point in the semiconductor in terms of its value at that point. This result reduces the task of solving Poisson's equation to one of finding the value of $\mathcal{E}_C$ at the physical junction for the pn-junction, or at the insulator–semiconductor interface for the MIS diode, for which the calculated displacement fields $D = \kappa \epsilon_0 E$ on the two sides of the interface are the same. Once this $\mathcal{E}_C$ at the interface is known, the energy bands can be easily integrated and graphed.

12.B.2   pn-junction

Quasi-exact solution For the 'exact' solution, "poisson" first needs to calculate the value of $\mathcal{E}_C$ at the interface between the two doped regions. To do this it starts with two bounds for the value of $\mathcal{E}_C$ at the interface, $\mathcal{E}_C(-\infty)$ and $\mathcal{E}_C(+\infty)$, the values of $\mathcal{E}_C$ in the bulk n- and p-regions respectively. It performs a binary search, starting with a trial $\mathcal{E}_C(0)$ which is the mean of the bounds and calculates, from Eq. (12.8), the slope $d\mathcal{E}_C/dx$ on both sides of the interface. If these are equal, then the trial $\mathcal{E}_C(0)$ was correct. If not, then a new trial value of $\mathcal{E}_C(0)$ is chosen which is the mean of the preceding trial value and one of the preceding bounds. The upper preceding bound is used if the slope on the n-side is less than that on the p-side; otherwise, the lower preceding bound is used. Using sixteen cycles of the binary search gives the value of $\mathcal{E}_C$ at the interface to nearly one part in 10^5 of the energy gap.

Having the value of $\mathcal{E}_C(0)$ at the interface, "poisson" numerically integrates the differential equation (12.8) using a simple, adaptive, step-sizing, first order integration routine. At each step the program uses a step size in x which is chosen to move the conduction band a given percentage (the STEP SIZE parameter in the CONFIGURE menu) of the current deviation from the bulk equilibrium value. The integration is terminated when the band energy gets to within a fraction, specified by the ENERGY CUTOFF in CONFIGURE, of the equilibrium value. At each point of the integration the charge densities are calculated from the relative positions of the band edges and chemical potential, and the energy bands and charge densities are plotted.

When a battery or some other source of chemical potential difference is applied across the *pn*-junction, then it is no longer in equilibrium and, in principle, the band bending will depend on the details of electron–hole recombination and carrier mobilities. Unless the *pn*-junction is strongly forward-biased, however, the redistribution of charge associated with the current flow is negligible and a quasi-equilibrium approximation can be adopted. (The entry box for VAPPLIED is instructed not to accept forward biases greater than half the energy gap.) In the *p*-region the chemical potential is displaced by eV_{applied} with respect to its value in the *n*-region. The algorithm then proceeds as described above, but using a different value of the chemical potential μ on the two sides of the junction.

Approximate solution   "poisson" also generates an APPROXIMATE solution for comparison with the 'EXACT' one. The concentration of electrons and holes on either side of the interface is taken to be entirely negligible in the region within a distance, the depletion length, of the interface. Outside the depletion layer the carrier density is taken to be its bulk value. The total charge density, therefore, has a square profile: within the depletion layer it is just that of the fixed charged impurities, and outside it is zero. The solution to Poisson's equation for a constant charge density is a parabola. The program constructs the solution from two parabolas with curvatures of $+eN_D/\kappa\epsilon_0$ and $-eN_A/\kappa\epsilon_0$ which satisfies the boundary conditions used above. See Ashcroft and Mermin [1, p. 595–596].

12.B.3   MIS diode

The MIS diode is a device in which layers of metal, insulator, and semiconductor are sandwiched together. By applying a voltage between the metal, called the gate, and the semiconductor, a screening charge density is built up in the semiconductor near the insulator. The conductivity of that region can therefore be regulated by the applied voltage. This effect forms the basis of operation of the MISFET which utilizes this change in conductivity to make an electronic switch or amplifier.

Finding $\mathcal{E}_C(x = 0)$ Again using Eq. (12.8), the algorithm employs a binary search to find the value of $\mathcal{E}_C$ at the insulator–semiconductor interface, and then calculates the energy bands in the semiconductor using the same integration routine as in the *pn*-junction. As with the *pn*-junction, $\mathcal{E}_C(0)$ is adjusted until the boundary conditions at the insulator–semiconductor interface, i.e., continuity of $\mathcal{E}_C(x)$ and the condition (12.11) on

$d\mathcal{E}_C(x)/dx$, are satisfied. The condition on $d\mathcal{E}_C(x)/dx$ is more complicated than for the pn-junction because of the discontinuity in dielectric
constant, and the inclusion in "poisson" of the possibility of interface
states, at the boundary.

"poisson" models the interface-state density as sheet charge density σ_{int}
which varies linearly with $\mathcal{E}_C$ and is zero when the chemical potential[1] is
in the middle of the band:

$$\sigma_{int} = -eN_{int}[\mathcal{E}_C(0) - \tfrac{1}{2}\mathcal{E}_{gap} - \mu]. \tag{12.9}$$

The constant N_{int} is the number of states per unit area per unit energy
and is specified with the N_{int} SLIDER on the main panel.

The boundary condition on the fields at the interface is that the discontinuity in displacement field,

$$D = \kappa\epsilon_0 E = \frac{\kappa\epsilon_0}{e}\frac{d\mathcal{E}_C}{dx}, \tag{12.10}$$

must equal the sheet charge density at the boundary. This translates to
an equation for the slope of the conduction band edge in the insulator, in
terms of its slope in the semiconductor and $\mathcal{E}_C(0)$,

$$\kappa_i\epsilon_0\left(\frac{d\mathcal{E}_C}{dx}\right)_{ins} = \kappa_s\epsilon_0\left(\frac{d\mathcal{E}_C}{dx}\right)_{semi} - e^2 N_{int}[\mathcal{E}_C(0) - \tfrac{1}{2}\mathcal{E}_{gap} - \mu]. \tag{12.11}$$

With the help of Figure 12.2, the slope $(d\mathcal{E}_C/dx)_{ins}$ on the insulator
side allows the calculation of $\mathcal{E}_C(0)$ as

$$\mathcal{E}_C(0) = \mu - eV_{applied} + A_{metal} - A_{semi} + D_{ins}(d\mathcal{E}_C/dx)_{ins}. \tag{12.12}$$

The binary search starts with the bounds $\pm 2\mathcal{E}_{gap}$ for $\mathcal{E}_C(0)$. These are
safe bounds since, for any plausible applied voltage, the energy bands
cannot be pulled far enough to carry the chemical potential more than
a few $k_B T$ into either band. The midpoint of the bounds is taken as
a trial value of $\mathcal{E}_C(0)$. From the trial value, "poisson" uses Eq. (12.8)
to calculate $[d\mathcal{E}_C(x)/dx]_{semi}$ on the semiconductor side of the interface.
Eq. (12.11) then gives $[d\mathcal{E}_C(x)/dx]_{ins}$ in the insulator, and from that,
Eq. (12.12) returns a value of $\mathcal{E}_C(0)$ to be compared with the trial value.
If the returned value is too high, the old trial is used as the new lower
bound and the cycle is repeated. If the returned value is too low, the old
trial is used as the new upper bound. After 16 iterations of this procedure,
the value for $\mathcal{E}_C(x = 0)$ is determined to better than 10^{-4} $\mathcal{E}_g$.

Integration of the energy bands   Once the value of $\mathcal{E}_C$ at the semiconductor–insulator interface is known, the energy bands can be integrated

---

[1] In the source code for "poisson", the chemical potential μ in the semiconductor is
taken as the origin of energy. It is included explicitly in this description for clarity.

directly. The integration routine used is the same as for the *pn*-junction, except that instead of the distance of the energy bands to equilibrium being used to control the STEP SIZE, the smaller of either this distance or of $3k_BT$ is used. This produces smoother charge density curves in the inversion layer where the charge density is exponentially dependent on the energy.

As for the *pn*-junction, once the energy of the band edge is known as a function of position, the several charge densities can be calculated and plotted in units of e/cm^3. In addition, four sheet charge densities (densities per unit area) are computed and given in the readouts: the charge on the metal gate Q_{gate}, the charge due to the interface states Q_{int}, the charge in the inversion layer Q_{inv}, and the depletion charge Q_{depl}. The densities Q_{gate} and Q_{int} can be found from the electric field in the insulator, using $Q_{gate} = \kappa\epsilon_0 E_{ins}$, and Eq. (12.9) respectively. The values of $Q_{\mathrm{inv}} = \pm \int_0^\infty m[\mathcal{E}_C(x)]\, dx$ and $Q_{\mathrm{depl}} = \pm \int_0^\infty \{N_i - M[\mathcal{E}_C(x)]\}\, dx$ are determined by integrating the appropriate number densities. The quantities $m(\mathcal{E}_C)$ and $M(\mathcal{E}_C)$ are, respectively, the minority and majority carrier densities and N_i is the acceptor or donor concentration as appropriate. The signs are chosen to be appropriate to the choice of N-TYPE or P-TYPE material. The integration routine follows that described for the *pn*-junction. Q_{tot} the sum of all of the densities should be zero and hence, when compared with the individual densities, serves as a measure of the accuracy of the integrations.

12.B.4 Displays

Densities The upper graph plots the concentration of electrons in the conduction band in red, the concentration of holes in the valence band in blue, and the net charge density divided by the electronic charge in black. The origin of the x-axis is at the *pn*-interface for the *pn*-junction, and at the insulator–semiconductor interface for the MIS diode. For the MIS diode, the surface charge densities at the two interfaces appear as spikes whose heights are the corresponding surface charge densities divided by the SPIKE WIDTH parameter available in the CONFIGURE menu. Thus, for the default value of 10^{-5} cm, a spike height of 10^{17} cm^{-3} would correspond to a surface density of 10^{12} cm^{-2}. There is *no* physical significance to the value of the SPIKE WIDTH: it is chosen to give a proportional representation of the surface densities which remains on scale in the density plots.

Energies The lower graph gives the conventional *energy band diagram*, showing the variation with position of the band edges, $\mathcal{E}_C$ and $\mathcal{E}_V$, the donor and acceptor energies, $\mathcal{E}_D$ and $\mathcal{E}_A$, and the chemical potential μ. For the MIS diode, only the conduction band edge is shown in the insulator, and only the Fermi level in the metal.

For both graphs, when looking for the effects of parameter changes, the autoscaling obscures the physics. Changes in width of the depletion layer, for example, appear only as changes in scale of the x-axis, leaving the plotted curve roughly independent of the parameter change. Turn off the autoscaling to get an intuitive grasp of what's happening.

Charges   The control panel on the right displays the sheet charge densities associated with: the depletion layer in the pn-junction Q_{depl}; and the depletion or accumulation layer Q_{depl}, the inversion layer Q_{inv}, the interface states Q_{int}, and the gate Q_{gate} in the MIS diode,. For the MIS diode, Q_{tot} the total of all four charge densities should be zero. The deviation from zero gives a measure of the accuracy of the integration routines. These charge densities are divided by the electronic charge e and given as particle densities.

12.B.5 *Bugs, problems, and solutions*

For the MIS DIODE, with high doping concentrations, the numerical accuracy is inadequate to give a smooth variation of surface charge densities close to the flatband, or Kelvin, condition. Be sure to compare the Q_{tot} values with the individual Q values to be warned of this problem.

13

"ising" – Ising model and ferromagnetism

Contents

13.1   Introduction

A collection of atoms, each having a permanent magnetic moment, gives a Curie law paramagnetic behavior, $\chi \propto 1/T$, as long as there are no interactions between the moments. At room temperature, achievable magnetic fields produce only a weak alignment of the spins. Such a system, though interesting for its physics, is not of technical importance. However, if there are interactions which favor a parallel alignment of neighboring moments or spins, then at high temperatures the paramagnetic susceptibility of the system is enhanced as each spin, which is on average partially aligned by the field, eggs on its neighbors to do the same. More interestingly, at low temperatures in the absence of a magnetic field, the cooperation among the spins gives a spontaneous parallel alignment of all the spins, the phenomenon of ferromagnetism, with its many implications for technology.

"ising" works with one of the classic models for ferromagnetic systems, the two-dimensional Ising model. This model's simplicity makes it easy to simulate with the computer, and the simulation illustrates a broad range of, not only magnetic phenomena, but features characteristic of many *phase transitions*. The results of the simulation will be compared with predictions of the mean field theory of ferromagnetism. Though unable to furnish a detailed understanding of ferromagnetism, the mean field theory provides a useful conceptual basis and is frequently a starting point for exploring the physics of novel systems.

13.2   Ising simulation

The Ising model is a system of N atoms, distributed on an $N_x \times N_y$ square lattice, with spin variables S_i which can take on only the values $+1$ or -1, corresponding to "up" and "down" orientations. The spins are assumed to interact with their nearest neighbors with an *exchange energy* $-J$ if the neighbor is *parallel*, i.e., it has the same value of S, and $+J$ if the neighbor is *antiparallel*, i.e., it has the opposite value of S. It interacts as well with an applied magnetic field with energy $-\mu S_i H$, where μ is the magnetic moment associated with each spin. The energy $\mathcal{E}$ of the full system is given by

$$\mathcal{E} = -\frac{1}{2} \sum_{ij} J_{ij} S_i S_j - \sum_i S_i H, \tag{13.1}$$

where J_{ij} is set equal to J if i and j are nearest neighbors, and zero otherwise. Note that these are not like real spins: if you like, they are "allowed" to point only along, or opposite to, a particular direction or axis. "ising" takes the applied magnetic field to be along the same special direction.

"ising" selects a spin at random and determines at the selected spin an effective field, which is the sum of the applied field and an *exchange field* from the neighboring spins. The exchange field is defined as

$$H_{\text{exch}} \equiv (n_+ - n_-)J, \qquad (13.2)$$

where n_+ and n_- are the numbers of near neighbors with up and down spins respectively. "ising" then assigns the chosen spin an up or down orientation on the basis of a coin toss with a coin that's loaded by a Boltzmann factor in favor of the orientation of the spin parallel to the effective field (page 270). It then selects another spin at random and repeats the process. One *sweep* by the program corresponds to sampling, on average, each of the spins once.

Exercise 13.1 (C) *When you open* PRESET 1 *you see the array of spins*   # 1
on the left, with red and white indicating respectively the up and down orientations of the spins. Click RUN *and drag the* EXTERNAL FIELD *slider slowly to both positive and negative values. What color of spin in the display corresponds to alignment parallel to a positive magnetic field? Leave the* EXTERNAL FIELD *at 1 J and* RUN, *varying the* TEMPERATURE *in the range* $T > 3\ J$. *Why does the magnetization increase as the temperature is lowered?* STOP *to see on the right a graphical history of the magnetization during your variation of the field and temperature parameters.*

As is typical with computers, everything is numbers and we've lost
all the units. The convention chosen for this simulation is to measure T and H in units of J. When the computer works with the number 3 for the variable T, we should think of $k_B T$ as being three times J. If the computer uses 0.5 for H, this corresponds to the interaction energy, $\pm \mu H$, of a spin of moment μ with the field being equal to $\pm 0.5\ J$. Also, the M in "ising" is not the magnetization, but rather the *normalized magnetization* $M = M_t/M_0$, where M_t is the true magnetization and $M_0 = N\mu$ is the saturation magnetization, obtained by aligning all of the spins parallel. Similarly, we introduce a normalized susceptibility $\chi_n \equiv M/H$, the ratio of the normalized magnetization to the field expressed in units of J.

The idea is to think of the simulation as an experiment in which you are to learn what you can about the behavior of this model system, and check out how well the mean field model predicts its behavior. After a brief look at the high temperature magnetization curve, we explore two related ideas with a single data set: the success of the mean field theory in its prediction of the *Curie–Weiss law* and the *fluctuation–response theorem* which relates the susceptibility to the equilibrium fluctuations of magnetization.

13.3   High temperature susceptibility

13.3.1   Paramagnetic response

Let's first check a couple of features of the behavior at high temperatures, $T \gg J$, where the exchange interaction among spins has little effect. Can we quantify the observation already made of the dependence of magnetization on field H?

#2   **Exercise 13.2 (M)** *Use* PRESET *2, with its rather unphysical initial condition, to obtain a quantitative plot of magnetization versus field at a temperature $T = 50$ J. RUN until the program stops automatically after the* NUMBER OF SWEEPS, *200. Increase the* EXTERNAL FIELD *by 10 J by clicking in the slider slot, and* RUN *again. Repeat up to a field of 100 J and the graph will give a stepwise plot of magnetization versus field.*

For the dimensionless H and T used by "ising", the standard expression for the magnetization of a paramagnet made from a collection of non-interacting spins becomes

$$M(T) = \tanh(H/T). \tag{13.3}$$

Exercise 13.3 (M) *Compare the experimental values determined in Exercise 13.2 with Eq. (13.3). Is the magnetization you measured more, or less, than predicted. Explain why. (Hint: remember that "ising" is working with interacting spins.)*

The initial slope of the magnetization curve, $\partial M(H,T)/\partial H$, is the *normalized susceptibility* χ_n, which will be the focus of our attention as we model analytically the consequences of including effects of the interactions.

13.3.2   Curie–Weiss law

The susceptibilities of magnetic materials, at temperatures well above any ordering temperature typically obey the *Curie–Weiss law*,

$$\chi_n = 1/(T - \theta), \tag{13.4}$$

where θ is called the *Curie–Weiss temperature* For non-interacting spins, θ is zero and Eq. (13.4) becomes the expression for the *Curie law* susceptibility. Note that a positive θ implies an enhancement of the high temperature susceptibility as was seen in Exercise 13.2. The form of the Curie–Weiss law (13.4) suggests plotting data as the inverse susceptibility χ_n^{-1} as a function of T. For a material obeying the Curie–Weiss law, the plot will be a straight line with a T-axis intercept at $T = 0$. Let's see how well the Ising system obeys this law.

Exercise 13.4 (M) *Use* PRESET 2 *and take a data series recording* T, H, $\langle M \rangle$, *and* $\langle (M - \langle M \rangle)^2 \rangle$, *starting at temperature* $T = 20\ J$. *Work down in* T *until you think you've reached a temperature at which the susceptibility has diverged, or until you feel the data no longer make sense. Choose H values large enough that the Ms are easily measurable, but small enough that M remains less than 0.25, to avoid non-linearities in the χ_n. You will need to decrease H as you reduce in T in order to continue to satisfy these conditions. It is useful to take smaller spacings in T as T is lowered. Also, repeat measurements frequently enough to have some sense of how well the results of successive runs agree. The readouts of $\langle M \rangle$ and $\langle (M - \langle M \rangle)^2 \rangle$ are average values over the time interval displayed in the graph. Be sure to* RESET *the graph before each change in temperature! Increase the* SPEED *to save time in taking data.*

Exercise 13.5 (M) *Plot the inverse susceptibility χ_n^{-1} versus T. From a distance it looks like a good Curie–Weiss law. Close examination of very good data, however, will show curvature of the plot at the lower temperatures. It may be impossible to see in your data. Fit the data two ways. For Fit A, fit with a straight line with both the slope, and x-intercept T_A, as free parameters. Since our aim is to find the temperature at which χ_n diverges, weight the low temperature points more heavily. For Fit B, weight the high temperature points more heavily, and force the slope to be unity, fitting with a single parameter, the x-intercept.*

Fit A gives a reasonable estimate of the *Curie temperature T_C or ferromagnetic transition temperature* for the "ising" simulation. We'll take $T_C = T_A$ as one result of the experiment. Fit A has the problem that the slope is *not* that implied by the Curie–Weiss law! The correct procedure to fit to the Curie–Weiss law is Fit B, and the single fitting parameter, the experimental Curie–Weiss temperature, is taken to be $\theta = T_B$.

What does the simplest of all theories of ferromagnetism, the *mean field theory*, predict for this system? The basic idea is to replace the magnetic field H in Eq. (13.3) by the sum of the magnetic field and an *exchange field* proportional to the magnetization of the sample. Note that this is *not* the exchange field based on the actual orientations of the neighbors, but rather its value *if* the neighbors had a magnetization equal to the sample average. In the units used by "ising", the mean field magnetization is given by the implicit equation for M,

$$M = \tanh\left(\frac{H + 4M}{T}\right). \qquad (13.5)$$

The high temperature susceptibility may be derived from Eq. (13.5) as $\chi_n = 1/(T - T_{\mathrm{mf}})$ with $T_{\mathrm{mf}} = 4\ J$. The mean field theory predicts a Curie–Weiss law behavior, and is the motivation for using the Curie–

Weiss law for data analysis. It predicts a common value T_{mf} for the Curie–Weiss temperature θ and the Curie temperature T_C.

Exercise 13.6 (M*) *Convert the mean field argument for the suscepti-bility as given in textbooks into the conventions for units used by "ising", and verify expressions (13.4), with $\theta = T_{mf}$, and (13.5). How well does the experimental Curie–Weiss constant $\theta = T_B$ agree with the mean field prediction $T_{mf} = 4\ J$?*

Exercise 13.7 (M*) *Show, using the mean field theory, that requiring $M < 0.25$ while taking the data insures the M/H values will be within 2% of the linear susceptibility. (Hint: what is the expansion of $\tanh(x)$ for small x?)*

So, the high temperature data are well fit by the mean field picture. What about its prediction of $T_C = T_{mf} = 4\ J$? Data from the susceptibil-ity is experimentally well defined and finite at $T = 4\ J$: the observed T_C is substantially lower. Thus the experimental transition temperature T_C and Curie–Weiss temperature θ are *not* the same. The mean field theory has its problems! Quantitatively, for this Ising system, the mean field pre-diction of the transition temperature is poor, off by nearly a factor of 2. The mean field theory does, however, give the high temperature limiting form and a good prediction of the associated Curie–Weiss constant.

13.3.3   Fluctuation–response theorem**

The fluctuation–response theorem of statistical mechanics leads to a very important result [53, Eq. (2.14)]: the thermal equilibrium *fluctuations* of some property of a system, e.g., its magnetization, are closely related to the magnitude of the *response* of the system to an applied driving force, e.g., its susceptibility. A suggestive analogy is the application of the equipartition theorem to a harmonic oscillator in thermal equilibrium with a heat bath. If its force constant, the analog of an inverse susceptibility, is κ then the equipartition theorem gives the mean square fluctuations in its displacement as

$$\tfrac{1}{2}\kappa\langle\delta x^2\rangle_{thermal} = \tfrac{1}{2}k_B T \quad \text{or} \quad \kappa^{-1} = \frac{\langle\delta x^2\rangle_{thermal}}{k_B T}. \tag{13.6}$$

The compliance κ^{-1}, or susceptibility, of the oscillator is equal to its thermal mean square displacement divided by $k_B T$. Similarly, a mag-netic system will have temporal fluctuations of its magnetization in zero magnetic field $\langle M^2\rangle_{H=0}$ which are related to the system's linear suscep-tibility. The appropriate relation for the magnetic system, expressed in the units used by "ising", is

$$\chi_n = M/H = N\langle (M - \langle M \rangle)^2 \rangle /T, \tag{13.7}$$

where N is the total number of spins, 900 for PRESET 2. A system which is "soft", i.e., which has a large response to an external driving force, will have correspondingly large thermal fluctuations.

Exercise 13.8 (M**) *Test "ising" for consistency with the fluctuation–response theorem. In terms of the dimensionless variables used in the simulation, plot $T/N\langle(M - \langle M \rangle)^2\rangle$ versus T on the Curie–Weiss plot you've already made. Is the fluctuation–response theorem obeyed by the results of this simulation?*

The good agreement between the fluctuation data and the susceptibility data gives added confidence to both and convincing evidence of the quantitative inadequacy of the mean field theory.

13.4 Ferromagnetic state

The Curie temperature of the infinite two-dimensional Ising system is 2.27 J. In PRESET 3 the temperature is below this value. The initial condition gives the spins random orientations and the EXTERNAL FIELD is zero. When you RUN, the spins quickly "condense" into a state with most of the spins in the same direction, or occasionally with two or a few large areas within which the spins are mostly parallel to one another. The temperature is not far below the Curie temperature and you see a few spins "testing the waters", flipping via a thermal fluctuation but quickly deciding it was a bad idea and accepting the majority decision. Increase both the NUMBER OF SWEEPS and the SPEED by a factor of 10. While RUNNING, click on INIT whenever the system has decided on a preferred spin orientation to see the result of repeating the experiment. Though indifferent as to whether the spins go up or down, "ising" clearly wants neighboring spins all to have the same orientation.

The mean field theory has some specific predictions about the temperature dependence of the magnitude of the spontaneous magnetization below the Curie temperature. We can look theoretically at the limiting behavior at low temperature and at temperatures very near the Curie temperature. The first corresponds to the limit $(1 - M) \ll 1$ and the second to $M \ll 1$.

Using PRESET 3, change the INITIAL CONDITIONS to $M = 1$, and RUN briefly at $T = 1.4$ J. You see the occasional reversed spin resulting from the occasional toss of the loaded coin that puts the spin in the less favorable energy state. It is these few reversed spins that give a magnetization less than the saturation value of 1.

Exercise 13.9 (M*) *Take $M(T)$ data in the temperature range where the magnetization is in the range $0.3 < M < 1$. You will be troubled by the fluctuations at the higher temperatures. Sometimes the trace will be relatively stable for a while and then have a large fluctuation or even change sign before stabilizing again for a little while. Try estimating the "stable" values from the graph rather than using the $\langle M \rangle$ readout. Demonstrate graphically from your data that the deviation of M from its saturation value is plausibly fit at the lower temperatures by a dependence $e^{-A/T}$ with A a constant.*

Exercise 13.10 (M*) *What does the mean field theory have to say about the temperature dependence of $(1 - M)$ in the limit $M \approx 1$? For a quantitative comparison add the prediction of the mean field theory to the plot of the preceding exercise.*

The mean field theory is moderately successful in predicting, for the Ising model, the low temperature approach of the magnetization to saturation. However, real ferromagnets typically show a $\frac{3}{2}$ power law dependence on temperature rather than an exponential dependence on $1/T$. Neither the *Ising* model nor the *mean field* treatment of more realistic models is able to give this power law dependence. The $\frac{3}{2}$ power law relates in real systems to the presence of transverse components of the spin, which do not exist in the Ising model, and the effects of which are ignored by the mean field treatment of more general models. As we needed the Debye model to explain the power law temperature dependence of the lattice specific heat, so we would need the spin wave model to understand the power law dependence of the low temperature magnetization.

What about the mean field prediction for the variation of M with T as we approach the Curie temperature?

Exercise 13.11 (M*) *From the data of Exercise 13.9, make a linear plot of M versus T and estimate a value for T_C, the temperature at which the spontaneous magnetization goes to zero. Does it agree within estimated error with T_A, the value determined by extrapolation of the plot of χ_n^{-1} versus T in Exercise 13.5?*

Exercise 13.12 (M*) *Use the mean field theory to determine the temperature dependence of M near T_C, or more precisely, the limiting behavior of $M(T)$ as $M \to 0$. (Hint: in the implicit equation (13.5) for $M(T)$, expand the tanh function keeping the two lowest orders in its argument.) Plot the result on the graph of the preceding exercise.*

The mean field does a poor job here as well; it doesn't even get the power law right. The qualitative picture it provides is a useful one but we should not expect much more.

13.5 Coarsening and nucleation

We saw that the system, when cooled below the transition temperature, orders into the ferromagnetic state with a large degree of spin alignment. How does a quenched sample progress from the initial random condition to the ordered state? By what sequence of steps do the spins in the ferromagnetic state reverse their magnetization when the sign of an applied magnetic field is reversed?

13.5.1 Coarsening

Exercise 13.13 (C) *In* PRESET 4 *the Ising system is initialized in a random* T=INF *array in zero field, but then runs at a temperature* $T = 1$ *J, well below the Curie temperature. With* RUN, *only a single sweep is performed. Already you see that the spins have, on average, each preferentially aligned in a direction determined by the orientation of the majority of its four neighbors. We might call a cluster of parallel spins a nucleus of the ferromagnetic phase. Click on* RUN *a number of times to see a coarsening of the grainy assembly of growth domains. For longer times the larger grains grow at the expense of the smaller ones. (As time increases, increase the* NUMBER OF SWEEPS *and* SPEED *to save a little time.) Watch the clusters as they become comparable in size with that of the array. Does "ising" use open or periodic boundary conditions?*   # 4

Exercise 13.14 (C**) *Watch the coarsening process carefully in its intermediate stages. It may be easier to follow if you use the* CONFIGURE *dialog box to make a smaller array of spins. Invent a microscopic argument for why the large clusters, on average, grow at the expense of the smaller ones. (Hint: focus your attention on a small cluster of oriented spins surrounded by a background of spins of the other orientation. By looking at the consequences of specific spin flips along the interface line, can you argue why the small cluster is more likely to shrink than to grow? And why the bias towards shrinking becomes stronger for the smaller clusters? Note that you can toggle spins up and down manually by* CLICKING *on them.)*

13.5.2 Nucleation

When the Ising model is quenched to a temperature below the Curie temperature, we have seen how small grains or *nuclei* of aligned spins form spontaneously as each spin tries to accommodate to its neighbors. We speak of no *nucleation barrier* to the formation of the ferromagnetic phase: the coarsening process starts from the atomic scale. PRESET 5   # 5

gives an example of nucleation which is quite different. Here we start
with a sample fully magnetized *up* which is in a magnetic field which is
down. Clearly the system could lower its energy by flipping all of the
spins, but the fundamental process for achieving equilibrium, both in real
Ising systems and in the simulation, involves independent fluctuations of
individual spins. However, for a field $H = 0.1 \, J$, an individual spin is
unlikely to flip parallel to the field (with a gain in energy $2 \, H$) because
that puts it in antiparallel alignment with its neighbors (at a cost of $8 \, J$).
How, then, can the system ever get to the lower energy state of reversed
magnetization when the fundamental step is so unfavorable? RUN and
watch. Thermal fluctuations do occasionally flip individual spins briefly,
sometimes a second will flip beside the first, or a third as well, to form a
pair or a triple but they soon "give it up as a bad job" and submit again
to majority rule. On very rare occasions the thermal fluctuations may
produce a much larger *nucleus* which, if large enough, can gain control
and the whole system ultimately inverts its magnetization.

Exercise 13.15 (C) *Leave* PRESET 5 *running while you work on the
next exercise. Keep one eye on the screen and, if you're lucky, you will
see a nucleus form which becomes "large enough" to convert the system
to the stable configuration. (Once you have a sense of the futile attempts
at nucleation, increase the* SPEED. *Or, if you don't want to wait, increase
the* EXTERNAL FIELD *or* TEMPERATURE *by a small amount.)*

How large is "large enough"? What is the energy cost to form a nucleus
or a cluster of N flipped spins? The gain in magnetic field energy is
proportional to NH. The cost in exchange energy is proportional to the
perimeter of the cluster, hence is proportional to $\sqrt{N}$. Can you argue for
the $\sqrt{N}$ dependence? For small N the exchange wins, and the nucleus is
unstable against decay. However, for a large enough cluster the external
field energy wins and the nucleus will grow. The critical size of the cluster
is the value of N for which the total energy is a maximum. A nucleus of
this size is called a *critical nucleus*, and the energy required to generate
a cluster of this size is the *nucleation energy*. One might expect the
nucleation rate to be proportional to a Boltzmann factor involving this
energy. A more careful argument shows the need to include the entropy
of formation of the cluster and to use the free energy of nucleation. The
entropy terms become particularly important as the Curie temperature
is approached. Let's first see what happens, keeping track only of the
internal energy.

Exercise 13.16 (M*) *Translate the argument above to a semiquantitative one by estimating the numerical coefficients. Knowing the energy
$\mathcal{E}(N)$, find the size of the critical nucleus by finding the value of N for*

which $\mathcal{E}(N)$ is a maximum. If a thermal fluctuation gives a nucleus this size, then it has a 50–50 chance of growing to invert the magnetization fully. The initial conditions of PRESET 6 have a nucleus that is approxi- # 6 mately the critical size. Run it a number of times and you will see that it sometimes decays, but sometimes grows to take over the whole sample. How does your estimate of the critical size compare with the size of the cluster in PRESET 6? For the temperature used in PRESET 6, the entropy correction is large, reducing the free energy per lattice constant of the perimeter to about 0.6 J/a [58]. Does this improve the agreement with the simulation?

The phenomena of nucleation, *coalescence*, and coarsening are common features of a wide variety of phase transitions. Often the ability to control the grain size through manipulation of the density of nucleation sites or the annealing procedures provides the ability to influence technically important parameters of the material.

13.6   Critical fluctuations*

There is interesting physics to be seen by comparing the time evolution of the magnetization in zero field for temperatures close to and far above the Curie temperature. Focus your attention on two issues: the *magnitude* of the fluctuations of the magnetization, and the *time* interval over which significant changes in the magnetization develop.

Exercise 13.17 (M*) RUN PRESET 7 *until "ising" shuts itself off. This*   # 7 *run is at high temperature and with the system initially fully magnetized. In the graph you see the approach of the magnetization to its equilibrium value of zero, and its fluctuations about that equilibrium value.* COPY GRAPH *this plot, click* INIT, *and then* RESET *the graph in that order.* *(The order is important!) Change the* TEMPERATURE *to 3 J and* RUN *again.* STEAL DATA *to compare the two plots. How do the magnitudes of the fluctuations compare? How do the characteristic times for the approach to equilibrium compare? (Alternatively,* RUN *at each of the temperatures with the T=INF initial condition. Use the $\langle (M - \langle M \rangle)^2 \rangle$* READOUT *to determine the magnitude of the fluctuations; and choose the* AUTOCORRELATION *function from the graph* MENU *to determine the times characteristic of the fluctuations. See page 272 for a description of the autocorrelation function.)*

Exercise 13.18 (C**) *Give qualitative physical reasons for the slowing down of the equilibration process and the increase in the magnitude of the fluctuations as the temperature approaches the Curie temperature. (Hint: remember that the Curie temperature is characterized by a divergence of* *the susceptibility, or by the development of spontaneous magnetization. Think about the driving force trying to return the system to equilibrium.)*

13.7   Correlation lengths*

The approach of the temperature to the Curie temperature not only enhances the amplitude of the fluctuations and slows their time variation, but also increases the length characterizing the spatial correlations in the orientations of the spins.

Exercise 13.19 (C*) *Compare the spin display after a brief* RUN *of* # 8   PRESET 8 *at high temperature with a second* RUN *with the temperature set to 2.8 J. Let this second run continue long enough for the pattern to stabilize in quality. Characterize the difference in appearance of the two patterns, one at high temperature, the other near the Curie temperature. Alternatively, set* NUMBER OF SWEEPS *to a large value and watch the character of the display vary as you drag the handle of the* TEMPERATURE *slider over a wide range of temperatures above the Curie temperature. At* $T = 2.8$ J, *how far, on average, can you move from a spin and still find spins which are more likely to be parallel than antiparallel to the one you started with? This distance, crudely speaking, is defined as the* correlation length. *How does the correlation length vary with temperature?*

At very high temperatures the neighbors of any spin are equally likely to be up or down. Close to the Curie temperature, however, the neighbors of an up spin are more likely to be up and those of a down spin more likely to be down, even though over the whole sample there are equal numbers up and down. We speak of the development of *short range order* as the temperature approaches the Curie temperature from above, and the correlation length is a measure of the spatial extent of these correlations.

The development of short range order, or the divergence of the correlation length with decreasing temperature, is another characteristic of the approach to the Curie temperature. This short range order is a principal source of the shortcomings of the mean field approximation, which takes no account of the obvious correlation of the orientation of one spin with that of its neighbors. It is not surprising, then, that the failures of the mean field method become most apparent for temperatures near the Curie point, where the correlations become most apparent.

13.8   Thermodynamic limit*

It is tempting, but soon becomes frustrating, to pin down the Curie temperature in "ising" by taking data at temperatures closer and closer to the transition temperature. There is a lesson to be learned from this frustration. Part of the problem is the *critical slowing down*: as we just saw in Section 13.6, thermal equilibrium is established only slowly when T is near T_C, and at thermal equilibrium the system undergoes large fluctuations over long time periods. These effects near T_C present a severe problem for careful measurements on real systems, as well as with simulated ones. It is more important to recognize that there *is* no sharp transition for a system with few degrees of freedom. Nine hundred spins is not enough to give a sharp transition. How would you define the transition anyway? Suppose you are at zero field and at a temperature you think is "just below the transition". The majority of the spins might be in spin-up clumps, but there would also be regions or clusters with spin-down. You watch for a while and the magnetization M stays positive though small. But soon there is a fluctuation, one of the spin-down clusters grows at the expense of the spin-up to the point that the total M becomes negative, and continues that way for a while. Was the system really below the transition temperature and then a fluctuation reversed the magnetization? Or was it above the transition and, because of the critical slowing down, the large amplitude, long time fluctuations just made it look, briefly, as if a permanent magnetization had been established? (Try running a 4×4 array of spins at $T = 2.2\ J$.) Don't bother deciding! The question is trying to draw a meaningless distinction: the idea of a "yes or no" transition is valid only in the limit of an infinite system, the *thermodynamic limit*.

13.9   Summary

The paramagnetism of a collection of non-interacting spins is easily described by assigning each spin an up or down orientation in accord with a Boltzmann factor in which the energy is determined by the interaction of that spin with an external field. With exchange interactions among the spins the situation is more complex. One can think of each spin as responding to an effective field which is the sum of the applied field and an exchange field (13.2), a measure of the exchange interaction with its neighbors. Simple analytic solution becomes impossible because we don't know what to do with any individual spin until we know what its neighbors are doing. The Monte–Carlo method used by "ising" allows a straightforward simulation of the Ising model and reveals a wealth of qualitative behavior illustrative of many phase transition phenomena: nucleation and growth, coarsening, critical fluctuations, critical slowing down, and divergence of the correlation length. The mean field method, the simplest approximate analytic solution, successfully gives the high temperature Curie–Weiss law, as well as a useful qualitative picture of the development of the spontaneous magnetization of the ferromagnet. It expresses some of the physics well, but, by ignoring correlations among neighboring spins, is unable to get the details correct.

13.A   Deeper exploration

Nucleation   The nucleation issue [57, 58] may be studied in more detail. One possible project is to determine how the size of the critical nucleus depends upon the field and temperature. Another would involve a study of the nucleation time and its field and temperature dependence. Just don't try to do everything!

Boundary effects   "ising" has used periodic boundary conditions: spins at the top of the array are coupled to the spins at the bottom, and those at the left to the ones at the right. As the correlation length becomes long, what happens at one site influences, indirectly, what happens at sites up to a correlation length away. When the correlation length becomes comparable with the sample size, a spin begins "to influence itself". Explore this issue in a systematic way by changing the aspect ratio N_x/N_y of the sample, keeping the total number of spins roughly constant. Do you expect a larger influence on the Curie–Weiss constant θ or the Curie temperature T_C? The question of the effect of sample size with fixed aspect ratio would also be relevant.

Noise characterization   There are many problems in physics requiring characterization of the time variation of statistical fluctuations. See Reif's discussion of Brownian motion [54, p. 582] for an example. "ising" computes two functions which characterize the time dependence of the fluctuations in magnetization. The power spectrum reveals the relative importance of the different frequencies required for a Fourier representation of the time dependence of $M(t)$. The autocorrelation function gives the degree to which the magnetization at one time is correlated with the magnetization at a different time. Read up on these two representations of random signals and the relationship between them. Use "ising" to illustrate what you have learned.

Quenched textures   Control of grain size in mixed phase alloys can be achieved in part by suitable quenching or annealing procedures [57]. This can be illustrated in "ising" by extending our introduction to nucleation. # 4   Starting from PRESET 4, RUN first with $H = -0.2\ J$, and then with $H = -1.5\ J$. In each run, hit the STOP button when the magnetization is roughly zero, i.e., when roughly half of the spins have flipped down.

COPY GRAPH the first and compare the textures in the two runs. Why is
the texture so much finer for the run at the larger field? Develop this into
a study of the dependence of the texture on the field and temperature, the
quench temperature, at which the program is run. Run with the largest
sample sizes you have the patience to work with to see the effects more
clearly.

Critical exponents An obvious exploration is an experimental study of
the critical exponents of this Ising system. The small system size and long
run times can make the exercise arduous, but nonetheless informative. It
should be undertaken only if the data taking can be automated.

13.B   "ising" – the program

"ising" simulates the two-dimensional Ising model on a square lattice.
The underlying algorithm is a simple Monte–Carlo algorithm in which
a randomly picked spin is equilibrated with its four nearest neighbors
according to Boltzmann weighting factors.

13.B.1   "ising" algorithm

"ising" simulates a spin system in which each spin interacts with both its
four nearest neighbors on a square lattice and the external magnetic field
H. The energy $\mathcal{E}_i$ of a spin S_i is given by

$$\mathcal{E}_i = -S_i \left(J \sum_{\langle j \rangle} S_j + H \right) \equiv -S_i H_i^{\text{eff}}, \tag{13.8}$$

where J is a coupling constant and the sum extends over the four nearest
neighbors (indicated by $\langle j \rangle$) of the spin S_i. Each spin is assumed to point
either up ($S_i = 1$) or down ($S_i = -1$).

Initial configuration   The initial spin configuration (INITIAL CONDITION)
can be chosen to be either random (T=INF), fully magnetized upward
(M=1), or a previously STORED configuration can be RECALLED. Note
that CLICKING with the mouse cursor on a lattice site flips the spin at
that site, which allows any spin configuration to be created and STORED.

Dynamics   According to statistical mechanics the probability of a spin
S_i pointing up or down at a temperature T is proportional to the Boltzmann weight $e^{-\mathcal{E}_i/T}$. The normalized probability that a spin points up

$(S_i = 1)$ is thus given by

$$p(S_i = 1) = \frac{e^{H_i^{\text{eff}}/T}}{e^{-H_i^{\text{eff}}/T} + e^{H_i^{\text{eff}}/T}}, \tag{13.9}$$

where H_i^{eff} is the effective field, defined in Eq. (13.8), at the lattice site of the spin S_i. (If you expected a k_B in Eq. (13.9), recall that "ising" measures both T and H in units of J.)

To simulate the dynamics of the spin lattice, "ising" randomly picks a spin S_i on the lattice and computes, for the chosen TEMPERATURE T and EXTERNAL FIELD H, the probability $p(S_i = 1)$ according to Eq. (13.9). Independent of the current spin orientation, that spin is left pointing upward with probability $p(S_i = 1)$ and left pointing downward with probability $1 - p(S_i = 1) = p(S_i = -1)$. A single *sweep* consists of $N_x \times N_y$ such attempts to flip randomly chosen spins where N_x and N_y are the LATTICE WIDTH and HEIGHT given in the CONFIGURE menu. The NUMBER OF SWEEPS for which "ising" runs can be specified on the panel. (You also can press STOP to stop a run at any time.) Since the graphics display of the spin lattice is the slowest part of the "ising" simulation, choose a SPEED> 1 to have "ising" run faster by displaying the spin lattice only every SPEEDth sweep.

Boundary conditions To minimize the effects of the finite lattice size, "ising" assumes periodic boundary conditions. For example, a spin at an "edge" of the lattice is coupled not only to its three immediate nearest neighbors but also to a spin at the opposite edge of the lattice. If you RUN "ising" at a TEMPERATURE below the ordering or Curie temperature $T_C \approx 2.27\ J$ and a magnetic field $H = 0$ then, starting from random initial conditions (T=INF), you will see spin-up and spin-down domains form which wrap around either or both sides of the lattice.

13.B.2   Displayed quantities

"ising" shows the spin lattice on the left. The graph on the right displays the MAGNETIZATION versus the number of sweeps ("time"), the POWER SPECTRUM of the time history of the magnetization, or its AUTOCORRELATION function. The time average magnetization and energy and their mean square fluctuations are given above the graph.

Spin lattice The display on the left shows the $N_x \times N_y$ lattice of spins. Spins pointing upward are indicated in red, spins pointing downward in white. The LATTICE HEIGHT and WIDTH and the spin COLORS can be changed in the CONFIGURE menu.

Pressing INIT initializes the spin lattice according to the specified INITIAL CONDITIONS. An individual spin can be flipped by clicking on its lattice site with the mouse cursor. Several spins can be flipped at once by holding down the left hand mouse button and dragging the mouse cursor across the lattice.

Magnetization graph Choose MAGNETIZATION in the right hand corner above the graph to see the magnetization versus time displayed. Note that the graph is only cleared if RESET is pressed, *not* when the lattice is INITIALIZED. This allows data from several RUNS to be accumulated in the graph, which is useful to improve the quality of the power spectrum discussed below.

Power spectrum Choose POWER SPECTRUM in the right hand corner above the graph to see the power spectrum of the magnetization data displayed in the graph.

To compute the power spectrum using a fast Fourier transform (FFT), let M_t, $t = 0, \ldots, N_s - 1$, be the normalized magnetization of the $N_x \times N_y$ spin lattice, sampled after the sweep t,

$$M_t = 1/(N_x N_y) \sum_{i=1}^{N_x N_y} S_i(t), \qquad (13.10)$$

and let $\langle M \rangle$ be the (time) average magnetization over all sweeps N_s since the last graph RESET,

$$\langle M \rangle = (1/N_s) \sum_{t=0}^{N_s-1} M_t. \qquad (13.11)$$

We are interested in "ising" in the discrete Fourier transform F_n of the deviation of the magnetization[1] $(M_t - \langle M \rangle)$ which is given by

$$F_n = \sum_{t=0}^{N_s-1} (M_t - \langle M \rangle)\, e^{2\pi i t n / N_s}. \qquad (13.12)$$

The power spectrum of the magnetization is then defined as

$$P_n = (1/N_s) |F_n|^2. \qquad (13.13)$$

The normalization of P_n is chosen so that the area under the P_n curve is just the mean square deviation of the magnetization $\langle (M - \langle M \rangle)^2 \rangle$.

---

[1]Taking the Fourier transform of $(M_t - \langle M \rangle)$ rather than of M_t, we make the zero-component of the power spectrum vanish, $P_0 = 0$. Otherwise, the power spectrum of a spin system with a large average magnetization would be dominated by the large intensity of the zero-component P_0.

Note that we display in "ising" only the positive frequency half of the (symmetric) power spectrum, with the area under the displayed curve equal to $\frac{1}{2}\langle (M - \langle M \rangle)^2 \rangle$.

"ising" allows the power spectrum P_n to be displayed either as computed according to Eq. (13.13) or averaged over l successive data points, with the AVERAGE LENGTH l chosen in the CONFIGURE menu. (The AVERAGE LENGTH $l = 1$ corresponds to *no* averaging.) The average $\bar{P}_n$ of P_n is defined to be

$$\bar{P}_n = \frac{1}{l} \left(\sum_{i=n-l'}^{n-1} \bar{P}_i + \sum_{i=n}^{n+l'} P_i \right), \tag{13.14}$$

with $l' = \frac{1}{2}(l-1)$ and minor modification for the end regions near $n = 0$ and $n = N_s/2$. Starting from P_0, "ising" averages the power spectrum according to Eq. (13.14). This procedure, though giving an asymmetric average, is calculationally convenient.

There is one more technical difficulty to overcome in the computation of the power spectrum (13.13) and the autocorrelation function in Eq. (13.16) below. Our FFT algorithm requires the number of data points to be a power of 2. The M_t array, however, can be of any length and we have to use a "fix" known as zero padding to be able to use this FFT routine: "ising" computes the FFT from an array of M_t, $t = 0, \ldots, N_s' - 1$, where N_s' is the next power of two higher than N_s and all M_t, $t \geq N_s$ are zero. The power spectrum (13.13) continues to be normalized by $1/N_s$.

Autocorrelation function   Choose AUTOCORRELATION for the right hand graph to display the autocorrelation function of the magnetization,

$$G(t') = (1/N_s) \sum_{t=0}^{N_s-1} (M_t - \langle M \rangle)(M_{t+t'} - \langle M \rangle). \tag{13.15}$$

An important relation in noise analysis is that the autocorrelation function is the Fourier transform of the power spectrum [54, p. 582],

$$G(t') = (1/N_s) \sum_{n=0}^{N_s-1} P_n e^{2\pi i t' n / N_s}. \tag{13.16}$$

This can easily be seen by expressing M_t as the inverse Fourier transform of F_n in Eq. (13.15) and noting that $F_n = F_{-n}^*$ since M_t is real. Here we have chosen the normalization of the autocorrelation function in Eq. (13.15) such that its zero-component is just the average mean square deviation of the magnetization, $G(0) = \langle (M - \langle M \rangle)^2 \rangle$.

Readouts   "ising" displays the (time) average magnetization and energy $\langle M \rangle$ and $\langle \mathcal{E} \rangle$ as well as their mean square fluctuations $\langle (M - \langle M \rangle)^2 \rangle$ and $\langle (\mathcal{E} - \langle \mathcal{E} \rangle)^2 \rangle$. The averages are taken over the accumulated N_s sweeps since the last graph RESET.

The time average magnetization $\langle M \rangle$ is obtained from Eq. (13.11). The average energy $\langle \mathcal{E} \rangle$ is given by

$$\langle \mathcal{E} \rangle = \frac{1}{N_s} \sum_{t=0}^{N_s-1} \left(\frac{-1}{N_x N_y} \frac{1}{2} \sum_{i=1}^{N_x N_y} S_i J \sum_{\langle j \rangle} S_j \right)_t , \qquad (13.17)$$

where the term in parenthesis is computed after every sweep t and the factor $\frac{1}{2}$ avoids double-counting of the exchange interaction. Note that we have chosen not to include the interaction energy of the magnetization with the external magnetic field, $-H\langle M \rangle$, in the energy (13.17).

13.B.3   Bugs, problems, and solutions

When the graph is RESET, the current magnetization is taken as the zeroth data point. Therefore always first INITIALIZE the lattice and *then* RESET the graph if you want the first data point to correspond to the initial conditions!

14

"néel" – Antiferromagnetism and magnetic domains

Contents

14.1 Introduction

"ising" provides us with a good sense of what goes on at the atomic scale in a ferromagnet, but gives no help in understanding those properties of ferromagnets that we most care about. What do you do with a piece of transformer iron to maximize its permeability and minimize its hysteresis? How do you make a permanent magnet stay magnetized, despite the fact that the magnetization M throughout most of the volume is opposed to, *not* parallel to, the magnetic field H? Recognizing the importance of the *dipolar interaction* among the spins is the key to understanding the technical properties of magnetic materials.

"néel", though similar to "ising", includes dipolar interactions in addition to the nearest-neighbor $-JS_iS_j$ exchange interaction. The strength of the dipole–dipole interaction is defined by the parameter D, the dipolar interaction energy of two spins at neighboring sites. The long range and the angular dependence of the dipolar interaction introduce important new physics into the problem. Though working with a two-dimensional model, "néel" allows simulation of a specific three-dimensional ferromagnet with a planar domain structure, illustrating a number of features of magnetic domains.

It also simulates the properties of a planar array of Ising spins with the moment direction either within or perpendicular to the plane of the array. These spins can interact with a tunable combination of exchange interaction with nearest neighbors only, and dipolar interaction with all of the other spins in the array. Thus "néel" is able to simulate quite a variety of ferro- and antiferromagnetic structures.

This chapter consists then of two main parts. The first, a generalization of ideas introduced already in "ising", discusses examples of antiferromagnetic ordering and the effects in antiferromagnets of competition between interactions favoring different types of order. The second addresses technically important questions about the properties of bulk ferromagnetic materials below their Curie temperature, properties which can be understood only in terms of domain structure.

You may be disturbed at first when using "néel" that none of the parameters T, H, D, or J is given units. This is the natural language of a computer, which always works just with numbers. In other programs we have artificially introduced numbers to connect the parameters more directly with real physical systems. "ising" changed things a little by expressing T and H relative to J since only ratios of pairs of these quantities were relevant to the physics, not their individual absolute values. The situation is similar in "néel", but here there is no obvious choice as to which parameter to make the "master". We adopt the computer's natural language and work with numbers alone, remembering that the statement by "néel" that $H/T = 3$ represents the physical statement that $\mu H/k_B T = 3$, etc.

Exercise 14.1 (M) *Suppose you do want to define units for each of the four parameters. If you wish to express the temperature T in units of kelvin, then by what conversion factors must you multiply the numerical results of "néel" in order to find the field H in tesla, the exchange interaction J in electron volts, and the nearest neighbor dipole interaction D in joules?*

14.2   Antiferromagnetism

For this section of the chapter we consider only the PERPENDICULAR choice of the EASY AXIS which implies an orientation of the Ising spins perpendicular to the plane of the array. The n choice in the POWER LAW menu will be 3 as appropriate for the dipolar interaction. (In Section 14.3 the power law is changed to $1/r^2$ for the simulation of a particular domain array in three dimensions. For now, be sure POWER LAW is set to $n = 3$.) For a planar array, with spins perpendicular to the plane, the dipolar magnetic field due to an up-spin at one site is directed down at all other sites in the array, favoring the antiparallel alignment of any pair of spins. The dipolar interaction energy for this geometry is $(Da^3/r^3)S_iS_j$ with r the distance between the spins and a the lattice constant of the array. The full energy of the system becomes

$$\mathcal{E} = -\tfrac{1}{2}\sum_{ij} J_{ij}S_iS_j + \tfrac{1}{2}\sum_{ij} \frac{Da^3}{r_{ij}^3}S_iS_j - \sum_i HS_i. \qquad (14.1)$$

"néel", like "ising", takes $J_{ij} = J$ for nearest neighbors and zero otherwise. Be careful with signs: spins like to align *parallel* for positive J, but *antiparallel* for positive D.

14.2.1   Antiferromagnetic domains

If you were adventurous in using "ising" you might have tried changing the sign of J. If you had, you would have seen the same thing as shown in PRESET 1. Here the dipole interaction has been turned off and the exchange has been set to $J = -1$. When you RUN this PRESET you see the system of 30×60 spins quickly transform into a new structure in which every spin is oriented antiparallel to each of its nearest neighbors. Such an ordered arrangement of antiparallel spins is called an *antiferromagnet*. The arrangement you see is not perfect, however. There is a set of *growth domains*, isolated regions of perfect antiferromagnetic order. The domains are separated by strange looking domain boundaries. Let the system anneal until there are only a few large domains. The domain structure becomes more obvious if you back away from the screen, or squint, to blur the image of the screen. Try it.

Exercise 14.2 (C) *Here is a puzzle. In the "ising" case the growth domains were clearly distinguished. Some, call them A domains, had all spins up and others, B domains, had all spins down. The boundary was simply the inevitable line dividing the two. For this antiferromagnetic case if you look within the domain on either side of a boundary it looks just the same. What is it that distinguishes A domains from B domains? There must be something different or the boundary between them would not be so obvious. In fact it is only because of the boundary that we even notice that there is a domain structure. (Hint: what is it in the arrangement of spins on the two sides of the domain wall that prevents the two domains from merging into a larger single domain?)*

Exercise 14.3 (M) *Raising the temperature slowly, determine roughly the temperature at which the antiferromagnetic phase "melts", i.e., the temperature of the antiferromagnetic transition, called the Néel temperature. The mean field theory gives $T_N = 4\,|J|$ for the Néel temperature. Does the mean field theory do any better here than it did for the ferromagnet?*

Exercise 14.4 (C) *Now set the* EXTERNAL FIELD *to 2 and the* TEMPERATURE *to 3. Change the* NUMBER OF SWEEPS *to 10^4, and while* RUNNING *the* PRESET *grab the handle of the* EXCHANGE J *slider and watch the screen to see the effect of the exchange on the magnetization. Drag the slider to both positive and negative values of J. Explain why the magnetization varies as it does with J.*

14.2.2 Magnetic susceptibility

The analysis of the antiferromagnet will turn out to be more interesting if the interactions go beyond nearest neighbors. PRESET 2 gives a purely dipolar antiferromagnet, $D = 1$ and $J = 0$, at a temperature below the Néel temperature. Be sure the INITIAL CONDITIONS are set to AFERRO to insure that the array will be a single domain. The temperature is high enough to see some thermal fluctuations, but the single domain structure is preserved.

2

Exercise 14.5 (M) *First, with $J = 0$ and $D = 1$, measure the magnetic susceptibility of the system, $\chi = M/H$, as a function of T, for temperatures below $T = 4$. (Remember to RESET the graph after each RUN of 100 sweeps to get meaningful values of $\langle M \rangle$ from the readout.) Start at a low*

temperature and work upwards to insure that the system remains single domain while below the Néel temperature T_N. Reinitialize with the INIT button if you notice that the system has broken into several domains and you wish to work again well below T_N. To insure that you stay within the range of a linear susceptibility you should keep $H < \frac{1}{4}T$. (T here is the dimensionless temperature, not tesla.) Plot the data and estimate the Néel temperature as the temperature of the poorly defined maximum of the susceptibility. What is the limiting temperature dependence of the susceptibility at the lowest temperatures?

The susceptibility is much less interesting than that of the ferromagnet. At high temperatures they are similar, but at or near the Néel temperature the antiferromagnetic susceptibility starts to fall with decreasing temperature, with no hint of any divergence.

3   **Exercise 14.6 (M**)** PRESET 3 *illustrates the need to work with a single domain to measure the susceptibility of the antiferromagnet. Here the initial condition freezes the system into the antiferromagnetic state with several domains. Check the value of the susceptibility at some temperature well below the Néel temperature to compare with a value you got from the single domain. Why is it so much higher than it was for the single domain? (Hint: watch the spin display carefully as you change the magnetic field.)*

Exercise 14.7 (M) *Still with $J = 0$ and $D = 1$, take more data working at temperatures $T = 4$ and above and make a Curie–Weiss plot χ_n^{-1} versus T. What is the Curie–Weiss temperature θ, the temperature-axis intercept, for this sample? What does the negative value of θ imply about the effect of the antiferromagnetic interaction on the high temperature susceptibility? Is it enhanced or suppressed relative to the Curie law of non-interacting spins?*

Although the Néel temperature cannot be determined very well, it is clear that it is quite different in magnitude from the Curie–Weiss temperature. The mean field theory can give us some guidance on this issue, as we'll see in the next section, but we should not expect too much from it.

14.2.3 Mean field analysis*

The mean field theory of the antiferromagnet starts with a supposition of what the ordered, low temperature arrangement of spins will be. For the simple square lattice in two dimensions the natural antiferromagnetic structure has the spins arranged in checkerboard fashion: an arrangement with each spin antiparallel to all of its nearest neighbors. Recalling the terminology of Chapter 2, you will recognize this as a centered square structure, rotated by 45° from the atomic lattice, with two sites per unit cell, an A site at the origin and a B site at the centered position. We think in terms of two *interpenetrating sublattices*, labeled A and B. With each sublattice we identify a normalized *sublattice magnetization* M_A or M_B which can take on values between -1 and $+1$. The normalized total magnetization, with this normalization convention, will then be $M = (M_A + M_B)/2$.

We evaluate the effective field at a given site as the applied field plus the sum of the dipole fields from the spins on all other sites. The essential assumption of the mean field approximation is to place an "average spin" on each site with spin value S equal to the sublattice magnetization appropriate to that site (*not* ± 1). This gives for the effective fields at the two kinds of sites,

$$H_A^{\text{eff}} = H + \mu M_A + \lambda M_B \qquad \text{and} \qquad H_B^{\text{eff}} = H + \lambda M_A + \mu M_B. \quad (14.2)$$

The constant μ is the effective field at an A (or B) site due all the other A (or B) spins when they are aligned in the positive direction. If the interaction is only dipolar, evaluation of the dipolar sum for an infinite lattice gives $\mu = -3.9\ D$. Similarly, λ is the effective field at an A (or B) site due all of the B (or A) spins when they are aligned in the positive direction. If the interaction is only dipolar, $\lambda = -6.5\ D$. If there is exchange interaction between nearest neighbors in addition, then $\lambda = -6.5\ D + 4\ J$ where we have continued the convention that positive J corresponds to a ferromagnetic interaction. For the finite lattice used by "néel", the cut-off of the long-range sums reduces the dipole mean field values to $\lambda = -5.36\ D$ and $\mu = -2.76\ D$.

Exercise 14.8 (M**) *Verify the numerical values of the constants λ and μ for the infinite array by working out the appropriate dipole sums numerically. (Because of the slow convergence of the sum, if you do this by hand you will want to replace the sum of contributions of far distant neighbors by an integral.)*

Exercise 14.9 (M*) *Work out the high temperature susceptibility and show that it gives the Curie–Weiss law with a Curie–Weiss temperature $\theta = (\lambda + \mu)$. How does this predicted value compare with the value determined from the Curie–Weiss plot? Remember the ambiguity in the dipole sums because of the finite sample size. (Hint: start with the effective fields (14.2) and use them in Eq. (13.3).)*

As with the ferromagnet, the mean field theory gives a good approximation to the limiting behavior of the system at high temperature. What about the ordering behavior below the transition?

The expression for the normalized susceptibility $\chi_n = 1/(T - \theta)$ gives no hint of a phase transition if θ is negative. On the other hand, it is evident from the freezing out of the susceptibility at low temperature and from the antiferromagnetic ordering of the spins that something is going on. Can the mean field picture say anything about that?

We can get some useful ideas by asking for the *alternating susceptibility*, $\chi_n^* \equiv (M_A - M_B)/H^*$. Suppose we could apply a positive field of magnitude H^* at the A sites and a negative field $-H^*$ at the B sites. We will speak of applying an *alternating field* H^* (alternating in space, *not* time). Now, *if* we could apply such a field it would drive the system towards the favored antiparallel arrangement. We expect the alternating magnetization which develops in response to such a driving force to *did* diverge at some temperature which we identify as the Néel temperature.

Exercise 14.10 (M**) *Continue the argument by asking for the alternating magnetization, $M^* \equiv (M_A - M_B)$ in the presence of the alternating field H^* and determine at what temperature the alternating susceptibility diverges. Show that the transition or Néel temperature, for $J = 0$, is given by $T_N = -\lambda + \mu$. How does this compare with the experimental data? Again, we're dealing with a finite, not an infinite sample.*

Exercise 14.11 (C**) *Explain physically why the magnitudes of the Néel and Curie–Weiss temperatures differ for the case with farther neighbor interactions, while they are the same with only nearest neighbor interactions (i.e., $\mu = 0$)? That is, why is it that with λ and μ both negative, they seem to work together to increase $|\theta|$, but work against each other to decrease T_N.*

Exercise 14.12 *Consider the case in which λ is negative (antiferromagnetic coupling) but μ is positive (ferromagnetic coupling). Now $T_N = |\lambda| + |\mu|$. How can it be that increasing the magnitude of μ, a ferromagnetic interaction within each of the sublattices, raises the temperature of the transition to the antiferromagnetic state? Further, explain why the antiferromagnetic interaction between the different sublattices λ can be arbitrarily small leaving the antiferromagnetic transition temperature at a finite value determined almost entirely by the ferromagnetic μ. (Hint: decide first the behavior of the system for a zero value of λ.) See if you can construct this situation in "néel". (Hint: remember that you may* *choose D or J either positive or negative. Be very careful with signs. It's easy to get them all mixed up if you're in a hurry.)*

14.2.4   Other phases

The structure we have been looking at is the most obvious antiferromagnetic arrangement of the spins in "néel", but there are other possibilities.

What happens if a magnetic field is applied or if the parameters D and J are changed? Other well-ordered phases may develop.

Exercise 14.13 (C*) *Using* PRESET 2 *again, change the temperature to*   # 2
$T = 0.2$, *the external field to* $H = 5$, *initialize to* AFERRO, *and* RUN. *A new phase develops which is reminiscent of the antiferromagnetic phase, except that now there are more spins up than down. We might make a distinction between the checkerboard phase at* $H = 0$ *and the tablecloth phase at* $H = 5$. *What are a primitive unit cell and basis for the tablecloth? Predict from the structure of the tablecloth phase, its net magnetization relative to the saturation magnetization.*

Exercise 14.13 considered the competition between the dipolar energy and the interaction energy with the external field. PRESET 4 explores   # 4
the effects of the competition between the dipolar antiferromagnetic coupling and a ferromagnetic nearest neighbor exchange. The PRESET opens with the dipolar antiferromagnetic state at the low temperature $T = 0.1$, and an exchange $J = 0.48$. (Nothing interesting happens for smaller J.) Increase J in steps of 0.01 by clicking on the small triangle at the right hand end of the J slider, waiting a little while at each value. By the time you reach $J = 0.52$ or 0.54 you should have seen the beginnings of a transition to a new stable phase, also antiferromagnetic but with a different spin structure. See what it anneals to.

Exercise 14.14 (C**) *Before reading further, construct your own argument to show that the lamellar phase which develops has a lower energy than either the ferromagnetic phase or the simple antiferromagnetic phase we saw before. (Hint: why is the ferromagnetic phase "bad news" for the dipolar interaction?) Check some of your ideas using the average energy* $<E>$ *readout at the bottom of the panel. (The energy here includes the* *exchange and dipolar energies, but not the interaction energy* $-MH$ *with the external field. For meaningful averages, you must reset the graph after each parameter change.) Why do the strips prefer to be perpendicular, rather than parallel, to the boundaries?*

The checkerboard phase is the worst possible arrangement of spins as far as the nearest neighbor ferromagnetic exchange is concerned. As J is gradually increased the exchange penalty becomes greater and greater. Finally the system looks for another arrangement with lower energy. In the lamellar phase each spin has two spin-up and two spin-down neighbors; hence the exchange energy of this phase remains zero. The system has avoided the exchange penalty of the checkerboard phase, though in the process it has ended up in a structure less energetically favorable for the dipolar interaction.

Exercise 14.15 (C) *In* PRESET 4, *with* $J = D = 1$, *gradually increase the magnetic field and watch the magnetic structure change. At a field not far below the one at which the whole system is driven to saturation, you see isolated pockets, one or two spins, of reverse magnetization which are energetically stable. They are basically the magnetic bubbles [59] that at one time were thought to be the key to bigger and better computer memories. Magnetic bubbles are now used in some special purpose devices. Why are there more reversed spins, or bubbles, in the interior of the array than near the boundary? (Hint: do you think "néel" uses free or periodic boundary conditions?)*

In considering the competition between exchange and dipolar energies, we started with large D/J and increased J until $J = D = 1$. We could also imagine starting from the limit of large exchange, a single-domain ferromagnetic phase, and considering the effect of increasing the dipolar interaction until it equals the exchange. For a single ferromagnetic domain, there is a large dipolar energy cost because every spin is unfavorably oriented with respect to the dipolar field of all the other spins. The lamellar arrangement which develops at $D = J = 1$ represents a compromise. The system has been clever enough to find a configuration which sacrifices some exchange energy in order to reduce the long range dipolar energy. Similar competition between the exchange and dipolar energies is the source of the equilibrium domain structures in three-dimensional ferromagnets, although the combination of the slow divergence of the dipole sums and the angular dependence of the dipolar interaction makes the three-dimensional problem a more subtle one.

Exercise 14.16 (C) *Start at the large J/D end of the range, with $J = 1$ and $D = 0$. Change the* INITIAL CONDITION *to* T=INF *and* RUN. *Recycle a couple of times by pressing* INIT *to get a feel for the condensation into the ferromagnetic phase with one or a few large domains. Next add a very small amount of* DIPOLAR *interaction, $D = 0.05$, and contrast the final configuration after annealing from* T=INF *with that for the case $D = 0$. Be sure to initiate each* RUN *from* T=INF. *Explain the qualitative changes, as D is gradually increased to $D = 1$, in the structure which develops after an anneal from the* T=INF *initial condition. How small a value of D/J is sufficient for the effects of the dipolar interaction to become evident?*

We have just seen how, for the planar ferromagnetic array of Ising spins, a weak, long range dipolar interaction among the spins breaks the uniform magnetization into a domain structure. This same physics, in more subtle form, drives the formation of a domain structure in macroscopic three-dimensional ferromagnetic systems.

14.3   Ferromagnetic domains

A nail from the hardware store is strongly magnetized if probed on a scale of 100 Å, but shows no macroscopic magnetization. Why not? What makes good material for permanent magnets, what makes good iron for transformers? Material designed for either purpose would be useless for the other (do you understand why?); yet on the microscopic scale, they seem to be quite similar, both showing locally a large, nearly saturated magnetization at room temperature.

We have seen part of the answer already on realizing that the dipolar interaction among spins favors the break-up of a uniformly magnetized sample into domains of smaller size. Though we must work a little further with this idea to see how it develops in a three-dimensional example, we have a sound argument for why many magnetic materials show no macroscopic magnetic moment. But we will soon be embarrassed by the implication that the formation of the domain structure will always force ferromagnets to have zero macroscopic magnetization in zero field. We've found good transformer iron, but haven't yet got a permanent magnet. Our second task will be to find how a system can maintain a substantial magnetization despite the efforts of the dipolar energy, and an opposed internal field, to destroy it. What is the source of magnetic hysteresis?

The effect of the dipolar interaction is more complicated in three dimensions than in the planar array of Section 14.2. Consider one spin in a uniformly magnetized sample. The dipolar field at that spin from other spins in a spherical shell centered on the first spin is zero because of the angular dependence of the dipolar interaction. Hence the dipolar interaction seems impotent in three dimensions. However, because it falls off as slowly as $1/r^3$, contributions from the far boundaries of the sample are significant, and the dipolar field at the chosen spin depends in magnitude and sign upon the shape of the sample.

In "néel", changing the allowed orientation, or EASY AXIS, of the spins from PERPENDICULAR to the array to parallel, either VERTICAL or HORIZONTAL, modifies the role of the dipolar interaction. (See Figure 14.2 on page 292.) Now it implies a ferromagnetic interaction with some of its neighbors, but antiferromagnetic with others. The VERTICAL choice allows "néel" to simulate the behavior of the three-dimensional ferromagnet shown in Figure 14.1. Think of the screen as giving a cross-sectional view of a bulk ferromagnetic sample L high, W wide and R (into the screen) deep. A three-dimensional analog of the lamellar structure we saw in PRESET 4 with $J = D = 1$ can develop, in which the domains of spins are planar slabs extending back from the screen with the normal to the slabs in the plane of the screen as illustrated. In this domain structure, all spins in a line perpendicular to the screen have the same orientation. "néel"

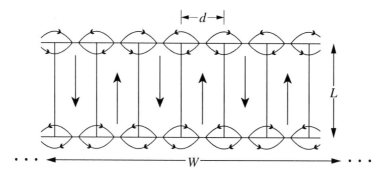

Figure 14.1: A lamellar domain structure with fringing magnetic fields near the ends of the domains.

shows the orientation of such a line of spins as up or down, according to the color in the display of the intersection of the line with the screen.

 # 5   **Exercise 14.17 (C)** PRESET 5 *shows the three-dimensional model at a low temperature with* T=INF *initial conditions. After a brief* RUN, *the domain structure suggested in Figure 14.1 becomes evident. What happens to the structure if you turn off the dipolar interaction* $(D = 0)$*? It really is the dipolar interaction that drives the formation of the domains.* RUN *with* $D = 0.5$ *and change the* EASY AXIS *back and forth a few times between* HORIZONTAL *and* VERTICAL *and compare the patterns that develop. Explain why the domain pattern changes as it does.*

14.3.1   Domain size

The quantitative conversion of "néel" to the description of the three-dimensional system requires two changes in the nature of the dipolar interaction. The $1/r^3$ dependence on the distance between spins is re-placed by a $1/r^2$ dependence on the distance between the lines of spins, or 'spins'. In the subsequent discussion, 'spins' is to be interpreted as 'lines of spins', and the orientation of a 'spin' refers to the orientation of the set of spins in the corresponding line. An angular dependence $\cos 2\phi$ is added where ϕ is the angle between the EASY AXIS, the allowed axis for the Ising spins, and the vector connecting the two 'spins'. The system energy, per lattice constant perpendicular to the screen, then becomes

$$\mathcal{E} = -\tfrac{1}{2}\sum_{ij} J_{ij} S_i S_j - \tfrac{1}{2}\sum_{ij} \frac{Da^2}{r_{ij}^2} S_i S_j \cos 2\phi_{ij} - \sum_i H S_i. \qquad (14.3)$$

Note that for two 'spins' separated by a vector *along* the EASY AXIS, the dipolar interaction is ferromagnetic; for two 'spins' with $\phi = 90°$, the dipolar interaction is antiferromagnetic.

Exercise 14.18 (M**) *Verify quantitatively the relationship between the "néel" simulation and the three-dimensional problem as described above. That is, show that the energy of interaction per unit length between two long parallel lines of dipoles, with the moments in each aligned perpendicular to the dipole lines, is proportional to $(-1/r^2)(\cos 2\phi)$ where r is the distance between the lines and ϕ the angle between the spin axis and the perpendicular vector connecting the lines.*

In the example of the spin array with the EASY AXIS perpendicular to the array, in Section 14.2, it was clear that the single-domain ferromagnet gave the maximum possible dipolar energy: every spin was unfavorably oriented with respect to every other spin. The angular dependence of the dipolar interaction complicates the treatment in three dimensions. The argument for domains is more conveniently given in terms of a field energy. Section 14.B.2 shows how appropriate manipulation of the magnetostatic equations can convert the expression for the dipole–dipole energy of a distribution of magnetization into a magnetic field energy, the volume integral of $\mu_0 H^2/2$. Note: this is a *different* result from the more familiar statement, for fields produced by a current distribution, that the magnetic energy density is given by $\mathbf{B}\cdot\mathbf{H}/2$.

To see the motivation for the formation of domains in three dimensions, refer to Figure 14.1 and consider the case $W \gg L$ and $R \gg L$. Suppose first that the magnetization is uniform. Then the H-field is zero outside the volume and equal to $-M$ within. The $\mu_0 H^2$ field energy is of the order of $\mu_0 M^2 LWR$, the energy density times the total volume. However, suppose we break the material into slab domains of thickness d. In Figure 14.1, the curved arrows at the top and bottom surfaces are to suggest lines of magnetic field running from the termination of lines of positive magnetization of one domain to the beginnings of lines of negative magnetization in the neighboring domains. The H-fields are now confined to within about d of the top and bottom surfaces, and again are of the order of M in magnitude. This gives a magnetic energy of the order of $\mu_0 M^2 dWR$, i.e., smaller than for the single domain by the factor d/L. The dipolar energy, though too small to be relevant to the development of the spontaneous microscopic magnetization, becomes the driving force for the development of the domain structure. The finer the size scale of the domains, the smaller is the dipolar energy. Expressed in terms of the strength of the dipolar coupling D instead of the magnetization, this energy may be written as $\mathcal{E}_{\mathrm{dip}} \sim DdWR/a^3$.

Exercise 14.19 (M*) *Show that the dipolar energy $\mathcal{E}_{\mathrm{dip}} \sim \mu_0 M^2 dWR$ can be converted to the form $\mathcal{E}_{\mathrm{dip}} \sim DdWR/a^3$. (Hint: both the magnetization and the dipolar constant D can be expressed in terms of the magnetic moment μ of the spins and the lattice constant a.)*

If a few domains are good, aren't more better? Not if a smaller d implies too great a cost in exchange energy. Along each domain boundary, the spins on either side see one neighbor with the "wrong" spin orientation which gives an energy cost of $2J$ for each a^2 area of boundary. You should be able to convince yourself that the total cost in exchange energy in this model is of the order of $\mathcal{E}_{exch} \sim JLRW/da^2$.

Exercise 14.20 (M*) *Assume that the system can equilibrate to the domain configuration of lowest energy, for given R, W, and L, by adjusting d. Use the energy expressions above for the dipolar and exchange energies to determine the d which minimizes the total energy. What is the predicted dependence of d on the ratio D/J and on the value of L?*

Exercise 14.21 (M*) *Check for the dependence on D/J using "néel". Work, as in PRESET 5, with the TEMPERATURE T at about 1.5 to give the system an opportunity to search for the minimum energy configuration, and leave the EXCHANGE $J = 1$. For each D/J, average data from a number of runs starting each run from the T=INF INITIAL CONDITION. It is clear from the results that the system is not reaching a cleanly defined equilibrium configuration; hence the need for repeated runs. (The easiest way to measure d is to count the number n of strip domains and, for the EASY AXIS set to VERTICAL, set $d = 60a/n$). For the EASY AXIS set to VERTICAL and $J = 1$, check the domain width for a few values of the DIPOLE constant D in the range 0.2–2.0. (You can increase the program SPEED, while losing time detail of the animation, from the CONFIGURE dialog.)*

14.3.2 Hysteresis

The energetics of the domain structure tell us that a material with large microscopic magnetization will have zero macroscopic magnetization in equilibrium in zero applied field. In this picture, what will be the magnetization in the presence of a magnetic field? One can extend the ideas of the previous section to include the energy of interaction of the spins with an external field, and determine an equilibrium $M(H)$.

\# 6 **Exercise 14.22 (C)** *PRESET 6 suggests the expected behavior. It opens with a domain structure corresponding to zero magnetization. RUN briefly and STOP to see the graph showing the fluctuations near the zero value. RUN again and click in the slot of the EXTERNAL FIELD slider to increase the EXTERNAL FIELD to 1. What happens to the number of domains? What happens to the relative sizes of the spin-up and spin-down domains? STOP and RUN again to update the graph of the magnetization. Has the system equilibrated or is it still changing? When the system has* *settled down, return to zero field and let it equilibrate again. Does the magnetization return to zero?*

One mechanism for changing the magnetization in a ferromagnet is the variation of the relative sizes of the domains of the two orientations. Changes in relative sizes are achieved by the motion of the domain boundaries. This process is moderately reversible in "néel" if the temperature is not too low. We see another mechanism, however, if the change in applied field is made larger. In this case the irreversibility of the magnetization, *magnetic hysteresis*, is substantial.

Exercise 14.23 (C) *Starting again with* PRESET 6, *raise the* EXTERNAL FIELD *to 2.5 and wait for equilibration. (To see what the magnetization has been doing,* STOP *and* RUN *to update the graph.) Now increase the* EXTERNAL FIELD *to 3.5, wait for the magnetization to settle down, and then return to 2.5. Does the magnetization return to its earlier value? Based on what you saw in the display of the spin array, explain why the hysteresis is so much larger here than in the field cycle $H = 0 \to 1 \to 0$. (Hint: were the changes in magnetization a consequence of changes only in the sizes of domains?)*

If you try these again at a lower TEMPERATURE, say $T = 1$ or 0.5, you will find the irreversibility to be greater. In "néel" the major source of irreversibility is the difficulty of nucleating new, reverse domains, and at lower temperatures of initiating new columns of reversed spins on the boundary of an existing domain.

Typical magnetic materials show hysteresis which is related not so much to thermal barriers to the nucleation of reverse domains as to the *pinning* of motion of the boundaries between domains, called *domain walls*. The walls are pinned by inhomogeneities in the microstructure of the material, which energetically favor certain positions of the domain walls with respect to others. The inhomogeneities may be randomly distributed impurities, grain boundaries, grain structure in mixed phase alloys, dislocation strain fields, and so on. As the magnetic field is changed, domain walls can remain pinned at inhomogeneities until the force on the domain wall is large enough to tear the domain wall away from the pinning site. When the field is returned to the original value, the wall is left pinned at some other site corresponding to a different magnetization. In an attempt to simulate this pinning, "néel" adds a term to the energy which puts an extra fixed pseudo-magnetic field or *pinning field* at each lattice site, a field which varies randomly from site to site with the rms value SIGMA.

Exercise 14.24 (M*) *Plot some hysteresis curves $M(H)$ using data from "néel" at* PRESET 7 *for two values of T, 0.5 and 1.5. Start each* #7 of the two data sets with a quench from T = INF at zero applied field. The coarse clicks of 1.0 in the EXTERNAL FIELD slider define a convenient set of fields for the plots. What magnitudes of reversed field are required to reduce the magnetization to zero at these different temperatures? Why are they so different for the two temperatures?

Exercise 14.25 (M**) *Develop some experiments and use them to determine whether the hysteresis observed in the preceding exercise is principally due to pinning by the inhomogeneous field or by the barrier to the nucleation of reverse domains.*

14.4   Summary

In the Ising model, changing the sign of the exchange interaction J gives an antiferromagnetic ground state in which neighboring spins are oriented antiparallel to one another. Returning J to a positive value and with the additional introduction, with varying strengths, of long range dipolar interactions and an external magnetic field, we find an interesting variety of arrangements of the spins, or phases, which minimize the energy of the system. As with the analysis of ferromagnetism, the mean field theory gives useful insight into the behavior of the antiferromagnet, but not quantitative accuracy.

   The Ising model, with only nearest neighbor exchange, elucidates the microscopic nature of ferromagnetism. It is unable, however, to predict the existence of domain structure beyond that associated with the nucleation and growth of the ferromagnetic phase. It is the long range dipole–dipole interactions among the spins which drive the formation of domains. Inclusion of the dipole energy leads to an equilibrium picture of a domain structure in a ferromagnet, with a magnetization which responds reversibly to changes in an applied field. To explain the irreversible, or hysteretic, response of ferromagnetic materials requires the introduction of the ideas of domain nucleation, spatial inhomogeneities, and the pinning of domain walls.

14.A   Deeper exploration

Decoupled sublattices   Exercise 14.12 addressed the mean field picture of the behavior of a system in which the interaction between the two sublattices is weak, but in which individual sublattices are ferromagnetically coupled. This may be achieved by taking both J and D negative. An interesting starting point is to take $T = 1$, $J = -1$, and $D = -0.75$. Explore this system in detail.

Tablecloth phase   Look at the tablecloth phase found in Exercise 14.13 in terms of the mean field model. Now there are four atoms in the primitive cell of the ordered structure. Generalize the scheme used for the two-sublattice antiferromagnetic case to the tablecloth phase. You will need more than two sublattices because, in the ordered configuration, different spins of the majority type have different configurations of neighbors. Explore the phase experimentally as well as in the mean field picture. Include a plot of $M(H)$ from $H = 0$ to $H = 10$, with better resolution in the range 4.5–6.5. Why is the slope of $M(H)$ less near $H = 5.5$ than the average slope over the full range?

Long range exchange   With choices of the POWER LAW of 2 or 3, the EASY AXIS PERPENDICULAR, zero EXCHANGE J, and a negative DIPOLE D, "néel" simulates the case of a long range $1/r^2$ or $1/r^3$ ferromagnetic interaction among the spins. The long range of the interaction [60] should make the mean field approximation work better than for the case of pure nearest neighbor exchange. Does it?

Phase diagrams   With the choice available of both positive and negative values of J and D, there's a lot of parameter space to explore. The most obvious project is to identify the nature of the stable low temperature phases as a function of location in the JD-plane. The mean field picture will allow some predictions of relative stability of different arrangements of spins. Remember, from Exercise 14.14, there is a range of J/D in which neither the ferromagnetic nor the checkerboard phase is the one of lowest energy. Adding T and H as parameters offers a wide range of potential exploration: too wide a range. Explore the terrain briefly and then narrow down your objectives to a more careful investigation of limited scope.

14.B   Dipolar and field energies

14.B.1   "Magnetic poles"

From dipole–dipole sums, it is very difficult to make even qualitative arguments about the dependence of the dipole–dipole interaction energy upon the arrangement of magnetic domains. If the magnetization varies slowly with position on the scale of the lattice constant, the magnetic moment on a lattice site may be replaced by the product of a macroscopic magnetization and the volume of the unit cell. The dipole–dipole energy becomes

$$\mathcal{E}_{\text{dip}} = -\frac{\mu_0}{8\pi} \iint d\mathbf{r}\, d\mathbf{r}' \mathbf{M}(\mathbf{r}) \cdot \frac{3(\mathbf{r} - \mathbf{r}')(\mathbf{r} - \mathbf{r}') - 1|\mathbf{r} - \mathbf{r}'|^2}{|\mathbf{r} - \mathbf{r}'|^5} \cdot \mathbf{M}(\mathbf{r}'), \quad (14.4)$$

where **1** is the unit dyadic and μ_0 the permeability of free space. Though the physical system is assumed to be of finite extent, *all* integrals in this section are to be taken over all space. The characteristic dependence of the dipole interaction on the relative positions of the moments may be expressed as a double gradient of $1/|\mathbf{r} - \mathbf{r}'|$ to give

$$\mathcal{E}_{\text{dip}} = +\frac{\mu_0}{8\pi} \iint d\mathbf{r} \, d\mathbf{r}' \mathbf{M}(\mathbf{r}) \cdot \nabla\nabla' \frac{1}{|\mathbf{r} - \mathbf{r}'|} \cdot \mathbf{M}(\mathbf{r}'). \tag{14.5}$$

Note that the gradients are taken once with respect to each of the position variables **r** and **r**'.

The next step is to integrate by parts twice, once with respect to each of **r** and **r**', and to discard the surface integrals at infinity where M is zero, giving

$$\mathcal{E}_{\text{dip}} = +\frac{\mu_0}{8\pi} \iint d\mathbf{r} \, d\mathbf{r}' \, \nabla \cdot \mathbf{M}(\mathbf{r}) \frac{1}{|\mathbf{r} - \mathbf{r}'|} \nabla' \cdot \mathbf{M}(\mathbf{r}'). \tag{14.6}$$

This form is suggestive of Coulomb's law in electrostatics, with the divergence of **M** playing the role of a charge density. It is common to introduce the term *magnetic pole density* for $\nabla \cdot \mathbf{M}(\mathbf{r})$. Alternative representations of this energy are often useful. If magnetization terminates at a surface, then the discontinuous normal component of **M** is similarly interpreted as a *surface pole density*. (Note that we are suggesting a way of thinking in terms of a magnetic pole density, *not* implying that any real pole density exists!) This transformation from Eq. (14.4) to Eq. (14.6) allows the interpretation of the dipole energy in terms of volume and surface pole densities for which intuition already exists because of the direct analog with energies of charge distributions.

14.B.2   Poles and fields

In the electrostatic problem we are familiar with expressing the interaction energy of charges as a field energy. Here there is a similar argument. To start, note that the relations

$$\nabla \cdot \mathbf{B} = \mu_0 \nabla \cdot (\mathbf{H} + \mathbf{M}) = 0, \tag{14.7}$$

allow us to replace $\nabla \cdot \mathbf{M}(\mathbf{r})$ in Eq (14.6) by $-\nabla \cdot \mathbf{H}(\mathbf{r})$. The dipolar energy is thus expressed in terms of the divergence of the **H**-field as

$$\mathcal{E}_{\text{dip}} = +\frac{\mu_0}{8\pi} \iint d\mathbf{r} \, d\mathbf{r}' \, \nabla \cdot \mathbf{H}(\mathbf{r}) \frac{1}{|\mathbf{r} - \mathbf{r}'|} \nabla' \cdot \mathbf{H}(\mathbf{r}'). \tag{14.8}$$

We will consider only the case of no applied fields: i.e., the current density is everywhere zero. (The linearity of Maxwell's equations makes generalization to cases with currents straightforward.) From Maxwell's equations we know that zero current implies zero curl of **H** and therefore

H may be written as the gradient of a scalar potential, $\mathbf{H} = -\nabla\phi(\mathbf{r})$. We express one of the magnetic field terms in (14.8) using the potential ϕ to give the dipole energy as

$$\mathcal{E}_{\text{dip}} = -\frac{\mu_0}{8\pi} \iint d\mathbf{r}\, d\mathbf{r}'\, \nabla^2\phi(\mathbf{r}) \frac{1}{|\mathbf{r}-\mathbf{r}'|} \nabla' \cdot \mathbf{H}(\mathbf{r}'). \tag{14.9}$$

Two successive integrations by parts take this to

$$\mathcal{E}_{\text{dip}} = -\frac{\mu_0}{8\pi} \iint d\mathbf{r}\, d\mathbf{r}'\, \phi(\mathbf{r}) \nabla^2 \frac{1}{|\mathbf{r}-\mathbf{r}'|} \nabla' \cdot \mathbf{H}(\mathbf{r}'). \tag{14.10}$$

Using

$$\nabla^2 \frac{1}{|\mathbf{r}-\mathbf{r}'|} = -4\pi\delta(|\mathbf{r}-\mathbf{r}'|) \tag{14.11}$$

gives us

$$\mathcal{E}_{\text{dip}} = +\frac{\mu_0}{2} \int d\mathbf{r}\, \phi(\mathbf{r})\nabla \cdot \mathbf{H}(\mathbf{r}). \tag{14.12}$$

One final integration by parts yields

$$\mathcal{E}_{\text{dip}} = -\frac{\mu_0}{2} \int d\mathbf{r}\, \nabla\phi(\mathbf{r}) \cdot \mathbf{H}(\mathbf{r}) \tag{14.13}$$

$$= \frac{\mu_0}{2} \int d\mathbf{r}\, |\mathbf{H}(\mathbf{r})|^2. \tag{14.14}$$

The energy of the dipole–dipole interaction among the spins may be written in any of three forms: a dipole–dipole sum, a Coulomb's law integral of interacting volume or surface magnetic-pole densities, or a volume integral over all space of the square of the magnetic intensity **H**. Do not confuse this result with the more familiar statement that for linear media (i.e. $\mathbf{B} = \mu\mathbf{H}$) the magnetic energy density associated with a current distribution is given by $\mathbf{B}\cdot\mathbf{H}/2$.

14.C   "néel" – the program

"néel" simulates a two-dimensional spin system on a square lattice. The interaction between spins can be the Ising exchange interaction and, additionally, a dipolar interaction. The underlying Monte–Carlo algorithm equilibrates randomly picked spins with the effective field at their lattice site according to the Boltzmann weight.

14.C.1   "néel" algorithm

"néel" is an extension of "ising". Every spin on the 30×60 lattice interacts not only with the external magnetic field and, via exchange, with its four nearest neighbors, but also with *every* other spin in the lattice through

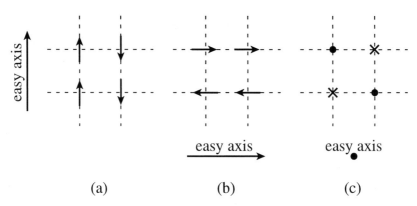

Figure 14.2: The allowed spin orientations for the three choices of EASY AXIS: (a) VERTICAL, (b) HORIZONTAL, and (c) PERPENDICULAR. The lattice is shown by the dashed lines, and the arrows give the spin orientation. (A • indicates that the spin is pointing out of the plane of paper, a × that it is pointing into that plane.) For $J = 0$, $H = 0$, and $D > 0$ the spin arrangement shown in each of the three cases is the ground state with the lowest possible energy.

a dipole–dipole interaction. Each spin can point either up $(S_i = 1)$ or down $(S_i = -1)$ along the specified EASY AXIS which can be either PERPENDICULAR to the lattice plane, or lie HORIZONTALLY or VERTICALLY within that plane. Figure 14.2 illustrates the three different choices for the EASY AXIS.

The energy $\mathcal{E}_i$ of a spin S_i is given by

$$\mathcal{E}_i = -S_i \left[J \sum_{\langle j \rangle} S_j + D \sum_{j \neq i} \frac{a^n}{r_{ij}^n} \cos(2\phi_{ij}) S_j + H + H_i \right] = -S_i H_i^{\text{eff}}.$$

(14.15)

Here the sum over $\langle j \rangle$ extends only over the four nearest neighbors of the spin S_i, while the second sum, over j, extends over all spins in the lattice except for S_i. The Ising and dipolar coupling constants are denoted by J and D. The distance between two lattice points i and j is r_{ij}, a is the lattice constant, and, depending on the choice of the POWER LAW, the power n of the dipolar interaction can be either 2 or 3. With ϕ_{ij} defined as the angle between the EASY AXIS and the vector connecting the lattice sites of the two spins i and j, $\cos(2\phi_{ij})$ is always -1 for the PERPENDICULAR choice of the EASY AXIS and the dipolar interaction favors antiparallel alignment of any pair of spins. For the HORIZONTAL or VERTICAL choices of the EASY AXIS, the dipolar interaction favors parallel alignment of two spins separated by a vector along the EASY AXIS or antiparallel alignment for a separation vector perpendicular to the EASY AXIS respectively, as indicated in Figure 14.2. In addition to the EXTERNAL FIELD H, a random magnetic field H_i along the EASY AXIS can be applied at each lattice site i. The H_i are randomly distributed according to a Gaussian distribution with a zero mean and a standard deviation SIGMA given on the panel. These random fields may be introduced to simulate system inhomogeneities.

Initial configuration   The INITIAL CONDITION for the spin configuration can be random (T=INF), fully magnetized upward (M=1), ordered in an alternating antiparallel manner (AFERRO), or a previously STORED configuration can be RECALLED. Note that CLICKING with the mouse cursor on a lattice site flips the spin at that site which allows any spin configuration to be created and STORED.

Dynamics   As in "ising", the probability for a spin to point up along the EASY AXIS ($S_i = 1$) is given by

$$p(S_i = 1) = \frac{e^{H_i^{\text{eff}}/T}}{e^{H_i^{\text{eff}}/T} + e^{-H_i^{\text{eff}}/T}}, \tag{14.16}$$

with the effective field H_i^{eff} at the lattice site i defined in Eq. (14.15).

To simulate the dynamics, "néel" computes the probability (14.16) for the given TEMPERATURE and effective field at a randomly chosen lattice site i. The spin at that site is left pointing upward with probability $p(S_i = 1)$ and downward with probability $[1 - p(S_i = 1)] = p(S_i = -1)$. One *sweep* consists of $N_x \times N_y$ such attempts to flip a randomly chosen spin, where N_x and N_y are the width and height of the spin lattice. The NUMBER OF SWEEPS for which "néel" runs can be specified on the panel. You also can press STOP to end a run at any time.

Every time a spin is flipped, "néel" updates the effective field of all other (1799) spins in the lattice. Because this requires a large amount of computation, "néel" will run more slowly when the spin lattice is out of equilibrium and many spins are flipped in every sweep. The speed of "néel" will increase near a low temperature equilibrium where only a few spins are flipped in each sweep, requiring fewer calculations. Away from equilibrium its speed is primarily limited by the computation rather than by the display. Hence, increasing the SPEED, in the CONFIGURE menu, to increase the number of sweeps per display, only gives a significant speed increase near a low temperature equilibrium.

Boundary conditions   In order to illustrate the effect of the dipolar energy in driving domain formation, "nèel" uses the straightforward *free boundary conditions* rather than the periodic boundary conditions used in "ising". For free boundary conditions, a spin S_i interacts only with

spins S_j *in* the lattice, *not* with their images which one gets from periodic translation of the lattice. For example, a spin in a corner of the
lattice has only two nearest neighbors (rather than four) for the exchange
interaction.

14.C.2 Displayed quantities

"néel" displays the spin lattice in the upper part of the panel. The graph
at the bottom shows the MAGNETIZATION versus time. The average magnetization and energy (dipolar and exchange contributions only) and their
mean-squared deviations are given below the control panel. These displayed quantities are the same as in "ising", see pages 270–273; and, as
in "ising", the averages are taken over the time interval since the most
recent RESET of the graph.

14.C.3 Bugs, problems, and solutions

When the graph is RESET, the current magnetization is taken as the zeroth
data point. Therefore, for the first point on the graph to be of significance,
 always first INITIALIZE the lattice and then RESET the graph!

15

"burgers" – Dislocations and plastic flow

Contents

15.1   Introduction

What determines the mechanical strength of materials? What is the microscopic mechanism that leads to the bridge collapse or the boiler explosion? We speak of *fracture* when something breaks catastrophically such as the breaking of a pane of glass. If there is deformation without breaking, as in the buckling of a bridge truss, we speak of *plastic flow*. In both cases the observed strength of materials is typically orders of magnitude lower than would be calculated from a perfect crystal model. Our concern in this chapter is only with plastic flow, permanent, stress-induced deformation of a crystal which leaves the crystal still intact.

Crystal imperfections and impurities frequently dominate the important properties of solid materials. We have already seen the critical role of impurities as a source of carriers in semiconductor materials; the active centers in solid state lasers are impurities or structural imperfections called color centers; and the hysteretic properties of magnetic materials are defined by impurities, grain boundaries, or precipitates. This chapter investigates dislocations, the structural defects which allow plastic flow of materials at stresses well below calculated yield strengths.

15.2   About the "burgers" model

Think of painting the atoms in the top half of a crystal blue and those in the bottom half red, the dividing boundary between the two being between two adjacent close-packed planes of the crystal. Failure by plastic flow is characterized by the shearing, across such a boundary plane, of the blue part of a crystal with respect to the red. We use the term *slip* to denote this type of deformation and the plane across which the deformation occurs is the *slip plane*.

"burgers" is designed to model many aspects of dislocation motion. It is difficult to illustrate three-dimensional crystals on a computer screen and time consuming for a program to execute the appropriate calculation. "burgers" compromises by modeling dislocation behavior in terms of only two planes, one on each side of the slip plane. The display in PRESET 1 shows the two planes of atoms adjoining the slip plane. The lower half of the crystal is represented by the single plane of red atoms, which are on a fixed infinite square lattice. The lattice constant of the red plane is a, and the red atoms are not allowed to move. The blue half of the crystal is replaced by a finite square $N \times N$ lattice of movable blue atoms, with the relaxed (unstrained) lattice constant c. The C/A RATIO slider sets the value of c relative to a. For most of this chapter the C/A RATIO is taken equal to 1.

1

The elastic rigidity of the blue half of the crystal is modeled by coupling the blue atoms to each other with harmonic nearest neighbor interactions with a nearest neighbor compressive force constant α and shear constant β. If **A** and **B** are the position vectors of two adjacent blue atoms, with atom B the right hand, nearest neighbor of atom A, then the harmonic interaction between them is written as

$$V_{A,B} = \tfrac{1}{2}\alpha(B_x - A_x - c)^2 + \tfrac{1}{2}\beta(B_y - A_y)^2, \qquad (15.1)$$

where α and β are given in units of (V_0/a^2) and V_0 is defined below.

Exercise 15.1 (M) *Suppose atom C is the nearest neighbor of A in the positive y-direction. What must be the form of the harmonic interaction between A and C in order to preserve the square symmetry of the problem?*

Note that this model for interactions within the blue plane is *not* equivalent to connecting the atoms together with relaxed springs. (Why not?) Finally, the interaction of each blue atom with the red plane is described by an *egg-carton potential*, which is, for the atom A,

$$V_{A,ec} = (V_0/2\pi)[-\cos(2\pi A_x/a) - \cos(2\pi A_y/a)], \qquad (15.2)$$

where V_0/π, the height of the saddle point separating two adjacent minima, defines the strength of the egg-carton potential.

There are some strange features of this model. First, the red substrate and the blue plane are treated very differently, while in a three-dimensional crystal there is an obvious symmetry between the two. Second, the interactions across the slip plane and within the upper plane are independently defined in "burgers", while in a real crystal they are, of course, closely related. Third, the picture of harmonic forces within a single plane of blue atoms does not properly represent the elastic energy of distortion of the three-dimensional half-crystal above the slip plane. These unrealistic features, though greatly simplifying the model, do not significantly alter most of our conclusions.

The model might look to you like a model for an atomic overlayer on the surface of a perfect crystal rather than a dislocation model. The resemblance will be even closer if you open PRESET 2 in which the LATTICE   # 2 CONSTANT c is set to give $c/a = 1.2$. Initially, with the perfect blue lattice set on top of the red one, many of the blue atoms are in positions of relatively high energy of interaction with the red atoms because of the misalignment forced by the disparity in lattice constants. When the program RUNS the blue atoms move to lower their energy of interaction with the red ones, but at some expense in strain energy within the blue plane. Ultimately the atoms find the best compromise, taking advantage of the egg-carton potential in which they rest while not straining the bonds of the blue structure too much. The final configuration may be described as an array of dislocations, but we're getting ahead of our story. You might enjoy changing the C/A RATIO to see what happens. From now on we will let $c = a$, and leave the overlayer problem for deeper exploration.

15.3   Perfect crystal

1   Returning to PRESET 1, we may try to shear the blue plane with respect to the red plane by applying a SHEAR STRESS σ across the pair of planes. Equivalently, it may be thought of as a force σa^2 on each of the blue atoms. A positive stress forces the blue atoms to the right. If you set the SHEAR STRESS to 0.7 V_0/a^3 and RUN, you see the blue atoms move slightly to the right. If you increase the SHEAR STRESS enough, however, the applied stress drags the blue atoms out of the red pockets, across the egg-carton and the perfect crystal is broken by shearing on the slip plane.

 Exercise 15.2 (M) *Find experimentally the value of the SHEAR STRESS σ which is just adequate to force the plane of blue atoms to move across the plane of red ones. This is the critical shear stress of the perfect crystal σ_0. Later, σ_0 will be a useful reference for comparison with the critical shear stress for a crystal containing a dislocation. Here, and later in this chapter, it is usually adequate to measure the critical stresses with an accuracy in the range of 5–10%. (In many of the exercises it helps to watch the <DELTA X> readout box in the middle of the panel. This indicates (see page 312 for details) the average displacement of the blue atoms in each step of the calculation and is a better indicator than the screen display of whether or not the system has really settled down.)*

Exercise 15.3 (M) *Show theoretically from Eq. (15.2) that the critical shear stress for the perfect crystal is $\sigma_0 = V_0/a^3$. (Hint: the quantity σa^2 is a force per square lattice constant or force per atom. How much force is required to move a blue atom out of a minimum of the egg-carton?) If $V_0 = 10$ eV and $a = 3$ Å, what is σ_0 in dynes/cm^2 (or gigapascals if you prefer)? Is this comparable with the values for the strength of perfect crystals as discussed in the textbooks?*

The puzzle is to understand why real crystals have critical shear stresses which are orders of magnitude smaller than this. The secret is to figure out how to distort the crystal locally in a few places so that there are atoms that are easier to push around *and* to arrange it so that by moving the location of this disturbed piece of the crystal the net effect is to shear the whole crystal. The dislocation is an ingenious structure which does just that. The introduction of the concept of dislocations and the

proposal of them as the source of low material strengths was a creative
and adventurous bit of physics.

15.4 Single dislocation

15.4.1 Dislocation geometry

PRESET 3 was constructed by moving the right hand half of the blue   # 3
plane to the left by one lattice constant without disturbing the left hand
half, except for some squeezing where they meet. The blue layer is everywhere in registry with the red except at the boundary between the
displaced and undisplaced portions of the blue plane. The line of misregistry between the red and the blue, or the boundary between the displaced
and undisplaced parts of the blue plane, is the *dislocation line*. This example, in which the displacement is perpendicular to the line separating
the displaced and undisplaced portions, shows the geometry of the *edge
dislocation*.

Exercise 15.4 (C) *Verify in* PRESET 3 *that there is an extra column of
atoms in the center of the blue plane relative to the red. In the three-
dimensional crystal this column corresponds to an extra half-plane of blue
atoms extending perpendicular to the screen, the more usual description
of the edge dislocation.*

Exercise 15.5 (M) *Determine the Burgers vector by taking a circuit,
called a Burgers circuit, enclosing the dislocation line: n steps of length a
in the blue plane across the dislocation line, down one step into the red, n
steps back across the line in the red, and then one step up into the blue.
By what vector did you miss the starting point? This, to within a sign,
is the Burgers vector* **b**.

The sign of the Burgers vector depends upon defining a convention
about whether to encircle the line in a clockwise or counterclockwise sense.
Relative signs are meaningful, absolute signs are not. For the dislocations
generated in "burgers" the magnitude of the Burgers vector happens to
be the lattice constant a, but this is not necessarily the case.

A warning: many pictures in books show cross-sectional pictures of
edge dislocations, pictures of bubble rafts, perspective drawings of screw
dislocations, etc., none of which correspond to the representation used by
"burgers". When you look at any of these, identify for yourself the three
entities: the dislocation line, the slip plane, and the Burgers vector. Be
sure you understand how they are oriented with respect to the plane of
the diagram. In "burgers", the slip plane is the plane between the blue
and red planes; the Burgers vector is one lattice constant in magnitude

and directed horizontally; and the dislocation line lies in the slip plane at a variety of angles depending upon the experiment. In many textbook pictures, we look *along* the dislocation line; in "burgers" we look *perpendicular* to it.

Exercise 15.6 (C) Run preset 3 *briefly to see the atoms settle from their initial positions into an equilibrium arrangement. Now adjust the* shear stress *(note that the* slider *has changed from a* coarse *to a* fine *control for this* preset*) and stress the crystal. You should be able to move the dislocation to the right or left by appropriately adjusting the* shear stress. *If a dislocation were to move from one side of the crystal to the other, by what amount would the blue half of the crystal have moved with respect to the red half?*

The motion of the dislocation in the slip plane is referred to as *dislocation glide*. When a dislocation glides completely through a crystal from one surface to the opposite, the crystal slips by one Burgers vector. If you lose the dislocation from the crystal you can write a new one by drawing a short horizontal arrow with the cursor with the right hand mouse button down. How does the dislocation structure differ when you draw the arrow to the left and to the right? We speak of dislocations of opposite sign.

\# 4 **Exercise 15.7 (C)** *The crystal in* preset 4 *contains a screw dislocation. Again, half of the blue plane has been displaced with respect to the other half. What is the direction of that displacement, the Burgers vector? What is the orientation of the boundary between the shifted and unshifted portions of the blue plane? Is the Burgers vector parallel to or perpendicular to the dislocation line? Decide which way the dislocation line will move when you apply a stress. Try it and see.*

Look again at the edge dislocation. Review the following imaginary prescription for its construction: slice the perfect crystal between the blue plane and the red substrate over the right hand half of the field. Leave the substrate fixed and push the right hand half of the blue plane one lattice constant to the left, squashing the atoms too close together in the middle. Then reattach the plane to the substrate as best you can. The region of poor matching is the edge dislocation.

Exercise 15.8 (C) *Reconstruct the preceding paragraph to make it appropriate for the screw dislocation: in translating, you must decide where to cut the crystal, how to distort it, and where to substitute top or bottom of the computer screen for right or left (or is it left or right?), in order to make the instructions appropriate for the screw dislocation. Note that there is one arbitrary choice of convention before the other answers are uniquely defined.*

The procedure we use for defining the edge dislocation differs from a common description involving the addition of an extra half-plane of atoms. Our picture of an incision, a relative displacement of the two sides of the cut, and reconnection has the advantage of working for edge, screw, and, in fact, for arbitrary orientations of the dislocation line with respect to the Burgers vector. Although we typically talk in terms of edge and screw dislocations, dislocation lines can run at any angle with respect to the Burgers vector.

Exercise 15.9 (C) *You can make dislocations yourself. With the right hand mouse button down, drag a short line in the middle of the display field at an angle θ with respect to the vertical. This will create a dislocation, perpendicular to the line that you dragged, oriented at θ with respect to the horizontal. Try a variety of orientations. Compare what happens if you drag to the left with what happens if you drag to the right. What is the difference between the dislocations you produce by these two methods? What about up versus down?*

We usually refer only to edges and screws for the simple reason that they are the two limiting cases that happen to be the easiest dislocations to visualize.

15.4.2 Dislocation width and energy

Figure 15.1 gives a cross-sectional view of the screen display, looking at columns of atoms, and the egg-carton potential in which they move. Though columns 1 and 7 are in good registry with the potential, because of an edge dislocation, which is roughly centered in the figure, there is one blue column too many to fit neatly into the intermediate wells. What determines the width of the dislocations made by "burgers"? That is, how far away from the core of the dislocation does the associated distortion extend, and how does this width depend upon system parameters? For the "burgers" model, for atoms away from the center of the dislocation, the displacements of the blue atoms from the minima in the egg-carton potential fall off exponentially with distance of the atom from the center of the dislocation.

Exercise 15.10 (M**) *Verify theoretically that the interplanar misregistry falls off exponentially with distance from the center of the dislocation. What is the characteristic decay length of that exponential? (Hint: write the condition for equilibrium of an atom in terms of its position relative to its neighbors and relative to the egg-carton. Show that an exponential solution satisfies this condition for all atoms as long as the displacements of the atoms from the egg-carton minima are small.)*

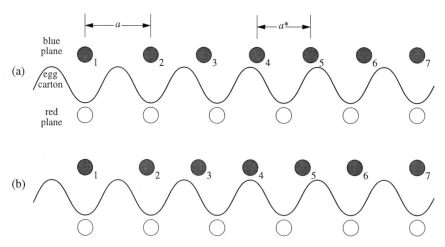

Figure 15.1: Cross-sectional view of an edge dislocation, looking along the dislocation. Each circle corresponds to a vertical column of blue atoms in the "burgers" display and the minima of the egg-carton potential are the positions of the red atoms. (a) The dislocation is centered half way between columns 4 and 5. (b) The dislocation, shifted a half lattice constant to the left, is now centered on column 4.

Note that this exponential decay is the most serious way in which the "burgers" model differs from the three-dimensional crystal. In a three-dimensional crystal there is no characteristic length defining the size of the dislocation: the associated strain field falls off as $1/r$, not exponentially. However, the exponential decay *does* give an appropriate description of the atom displacements for the misfit dislocations illustrated in the overlayer problem by PRESET 2.

The calculated exponential decay length gives a reasonable measure of the width for the "burgers" dislocation. An alternative approach, which should remind you of the treatment in Kittel [6, p. 472] of the Bloch domain wall, focuses attention on the energies involved in the distortions produced by the dislocation. With PRESET 3 watch the screen with "burgers" running as you vary the force constants ALPHA or BETA. Do the same for the screw dislocation using PRESET 4. We see an obvious correlation between the width of the dislocation and the strength of the coupling between atoms in the blue plane. Can we convert that observation into a semiquantitative argument?

3

4

In Figure 15.1(a), all columns to the left of number 3 and to the right of number 6 have atoms at the normal lattice sites. Between 2 and 7 the columns are crowded with nearest neighbor distance $a^* < a$ to accommodate the "extra plane of atoms" (an extra column of atoms in "burgers"). This configuration has an energy higher than that of the perfect crystal for two reasons. First, the five springs (force constant α) from 2 to 7 are compressed to give a strain energy within the blue plane. Second, the atoms 3–6 are displaced from the minima in the egg-carton, giving an increase in the potential energy associated with the binding between the planes.

Exercise 15.11 (M*) *Estimate roughly the magnitudes of these two energies, without regard to numerical coefficients, in terms of α and V_0, and in particular how each would depend upon n, the number of atoms in the disturbed region. Minimize this energy with respect to n to predict the dependence of the width on the force constant α, and to get an estimate of the energy per unit length of dislocation line, which we will call γ.*

Both methods give for the number of atoms within the width of the dislocation

$$n \approx \sqrt{\pi \alpha a^2 / V_0}. \tag{15.3}$$

Apart from a numerical factor, the *energy per unit length of line γ* is given in terms of the stiffness of the blue layer *alpha* and the strength of the coupling to the red layer V_0 by

$$\gamma \approx \sqrt{\alpha V_0 / \pi}. \tag{15.4}$$

It is useful to interpret γ also as a *line tension*, just as an energy per unit area of a surface or interface is a *surface* or *interface tension*.

Exercise 15.12 (C) *Explain qualitatively why increasing the force constants within the blue plane should increase γ, the energy per unit length. Explain why* ALPHA *is important for the edge dislocation and* BETA *for the screw.*

Exercise 15.13 (M*) *Measure the width of the edge dislocation for several values of* ALPHA *and confirm the predicted square root dependence (15.3) of n upon α.*

Exercise 15.14 (C*) *Open* PRESET 5 *which gives a dislocation oriented*   # 5
at 45° with respect to the crystal axes. What happens when you RUN *the program? Why does the dislocation move in the absence of any applied stress? Interpret the behavior in terms of the line tension γ.*

15.5   Critical shear stress with dislocations

You have already seen that application of a shear stress to a crystal containing a dislocation induces motion of the dislocation, and that the motion of the dislocation results in a displacement of the two halves of the crystal with respect to one another. How much stress does it take to get a dislocation to glide in its slip plane?

Exercise 15.15 (M) *With an edge dislocation in the crystal, as in* PRE-SET *3 but with different values of* ALPHA, *find experimentally the yield stress* σ_d, *the minimum stress at which the dislocation moves, for the three cases,* ALPHA $= 2, 5$, *and* 8. *Just above the yield stress the dislocation moves slowly: it takes a number of computer cycles before the displacement is large enough to displace an atom by a single pixel. (The behavior of the* <DELTA X> *readout is a better indicator of whether the system is above or below the critical shear stress. Have patience and try to get data to an accuracy of the order of* 5%.)

3

It's not obvious that it should take any stress at all to move the dislocation. To move the dislocation one half lattice constant to the left, individual atoms need move only a very small fraction of a lattice constant. Figures 15.1(a) and (b) show the column arrangements for two positions of the dislocation separated by $a/2$. In Figure 15.1(a), the center of the dislocation is the midpoint between columns 4 and 5 and is symmetrically disposed about a valley of the potential. In Figure 15.1(b), the dislocation is centered on column 4 and is centered on the top of the potential barrier. The configuration of the column with respect to the substrate potential has changed, implying a minor change in the magnitude of the line energy γ. Thus the energy of the dislocation varies periodically by a small amount as the dislocation moves through the lattice. Estimation of this energy variation requires a much more careful analysis than our estimate of the line energy γ. This periodic potential for dislocation motion, with the periodicity of the lattice, is called the *Peierls potential*.

Exercise 15.16 (C*) *Running* PRESET *3 you can see qualitative evidence for the Peierls potential. The* PRESET *opens with the dislocation in the arrangement of Figure 15.1(b). Describe what happens as the program is allowed to* RUN *in the absence of any applied stress, and interpret in terms of the Peierls potential. (To reinitiate the sequence, simply load the* PRESET *again.)*

There are two ways to approach the problem of estimating the shear stress σ_d required to move a dislocation through the crystal. One involves analyzing all the forces on all of the atoms and is not very practical. A simple alternative is to think in more macroscopic terms, with focus on the dislocation. To move the dislocation through the crystal requires a force acting on the dislocation sufficient to move it past the point of steepest slope of the Peierls potential, the point at which the Peierls potential provides the greatest force on the dislocation line. For the special case considered by "burgers", an applied stress parallel to the Burgers vector, the interaction of the applied stress with the dislocation line is conveniently expressed by the statement: the applied stress gives a force per unit length of line, directed perpendicular to the line, of magnitude σb. To show this, the key idea is to relate the work done by the externally applied stress in shearing the crystal by a distance b, the Burgers vector, to the work done by the force F on the dislocation line, as the dislocation glides from one side of the crystal to the other.

Exercise 15.17 (M) *Complete this energy argument to show that an applied stress σ acting on a macroscopic crystal implies an effective force per unit length, or two-dimensional pressure, acting on a dislocation line within the crystal of magnitude $F/L = \sigma b$.*

At the critical shear stress for dislocation motion σ_d the force per unit length, $F/L = \sigma b$, is just equal to the maximum force provided by the Peierls potential.

Exercise 15.18 (C**) *Construct a qualitative argument to explain why the yield stress for motion of an edge dislocation decreases as the value of the force constant α increases. (Hint: recall the variation with force constant of the width of the dislocation and stare at Figure 15.1.)*

There is an even easier way to move a dislocation. PRESET 6 contains a dislocation written at an angle of about 5° with respect to the vertical. The initial configuration looks more or less like such a line. After the line anneals at zero stress, it looks more like two segments of edge dislocation along two neighboring minima in the Peierls potential, with a *kink* in the middle connecting them. (You may be tempted to use the word *jog* instead of *kink*, but don't. A *jog* refers to a step in a dislocation line as the line moves from one slip plane to another; a *kink* refers to a step in the line which leaves the line in the same slip plane. The distinction between the two is important.)

6

Exercise 15.19 (M) *For the same set of ALPHA values 2, 5, and 8 used in Exercise 15.15, find experimentally the critical shear stress σ_k for the motion of the kink. Explain qualitatively why it is much smaller than the stress required to move a clean edge dislocation.*

The discovery of the dislocation was critical to the understanding of the strength of solids, or rather of the weakness of solids. It gave a way to understand why material mechanical strengths were so far below theoretical estimates.

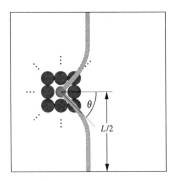

Figure 15.2: A dislocation with an applied stress pinned at an impurity.

15.6   Impurity pinning

Unfortunately we've overshot the mark. If kinky dislocations were the whole story, we'd have the new problem that most materials are much *stronger* than predicted by the models. To rescue the theory we must introduce a mechanism to impede the motion of the dislocations. In PRESET 7 the edge dislocation runs through an impurity atom which is smaller than the atoms of the host crystal. Vary the applied stress to see the pinning of the dislocation by the impurity. With enough stress the dislocation ultimately breaks away from the impurity.

7

Exercise 15.20 (M) *Be sure* ALPHA *is set to* 8 V_0/a^2 *and for* PRESET 7 *determine experimentally the critical shear stress* σ_p *for pulling the dislocation off the impurity. How does this compare with the critical shear stress* σ_d *for dislocation motion in the pure crystal as determined earlier, Exercise 15.15, for the same value of* ALPHA*?*

Exercise 15.21 (C**) *Note that the only effect of the impurity was to distort the arrangement of nearby atoms in the blue plane; its interaction with the substrate is the same as that of the blue atoms. Explain on energetic grounds why this still results in pinning of the dislocation.*

Figure 15.2 illustrates the "burgers" display with an edge dislocation and an applied stress. In Eq. (15.4) we introduced the line tension, or energy per unit length of the dislocation line γ. In the configuration illustrated, there is a line tension γ in each of the two segments of line; together they exert on the pin a force to the right of magnitude $2\gamma\cos\theta$. The applied stress, remember, exerts a force per unit length on the line, perpendicular to the line, of magnitude σb. Integrated over the length of the two line segments, the horizontal component of the stress driven force on the full line is σbL where L is the length of the line (the height of the display on the screen).

Exercise 15.22 (M*) *Complete the argument to show how to deter-
mine the line energy* γ *from a measurement of* θ *in a known stress* σ. *For*
PRESET 7, *with* IMPURITY SIZE $= -1$ *and* ALPHA $=$ BETA $= 8\ V_0/a^2$, *find
the* SHEAR STRESS *that gives the angle* θ *in Figure 15.2 to be about 60°.
From this determine the line tension* γ. *Compare this experimentally determined value with the value* $\gamma \approx (\alpha V_0/\pi)^{1/2}$ *calculated earlier from our
theoretical argument. (Please forgive the strange* IMPURITY SIZE *of* -1:
it happens to give a useful pinning strength.)

Not surprisingly, different impurities have different pinning strengths
depending upon their degree of misfit with the host crystal. We can
define the pinning strength F_p of an impurity in terms of the experiment
of Figure 15.2 by the relation $F_p = 2\gamma \cos \theta_{\max}$ where $\theta_{\max}$ is the value of
θ at which the dislocation is just able to break away from the impurity.

Exercise 15.23 (M*) *Find experimentally the pinning strengths for a
few* IMPURITY SIZES *in the range* -2.5 *to* 1 *with* ALPHA $=$ BETA $= 8\ V_0/a^2$.
*Explain why it is not possible for an impurity to have a pinning strength
greater than* 2γ.

Exercise 15.24 (C) *Change the* IMPURITY SIZE *to* 3, *an impurity larger
than the host. What would you predict for the effect on the pinning
strength of changing the nature of the impurity from pulling in its neighbors to pushing them away? Will that change the interaction of the edge
dislocation with the impurity from an attractive interaction to a repulsive
one, or not? Try it and see. What happens if you put in a dislocation of
the opposite sign? Be sure you can argue for whatever it is that happens.*

The ability of impurities to pin dislocations reflects both their pinning
strength and their concentration, since putting the impurities closer together along the line reduces the length of line L that any particular
impurity must pin.

Exercise 15.25 (C) *Add a dozen or so impurities at random by clicking
on the blue atoms. (A second click converts the impurity to a vacancy.
A click on the perimeter, but not the interior, of the vacancy converts it
back to a normal blue atom.) What shear stress is required to move a
dislocation across the array of impurities relative to the critical stress for
depinning from a single impurity of the same strength?*

Thus we begin to see what goes on in real materials and the importance
of impurities or precipitates or entanglements of dislocations or other
disturbances to the nearly perfect crystal, which can pin dislocations and
strengthen the material. Much of the art of controlling the mechanical
properties of materials concerns developing procedures to control densities
and distributions of dislocations and dislocation pinning sites.

15.7   Dislocation interactions

Dislocations interact not only with impurities and other crystal imperfections but also with each other. This is a source of the work hardening of materials. We can see one example of the interactions here.

8   **Exercise 15.26 (M)** PRESET 8 *opens with two edge dislocations of the same sign. Alternatively you can introduce a pair of dislocations of opposite sign by dragging once to the right with the right hand mouse button and once to the left. How does the attraction/repulsion depend upon the relative signs of the two dislocations? Postdict this behavior on energetic grounds. Follow a Burgers circuit, as described in Exercise 15.5, which encloses two dislocation lines of opposite sign. What is the Burgers vector of the pair? What if the two dislocations have the same sign? Do you expect dislocations to be attracted to or repelled by the surface of the crystal? What is the evidence for this?*

15.8   Frank–Read source

Our argument about the strength of materials has a gaping hole in it. Why doesn't a crystal quickly get stronger as dislocations left in it from the growth process are swept out by the applied stress. Clearly the full story requires the presence in the crystal of sources of new dislocations, and there are quite a few. "burgers" gives the opportunity to illustrate a particularly ingenious idea, the *Frank–Read source*.

9   In PRESET 9 a short row of vacancies has been inserted into the blue plane. It is equivalent to a short length of an edge dislocation which is firmly pinned at the two ends of the vacancy string. Increase the stress and the dislocation bows out in response to the stress. Again, think of the applied stress as equivalent to a force per unit length, or two-dimensional pressure σb. Remember the plastic ring you dipped into the soap solution when you were a child? When you blew gently on the film, it bowed out, the first step in making a soap bubble. This is the same thing as applying a stress to the dislocation line, pinned at each end, but in one more dimension. The radius of curvature of the dislocation line (the soap film) is determined by a balance of the force per unit length (pressure) tending to bow out the line (film) and the line tension (surface tension) tending to keep it straight (flat).

Exercise 15.27 (M**) *Show that the line tension γ and the force per unit length σb are related by $\gamma = \sigma b R$ where R is the radius of curvature of the dislocation line. (Hint: see if your freshman physics book talks about the pressure inside a soap bubble of radius R and its relation to*

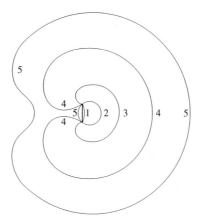

Figure 15.3: The Frank–Read source showing five successive stages of the generation of a dislocation loop.

the surface tension of the soap film, the law of Laplace. Or think of a dislocation ring of radius R as two semicircles and look at the two points at opposite ends of a diameter where they are tied together. The line tension is pulling the two semicircles together with a force related to γ. The applied stress is pushing them apart with a force related to σb.)

Exercise 15.28 (M*) *Determine γ from the radius of a stable bow you generate with* PRESET 9 *of "burgers" and the relation* $\gamma = \sigma bR$. *Be sure to use a stress small enough to leave the system in equilibrium. Check that* ALPHA *is equal to* $8\,V_0/a^2$ *to allow a valid comparison with the experimental value of γ determined from the contact angle with the pinning impurity, Exercise 15.22. How does the value compare with the two estimates you might have made of γ from Eq. (15.4) and from Exercise 15.22?*

With large enough stress, the bowed loop breaks away from the line of vacancies which can then serve to initiate a new loop. A Frank–Read source can continue generating and sending out *dislocation loops* indefinitely. Figure 15.3 is a sketch of five stages of the generation of a loop by the source. Check the illustration in references [6, p. 601] or [63, p.170] for a beautiful picture of the real thing! STOP the expansion at several different times in the cycle and establish for yourself that the material inside the loop has sheared by one burgers vector with respect to that outside the loop. Be sure you understand how the system passes from stage 4 to stage 5 in Figure 15.3. The key is to determine the sign of the two dislocation segments of stage 4 which are about to come together to the left of the source. See if you can STOP the simulation just before this point and determine the signs of the two screw dislocations segments that are about to annihilate one another.

15.9   Summary

Most materials are weaker than suggested by naive calculations based on
microscopic modeling of a perfect crystal. One mode of failure is plastic flow in which the material deforms without actually breaking. The
dislocation is an interesting structural defect, a line of disturbed material in an otherwise perfect crystal. It is constructed in such a fashion
that in moving through the crystal it leaves in its wake a perfect crystal
which has been sheared by one Burgers vector with respect to the original.
Since dislocation motion provides the microscopic mechanism for plastic
flow, the key to developing stronger materials is to develop well-controlled
methods to introduce defect structures which can pin the dislocations and
prevent their motion.

15.A   Deeper exploration

Dislocation loop   Explore stability issues with a single dislocation loop.
With a large array, perhaps 30×30 or larger with a fast computer, make
a single dislocation loop. Either use a Frank–Read source to make a
loop, STOP the program, replace the source vacancies by atoms, and then
proceed; or make a tic-tac-toe pattern of four dislocation lines, with appropriate signs, and anneal briefly. Find a stress for which the circular
loop neither grows nor collapses. How does the stress for a static loop depend upon the loop radius? Is this condition stable against small changes?
Discuss what happens if you make ALPHA and BETA different, and why.

Better value of γ Program a computer to improve upon the estimates
of Exercise 15.11. Using a variational approach, guess a dependence on
position of the displacements of the columns from the minima: possible
guesses could be $a[(1/2) + (1/\pi)\tan^{-1}(\lambda n)]$ or $(a/2)[1 + \tanh(\lambda n)]$ where
n is a column index measured from the center of the dislocation, and λ
a constant chosen to minimize the energy. Use a slight generalization to
determine the amplitude of the Peierls potential and try to predict the
critical shear stress σ_d for dislocation motion in "burgers".

More on kinks   A convenient way to explore the behavior of kinks is to
pin a dislocation at the top and bottom boundaries of the display by
adding an impurity or two at each edge. With one directly above the
other, the trapped dislocation is pure edge. If one is displaced laterally
with respect to the other the dislocation makes an angle with respect
to the vertical and, with small ALPHA and hence a high Peierls barrier,
appears as edge segments and kinks. There is now a range of SHEAR

STRESS for which the kinks move but the dislocation remains pinned. This behavior is referred to as *preyield microplasticity* [63, p.218]. Find out more about it.

15.B "burgers" – the program

"burgers" simulates a model system which is designed primarily to illustrate many of the characteristics of dislocations in solids, and which incidentally displays features of surface overlayers. It makes no attempt at quantitative resemblance to real systems, but has the flexibility to illustrate a variety of physical ideas.

15.B.1 Model system for dislocations

The "burgers" model represents two simple square arrays of atoms interacting across the plane that separates them. The blue atoms in the upper plane interact with their four nearest neighbors through harmonic potentials with an equilibrium spacing of a lattice constant c. The bottom plane of atoms is rigidly fixed in space with lattice constant a. The two planes interact via an *egg-carton potential* defined below.

The forces connecting the blue atoms have not only a longitudinal force constant α, but also a transverse force constant β. For example, the interaction energy of an atom A with its neighbor to the right B is given by Eq. (15.1) on page 297. The constants α and β can be thought of in terms of bulk and shear moduli, respectively. In the dimensionless form used by the computer, the equation takes a form similar to Eq. (15.1),

$$V_{A,B} = \tfrac{1}{2}\alpha(B_x - A_x - c)^2 + \tfrac{1}{2}\beta(B_y - A_y)^2. \tag{15.5}$$

The interaction of the atoms in the top layer with the substrate layer is modeled as a sinusoidal *egg-carton potential*, in which the top layer of atoms move in a fixed periodic potential with minima over the positions of the red atoms. The energy of an atom A moving in the egg-carton is

$$V_{A,\mathrm{ec}} = -\frac{1}{2\pi}\left[\cos(2\pi A_x) + \cos(2\pi A_y)\right]. \tag{15.6}$$

Eq. (15.6) differs from Eq. (15.2) by the deletion of the potential scale factor V_0 and by the use of a dimensionless position vector $\mathbf{A}$.

Finally a shear stress σ between the top layer and the substrate layer can be applied to shift the atoms in the positive x-direction. The shear force F on atom A may be expressed in terms of an energy,

$$V_{A,\sigma} = -FA_x. \tag{15.7}$$

The normalizations of the dimensionless egg-carton and shear force energies are chosen such that the critical shear stress for the perfect crystal corresponds to unit force $F = 1$.

The dimensionless forms are natural to the numerical calculations of the computer. Physical values are recovered by assigning the units V_0/a^2 to the force constants α and β, a to the position components, $A_x, A_y, ...,$ and relating the physical shear stress σ to the dimensionless force F by $\sigma = (V_0/a^3)F$.

The evolution of the atom positions is developed by an iterative energy minimization procedure. One cycle of the iteration consists of the following steps for each of the atoms in turn. With atom A as an example:

1. Use the current positions of the neighbors of A to calculate the sum of the four near-neighbor $V_{A,B}$-like terms. Determine the value $\mathbf{A}_m$ of the position vector $\mathbf{A}$ which minimizes the sum, and of the curvatures k_x and k_y of the sum of the $V_{A,B}$s.

2. Calculate the force $\mathbf{f}$ on atom A from the gradient with respect to $\mathbf{A}$ of the egg-crate potential $V_{A,\text{ec}}$ evaluated at its current position $\mathbf{A}_{\text{old}}$, plus the force F in the x-direction due to the applied stress.

3. Calculate a target position with the values $A_{x\text{tar}} = A_{xm} + f_x/k_x$ and $A_{y\text{tar}} = A_{ym} + f_y/k_y$.

4. Update the position vector to $\mathbf{A} = (1 - S)\mathbf{A}_m + S\mathbf{A}_{\text{old}}$, where S is the STEPSIZE FRACTION available in the CONFIGURE dialog box.

Increasing the STEPSIZE FRACTION speeds up the computation; decreasing it improves its stability. The iteration continues until STOP is clicked.

The finite screen resolution often can obscure slow changes in a configuration. The average of the atom displacements in the x-direction is computed for each cycle, and given in the <DELTA X> readout. This readout provides a means, more sensitive than the display of the atom array, to follow the progress of the simulation.

15.B.2 Impurities and vacancies

Impurity atoms may be substituted into the blue plane in "burgers", by CLICKING on the blue atoms. The impurity is coupled to its neighbors with the same force constants α and β, but the equilibrium distance to the neighbors is changed to the value $\frac{1}{2}[c + (\text{IMPURITY SIZE}) a]$. For an IMPURITY SIZE of 1 (and $c = a$) the impurity is identical to the host atoms. The program will accept negative values of IMPURITY SIZE if you wish to construct stronger pinning. Finally, the egg-carton potential is the same for the impurity as for the blue host.

The modeling of vacancies is slightly more complex. If one of the neighbors of an atom is found to be a vacancy, then the spring is attached not to the vacancy, but to the atom just beyond the vacancy. This simulates a short range attractive force. If the atom just beyond is also a vacancy, then the farther atoms are assumed to be out of range, and the spring is ignored for that atom.

15.B.3 Displayed quantities and controls

"burgers" displays a lattice of circles representing the positions of the atoms. Normal atoms in the plane above the slip plane are blue, impurities are yellow, and vacancies are transparent with an outline. The positions of the minima of the egg-carton potential are indicated by a lower layer of red atoms.

Sliders allow the user to change ALPHA, BETA, the C/A RATIO, the IMPURITY SIZE, and the SHEAR STRESS in the x-direction. A selection menu allows the sensitivity of the SHEAR STRESS slider to be changed conveniently. The SPEED slider controls the number of numerical cycles per display cycle. A low value for SPEED will result in more display cycles, giving a smoother motion of the atoms but a slower simulation speed.

The RUN/STOP button starts and stops the simulation, the atoms can be set back to their initial starting points by clicking on the appropriate PRESET number, and all dislocations may be removed from the array with the REMOVE DISLOCATIONS button. The size of the lattice and the STEPSIZE FRACTION can be controlled via the CONFIGURE menu.

Atoms can be changed from normal to impurity to vacancy and back to normal by consecutively clicking on them with the left hand mouse button. To change a vacancy to a normal atom it may be necessary to click near the outline of the vacancy.

Dislocations can be created by clicking with the right hand mouse button and dragging to form an arrow on the screen. The dislocation is formed perpendicular to the arrow. The initial width of the region of distortion is inversely proportional to the length of the arrow, hence a short arrow makes a wide dislocation. Drawing a vertical arrow creates a screw dislocation; drawing a horizontal arrow creates an edge dislocation. Dragging the arrow in the opposite direction, or in the same direction with the shift key down, creates a dislocation of opposite sign.

15.B.4 Bugs, problems, and solutions

The help system will not work while the program is still running. Try to avoid calling on HELP without first STOPPING the program. If you forget, simply STOP the program, click on the HOME button on any HELP page, and the system will work again.

Epilogue

Eight of the fourteen Solid State Simulations use random numbers generated by `ranmar`, a program given to us a few years back by Tom, then a graduate student at Cornell. Before distributing `ranmar` with the SSS package, it only seemed fair to check with him whether others could use his truly random random numbers.

One Friday morning we sat down to hunt for Tom, knowing only that he had moved to Munich, Germany. Neither Mark nor Badri, both of whom use `ranmar`'s random numbers on a daily basis, had his email address. It was Tatjana (who seems to have no need for random numbers) who knew Tom's electronic whereabouts. So we e-mailed Tom asking for his permission to distribute `ranmar`. Minutes later we had his reply:

> To tell you the truth, I didn't write ranmar, and though I'd be more than delighted to take credit for it, I got the ranmar source code years ago from Gerard (who I'm fairly sure grabbed it from somewhere else, etc). You could email him about it.

Tom also referred us to Paul, who used `ranmar` in one of his programs and therefore should surely know who the original author was. With two hot leads, we began to enjoy our new role as internet detectives. We e-mailed Gerard, asking if he knew the creator of `ranmar`, and before we could even finish our internet search for Paul, we had Gerard's reply:

> I got the code of ranmar from a CERN data division report, sometime in 1990 or so. I think that the author of this report was F. James. This report was a review of a couple of random number generators.
>
> My strong guess is however that F. James is not the author of ranmar, but George Marsaglia; the reason for this is the last three characters of the name of this random-number generator. Email him. He will surely know whether he is the author or not, and otherwise probably know the author.

Gerard turned out to be more of a detective than we. His guess that the author might have incorporated his last name into the code seemed

314

cunning. We dropped George a short email inquiry. While waiting for his reply, our own internet search succeeded: the program library at the European high energy physics center CERN revealed under the unsuspicious internet address

`http://www.hep.net/wwwmirrors/cernlib/docs/www.dir/shortwrups/`

the document `H2shortwrupsV113.html`, which stated that, on June 8 1989, F. Carminati and F. James had submitted a program called `ranmar`, written by G. Marsaglia and A. Zaman, to the program library. We were close. Would Marsaglia admit to having written the code?

4:34 pm that Friday afternoon the email beeped again. Marsaglia had replied. Had our quest come to an end?

> Yes, I created ranmar, and it is good, but generators based on my multiply-with-carry are better.
>
> If you send a mailing address I will send The Marsaglia Random Number CDROM which contains 5 billion random bits as well as the DIEHARD battery of tests of randomness.

We certainly had found the right guy. But had we found the right random number generator?

This spirit of collaboration and generosity has characterized every stage of the Solid State Simulation project. We would like to thank all those who have, directly and indirectly, contributed to its success.

Appendices

Appendix A

How to run and install the SSS programs

Contents

A.1   Information for all platforms

The SSS are distributed on a CD-ROM. For MacOS and Windows platforms you can run the simulations directly from the CD-ROM; you also can install the simulations on a hard disk drive for increased performance if you wish. For versions of Unix, the simulations need to be installed on a hard disk.

For all platforms, files and instructions are provided to allow modification of the programs and recompiling if you have available an appropriate compiler for your platform.

Many references in the HELP files and in this guidebook are made to the three (LEFT, MIDDLE, and RIGHT) mouse buttons which are standard for Unix systems. If your mouse has only a single button, hold down the CONTROL key when you press the mouse button when asked to click the RIGHT button. If instructed to click with the MIDDLE mouse button, for a single- or two-button mouse, hold down the SHIFT key when you press the single or left mouse button.

For additional information about the installation procedure, or to resolve installation problems, consult our Web pages:

$$\texttt{http://www.ruph.cornell.edu/sss/sss.html}$$

A.2   Microsoft Windows

If you have Windows 3.1 or Windows for Workgroups 3.1 *without* the win32s subsystem, go *first* to Section A.2.4 to install win32s.

A.2.1   Windows 3.1 with win32s, Windows for Workgroups 3.11 with win32s, or Windows NT-3.51

To run the SSS from the CD-ROM:

1. Load the SSS CD-ROM into the CD-ROM drive. Start the FILE MANAGER by double-clicking on its icon in the MAIN group in the PROGRAM MANAGER.

2. Open a FILE MANAGER subwindow for the CD-ROM by double-clicking on the CD-ROM drive's icon at the top of the FILE MANAGER window. You should see a top level subdirectory on the CD-ROM called `sss`.

3. Change to the `sss` directory by double-clicking on it. You should see two subdirectories to `sss`, namely `winbin` and `lib`.

4. Change to the `winbin` subdirectory by double-clicking on it. You should see the 14 SSS such as `bloch.exe` or `born.exe`. To run one of the simulations, double-click on it.

To install the SSS on your hard disk and run them from there:

1. Follow steps 1 and 2 above so that you see a top level subdirectory on the CD-ROM called `sss`.

2. Open a second FILE MANAGER subwindow for the hard disk partition where you wish to install the simulations by double-clicking on that drive's icon at the top of the FILE MANAGER window.

3. Drag the `sss` directory from the CD-ROM window to the directory in the hard disk window where you wish to install the Solid State Simulations. You may have to confirm your choices.

4. To run the simulations, just open the `winbin` subdirectory of the `sss` directory on your hard disk and double-click on the desired simulation program.

A.2.2   Windows 95 or Windows NT4

To run the SSS from the CD-ROM:

1. Load the SSS CD-ROM into the CD-ROM drive. Start the (NT) EXPLORER by selecting it from the START menu.

2. Change to the CD-ROM by double-clicking on its icon in the EXPLORERand find the top level subdirectory on the CD-ROM called `sss`.

3. Change to the `sss` directory by double-clicking on it. You will see two subdirectories to `sss`, namely `winbin` the `lib`.

4. Change to the `winbin` subdirectory by double-clicking on it. You should see the 14 SSS such as `bloch.exe` or `born.exe`. To run one of the simulations, double-click on it.

To install the SSS on your hard disk and run them from there:

1. Follow steps 1 and 2 above. You should see a top level subdirectory on the CD-ROM called `sss`.

2. Open a second (NT) EXPLORER subwindow for the hard disk partition where you wish to install the simulations by double-clicking on that drive's icon at the left of the (NT) EXPLORER window.

3. Drag the `sss` directory from the CD-ROM window to the directory in the hard disk window where you wish to install the SSS. You may have to confirm your choices.

4. To run the simulations, just open the `winbin` subdirectory of the `sss` directory on your hard disk and double-click on the desired simulation program.

A.2.3   Installation and running from a hard disk for all versions of Windows

The simulations rely on the `lib` and `winbin` directories being subdirectories of `sss`. If you rename `sss` or `lib` the simulations will probably cease to work. If you wish to move the executables out of the `winbin` directory, you must set the environment variable `SSS_DIR` to the full path of the `sss` directory in order for the simulations to run.

The minimal installation will require 10-15 MB of disk space. Alternatively, you may instead move the `compile` directory from the CD-ROM to your hard disk, which requires an extra 6 MB of disk space. `compile` contains, in addition to the files in the top-level `sss` directory, resources needed for modifying and recompiling the programs including the `src` directory which contains files with the source code for the programs, of interest to those concerned with the details of the algorithms used in the simulations. The additional files are *not* required to run the current versions.

 Note that some of the simulations use a MIDDLE mouse button. Under Windows, holding down the SHIFT key together with using the left button simulates the MIDDLE mouse button.

A.2.4   Installing win32s: Windows 3.1 or Windows for Workgroups without win32s

If you have a Windows 3.1 or Windows for Workgroups 3.11 system, and if you have *not* installed the win32s subsystem on your computer, then you will need to install the win32s subsystem before you can run the simulations. You only need to do this installation once. It should take about 2 MB of hard disk space.

Version 1.3 of the win32s subsystem, the latest version available at press time, is included on the CD. If you wish to install a later version of win32s, you will need to obtain it from Microsoft. You can find information about this by searching for win32s on Microsoft's web site: `http://www.microsoft.com/`.

To install win32s from the SSS CD-ROM:

1. Load the SSS CD-ROM into your CD-ROM drive. Open the FILE MANAGER which should be located in the MAIN program group in the PROGRAM MANAGER.

2. Open a FILE MANAGER window to the CD-ROM drive by double-clicking on the CD icon in the FILE MANAGER. You should see a top level subdirectory called `win32s`. Change to this subdirectory by double-clicking on it.

3. Within the `win32s` subdirectory you will find a `setup.exe` program. Double-click on this to start the win32s installation. Follow the instructions on the screen. We recommend that you install the Freecell game to test win32s.

4. After you have restarted Windows, run the FREECELL program to test the win32s installation. If it seems to work well, you can go to Section A.2.1 on running the simulations. If you notice screen corruption, or if the program stops responding to the system then win32s is almost certainly not working correctly.

5. Troubleshooting win32s: Your best option is to consult the Microsoft web site for information on win32s installation problems. We have found the video driver to be the most likely cause of win32s failure. You can test for video driver incompatibility by switching your Windows video driver to standard VGA using WINDOWS SETUP in the MAIN program group. If the FREECELL game works correctly under plain VGA, then you should obtain a more recent Windows video driver for your video adapter.

A.3 Apple Power Macintosh

A.3.1 Installation

To run the SSS from the CD-ROM:

1. Insert the CD-ROM into the CD-ROM drive. The MAC SSS volume should appear on your desktop. Opening the volume should reveal two folders, `sss` and `compile`, and a readme.

2. Open the `sss` folder to get to two sub-folders, `macPPCbin` and `lib`, and open the `macPPCbin` folder. It contains the fourteen Solid State Simulation applications such as `bloch.ppc` or `born.ppc`. Double-click on the desired application to run it.

To install the SSS on your hard disk and run them from there:

1. Insert the CD-ROM into the CD-ROM drive. The MAC SSS volume should appear on your desktop. Opening the volume should reveal two folders, `sss` and `compile`, and a readme.

2. Open the volume and subfolder on your hard disk where you wish to install the simulations. Drag the `sss` folder to the desired sub-folder to copy the entire contents of `sss` to your hard disk. You may need to confirm your choice.

3. To run the simulations from your hard drive, open the `sss` folder and the `macPPCbin` subfolder. Double-click on the desired application to run it.

A.3.2   Notes for Macintosh

Note that some of the simulations use the RIGHT or MIDDLE mouse button. Under the MacOS, hold down the CONTROL key and click with the mouse button to simulate the RIGHT mouse button; similarly, the SHIFT key together with a mouse click simulates the MIDDLE mouse button.

The minimal installation requires 25-30 MB of disk space. Alternatively, you may instead move the `compile` folder from the CD-ROM to your hard disk, which requires an extra 12 MB of disk space. `compile` contains, in addition to the files in the top-level `sss` folder, resources needed for modifying and recompiling the programs including the `src` folder which contains files with the source code for the programs, of interest to those concerned with the details of the algorithms used in the simulations. The additional files are *not* required to run the current versions.

The simulations rely on the applications being in a subfolder of a folder called `sss`, and there being a `lib` folder also under that `sss` folder. Therefore, if you rename the `sss` or `lib` folders, or if you move the `lib` folder, one of the `simulations`, or the `macPPCbin` folder to a different place, the simulations will probably not run. You may make aliases for the simulations and move those aliases wherever you wish. You may also move the `sss` folder in its entirety to a different location.

A.4   Versions of Unix

The SSS cannot be run from the CD-ROM. To install the simulations to a hard disk file system:

1. Insert the CD-ROM into the CD-ROM drive. Mount the CD-ROM at an appropriate mount point. You should see a top level directory called `unix` on the CD. Change directory to that subdirectory. A listing of this directory will show you many `tar` files. To install the executables you will only need the appropriate `xxxbin.tar` file (e.g., `sunbin.tar` for SunOS), where Table A.1 gives the translation key from `xxx` to the different operating systems.

2. Change directory to the subdirectory under which you wish to install the simulations – `/usr/local` is a good choice. Extract files from the tar file via something like the following (for AIX 3.2 with the CD mounted on `/cdrom`): `tar -xvf /cdrom/unix/aix3bin.tar`

3. If you wish to have the executables in the default path, please create symbolic links in a directory in the path that point to binaries under the `sss` directory. If the binaries do not find `../../sss/lib` they will not run correctly. The programs will detect invocation via a symbolic link and find the binary.

Table A.1: Key for connecting file names xxx to Unix operating systems.

xxx	Unix operating system
sun	SunOS 4.1.3 or later
aix3	AIX 3.2.5 or later
aix4	AIX 4.1.3 or later
linux	Linux, kernel 2.1.3 for Intel X86 or later

There is an additional tar file, of the form xxxall.tar, for each Unix platform. These contain the same files as the xxxbin.tar files, as well as the additional files, including the source code, needed to modify and recompile the programs.

A.5 Compiling the simulations for yourself

We have provided the source code for the simulations on the CD-ROM: in the compile directory for Windows, in the compile folder for the Macintosh, and in the xxxall.tar files for Unix. You can, if you wish, compile the source code yourself if you have an appropriate compiler. Please consult the README files on the SSS CD-ROM for more detailed instructions for recompiling.

Appendix B

Detailed system requirements

Contents

B.1   Introduction

The SSS were written in a combination of the C programming language for efficient numerical computations, and the tool control language Tcl for efficient graphical user-interface design. Tcl, developed by John Ousterhout, is itself written in C. The source code and some binaries are freely available to everyone [66].

Our simulations use the Tcl 7.6 and Tk 4.2 libraries and a variant of the BLT package for Tcl/Tk written by George Howlett. Both C and Tcl 7.6/Tk 4.2 work on many different types of Unix platforms running X, computers running Apple's MacOS, and 32-bit versions of Microsoft's Windows. At the time of press we have compiled versions of the simulations that are designed to work under Windows 3.1 or later, MacOS system 7.5 or later for the Power Macintosh, IBM's AIX versions 3.2.5 and 4.1.3 or later, Sun Microsystems' SunOS version 4.1, and Linux 1.2.13 on Intel. It should be possible to recompile the software for other versions of Unix where Tcl/Tk and BLT have been successfully built. What follows in Sections B.2–B.5 is a detailed description of the hardware and software platforms on which we have tested the simulations.

BLT is a very nice package written by George Howlett of Bell Laboratories. It is an addition to Tcl/Tk that provides a rich set of routines for many things from drag-and-drop to debugging. In particular, it contains routines for drawing scientific style xy-graphs with rotated y-axis labels. Such routines are essential to SSS, and are in fact the only features of BLT utilized in our simulations. We have used the semiofficial version 1.9 of BLT, which is an upgrade of BLT 1.8 by Jeffrey Hobbes of the University of Oregon. For the Unix platforms, only minor modifications to BLT 1.9 were necessary. To make its xy-graphs work under Windows and MacOS, however, we had to make extensive modifications to the limited portion of BLT 1.9 that we used.

Please consult our web site

                   http://www.ruph.cornell.edu/sss/sss.html

for more information on the SSS programs, possibly additional material, bug reports and fixes.

B.2   For all platforms

There are several system requirements that are essentially independent of platform.

CD-ROM: A CD-ROM drive is, of course, required to install the software on a hard disk drive or to run the software directly from the CD. If you don't have access to a CD-ROM drive for your system but do have access to the internet, you can obtain the essential parts of the software from our world wide web site given above.

Disk space: No hard disk space is required to run the software from the CD-ROM (Windows and MacOS only). For a minimum installation of the simulations on a hard disk drive, the amount of space required is 30 MB for the MacOS and 10 MB for Windows.

Screen resolution: The current version of the simulations requires a screen with at least 1024×768 pixels and 256 or more colors. Hopefully we will be able to decrease the number of pixels to 800×600 or fewer in the future. Check our web pages for possible updates.

Memory: Most of the simulations require about 4 MB of free RAM to run well, and a few need as much as 8 MB of free RAM in certain situations. By free RAM we mean the amount of memory available to run an application program after the operating system and other essential programs have been loaded. The total amount of RAM that must be installed in a computer in order to have 4–8 MB of free RAM varies widely depending on what platform you are using and how it has been configured. We provide some rules of thumb for particular platforms below.

Processor: The simulations do a great deal of calculation in general and a large number of floating point calculations in particular. The faster the processor the better, and a floating point coprocessor is required for an acceptable speed of operation. Specific processor recommendations for particular platforms are given below.

B.3   Microsoft Windows

The simulations are designed to run under Microsoft Windows. Microsoft has defined a common 32-bit application programming interface (API) called win32s which Tcl and hence the SSS programs require. The win32s API is supported by Microsoft Windows 3.1, Windows 95, Windows NT, and, to a limited extent, by the windows subsystem of IBM OS/2 Warp 3. The simulations were compiled for version 1.25 of the win32s API, and should be compatible with later versions.

Windows NT and Windows 95: The simulations should require no special setup under Windows NT for Intel x86 processors or Windows 95.

Windows 3.1 and Windows for Workgroups 3.1: The simulations are designed to run under Windows 3.1 in 386 enhanced mode or Windows for Workgroups 3.1. You will need to have version 1.25 or later of the win32s subsystem working. Version 1.25 of win32s is provided on the CD, and newer versions can be obtained from Microsoft's down-load sites. You may, however, need to update your Windows driver for your video card in order for win32s to function properly (see the installation notes in Appendix A for detailed instructions).

For the windows simulations a 486DX2-66 or faster processor is required. Processors without a math coprocessor such as the 486SX will not work acceptably. A Pentium processor is recommended. As a rule of thumb, Windows 3.1 or Windows 95 systems will need 12 MB of RAM, and Windows NT systems will need 16 MB of RAM in order to have 4 MB of free RAM available for the simulations. To install all of the Windows software on your hard disk , including the source code and resources for recompiling, requires approximately 15 MB of disk space. 10 MB is sufficient simply to run the programs.

B.4   Apple MacOS

The SSS are designed to run on Apple Power Macintosh computers. The Power Macintoshes should be running system 7.5 or later; system 7.5.5 is recommended. The simulations should run on any Power Macintosh with at least 16 MB of RAM. To install the full set of software, including resources for recompiling, requires about 42 MB of disk space. 30 MB is sufficient simply to run the programs.

B.5   Versions of Unix

For all flavors of Unix, the SSS require X11 R5 or later and a suitable X display. For any Unix platform, the source code will require 4 MB of hard disk space additionally to the disk space used for the binaries.

AIX: There are separate compiled versions of the SSS on the CD that are designed to run under AIX 3.2.5 and AIX 4.1.3 or later. For AIX, 32 MB of system RAM is generally recommended, although we have not tested in lower memory configurations. The slowest machines we have run the simulations on were RS/6000 models 320 and 40P for AIX 3.2.5 and 4.1.4 respectively, and even the 320 was acceptably fast. About 23 MB of disk space is needed to install one version of the executables.

SunOS: There is a compiled version of the SSS designed to run on Sun SparcStations running SunOS 4.1/Solaris 1.0. They are a bit slow on a SparcStation 1, and a faster processor is recommended. For a networked workstation 24 MB of RAM or more is recommended.

Approximately 25 MB of disk space is needed to install the simulations.

Linux: The compiled version of the SSS is designed to run on Linux 1.2.13 or later, on Intel x86 based systems. They are ELF binaries compiled with gcc 2.7.2. As with other x86 operating systems a 486DX2-66 processor or faster is required, and a Pentium or better is recommended. About 16 MB of RAM is needed to provide the 4–8 MB of free RAM. The Linux simulations require about 13 MB of disk space.

Other Unix: In principal, one should be able to compile the simulations on other versions of Unix. See the README files on the CD-ROM and our web page for instructions on compilation.

Appendix C

Presets and file editing

Contents

The PRESET system used in the SSS allows quick access to the physically interesting regions of the large parameter space. Most, though not all, of the parameters available on the main panels and in the CONFIGURE dialog boxes can be defined through the PRESET system. The system was designed to allow users to modify the provided presets, to define their own presets, and to add or edit the annotations. Thus, students and instructors can save their own presets for demonstrations and class work. This appendix gives instructions for editing and adding PRESETS and annotations. Similar procedures may be used to edit or add to the existing HELP files or ADDITIONAL EXERCISES.

C.1 Presets provided with SSS package

The presets provided on the SSS distribution CD can be loaded either through the hypertext (html) based help menu or directly from the main panel. To browse the simulations and to work with this text in conjunction with the simulation, we suggest using the HELP system to access the presets since only the HELP system gives a descriptive title and explanation for each preset. For lecture demonstrations the PRESET button on the main panel gives faster access to the presets.

The display rates for the programs containing animations will vary considerably from platform to platform. The SPEED and PARTICLE NUMBER choices in the SSS PRESETS were chosen appropriate to the machines on which the programs were developed: they may be unsatisfactory for your use. For class use in a computer lab setting, it would be desirable to tune the settings in the PRESETS to match the capabilities of the available computers. Too slow an animation is soporific; too fast and the physics is hidden.

C.1.1 *Loading SSS presets though the help system*

Note first that all "click-able" choices in the html menus are shown in blue. To load a PRESET through the html help system:

1. Click on the HELP button on the main panel. This opens the html browser.

2. Select PRESETS FOR BROWSING to get to a preset page specific to the program from which HELP was called. This page lists the available presets, giving both their NUMBERS and TITLES. Next, either

    (a) click on the PRESET NUMBER to load the parameters appropriate for this preset into the program. These numbers correspond to the preset numbers such as PRESET 5, used throughout the text; or

(b) click on the PRESET TITLE to get to an html page with detailed descriptions, hints, and questions for that preset. Finally, click LOAD THIS PRESET, highlighted in blue, on that page to load the preset into the program.

C.1.2 Loading SSS presets from the main panel

To load a preset directly from the main panel, without the preset titles and annotations:

1. Click on the PRESET button on the main panel to pull down a menu with the PRESET NUMBERS, e.g., ISING05. These numbers correspond to the preset numbers used throughout the text.

2. Select one of the PRESET NUMBERS to load the parameters of that preset into the program.

Hint: You also can "tear off" the preset menu at the dashed line above the LOAD option. The preset menu then stays on the screen even after you have selected a preset so that to select the next preset you only need to click on its number in that menu.

C.2 Customized presets

The PRESET system in the SSS has been designed such that instructors and students can SAVE and LOAD their own presets. This is useful for demonstrating an interesting set of parameters in a lecture, preparing a problem set based on presets different from the ones we provide, or saving a particularly interesting set of parameters to recall for later investigation.

There are two kinds of files important for writing your own presets: ".pre" files and ".htm" files. The ".pre" files contain values for most of the program parameters and are used by the *simulations* to set the sliders and buttons to the desired values. The ".htm" files contain the description of the corresponding preset and are needed by the *html browser* to display the annotations in readable format. Section C.2.1 describes how to save your own presets and to create the ".pre" file; Section C.2.2 gives the procedure to load customized sets while running a program; and Section C.2.3 shows how to access your own presets through the HELP menu and how to add annotations to the presets. Hint: if all you want is to store the parameters, skip Section C.2.3 and forget about ".htm" files.

Warning! Using somebody else's presets might cause damage to your computer system. A preset file consists of a set of commands which are executed when the preset is loaded. These commands set the parameters of the simulation such as lattice size, atomic radius, and charge. There is, however, no protection for damaging commands that might have been added to the preset file. On Unix systems, make sure that any preset files for shared use have their file permissions set such that the files cannot be modified by unauthorized people. Check with your system administrator for details.

C.2.1   Saving your own customized presets

To save your own presets:

1. Set all the SLIDERS, BUTTONS, and CONFIGURE and MENU CHOICES in the program to the desired parameter values. (Note that you may occasionally find a parameter, usually related to the displays, which is not included in the preset list.)
2. Click on the PRESET button on the main panel.
3. Select SAVE from the menu to get a FILE SELECTION menu.
4. Choose the DIRECTORY or FOLDER in which you want to save your preset by double-clicking on the NAME of the directory (folder) in the menu list. Note that double-clicking on the symbol ".." (":":" for MacOS) moves you "up" one level. The default directory for the preset SAVE is the directory from which you started the simulation. Once you have chosen a new directory, it will remain the default as long as the program is running.
5. Type into the SELECTION box the file name for the preset. It is recommended to use the extension ".pre" for the name of your preset. Press ENTER in the SELECTION box to save the preset.

C.2.2   Loading customized presets: main panel

To load customized presets from the main panel:

1. Click on the PRESET button on the main panel.
2. Select LOAD to get the FILE SELECTION menu.
3. Choose the DIRECTORY where the preset file is located by double-clicking on the NAME of the directory in the DIRECTORY box. Note that double-clicking on the symbol ".." (":":" for MacOS) moves you "up" one directory. The directory from which you started the simulation will be the default directory which can save you a lot of mouse-clicking. Once you have chosen a new directory, it will remain the default as long as the program is running.
4. If the preset file has the ".pre" extension, its name will appear in the FILES box. Double–click on the name to load it. If it does not have the ".pre" extension, either change the FILTER to reflect the correct file name extension or type in the preset file name in the SELECTION box and press ENTER. Once a preset is loaded, it will

appear in the preset menu the next time it is opened. (It will not, however, appear in a torn-off list which is already open. A double click in the upper left hand corner of the torn-off list will close it.)

C.2.3   Annotations and presets via the help system

To edit the SSS preset annotations, to write html annotations for your own presets, and to have access to them from the HELP system, the html browser provided with the SSS programs allows you to EDIT any page you have called up, provided you have the "right" file permissions set for accessing that file on your system.[1] Be warned, however, that this built-in editor has only the most basic functionality.

For example, to add a PRESET 9 to the presets for "ising":

1. Create and save a preset named `ising09.pre` using the procedure described in Section C.2.1 above.
2. Click HELP on the "ising" main panel.
3. Select PRESETS FOR BROWSING.
4. Click EDIT on the html page ISING PRESETS. You are now editing the file `presets.htm` in the directory `sss/lib/ising`.
5. Use the current entities as a model. Add to that file a line such as

   `<li><a href="ising09.pre">09</a> --`
   `<a href="ising09.htm"> My new title  </a>.`

   (If the preset file has been saved in a directory (folder) other than `lib/ising` (or :lib:ising for the Mac), a complete pathname must be defined.) Next you should click on SAVE to save the changes you made to `ising.htm`. When you later click on the number 09 in the list, PRESET 9 will be loaded into "ising"; and once you have constructed the annotation file `ising09.htm`, clicking on the MY NEW TITLE will load the file `ising09.htm` into the html browser.
6. Create a new annotation page in one of two ways. One is to change the pathname in the window in the upper right hand corner of the html page to the name of the annotation file you want to create, possibly `ising09.htm`. With a RETURN you will be asked if you wish to create the new file which is then opened with a click on the OK; finally, a click on the EDIT button lets you write an appropriate title and the annotation. You may find it useful to look at an existing annotation file as a model. Click the SAVE button and you have the annotation added. An alternative is to make a copy of `ising08.htm` and rename it `ising09.htm`. Either edit it with any text editor, or return to the PRESET LIST and click on MY NEW TITLE to open the new file which may then be edited.

---

[1]For Unix systems, check with your system administrator for details.

C.3   Exercises and other editing

You may create additional help files in a similar fashion. The procedure is to establish a new file link on an existing html page using the editor for that page and then to SAVE. Next create the new page by defining its pathname in the pathname window of the current html page and hitting the RETURN key. After agreeing with an OK, the new page appears, ready to edit with a click on the EDIT button. SAVE the edit and the job is done.

For example, to add a new exercise to "ising" entitled `My exercise`:

1. Open the main help file and click on the EDIT button.

2. Add a link to your `my_exercise` page using the links to the additional exercises as a model; for example, you would add the link

   `<li><a href="my_exercise.htm"> My exercise</a>`.

   If the `my_exercise.htm` file were to be saved in a directory other than `sss/lib/ising` a longer pathname would be required.

3. SAVE the edited help file.

4. Type the pathname, `my_exercise.htm`, of the link just created into the pathname window, ENTER, and OK the creation of the new file.

5. Click EDIT, and replace the existing pieces of text with a title for the window border, a title for the page, and finally the question. Leave in the `<h1>`, `</h1>`, etc. commands which define the formatting of the page.

6. If the question is to include a figure, the figure is accessed with a link of the form `<img src="filename">`, with `filename` the pathname of the GIF file with the desired figure. The exercises provided on the SSS disk involve the use of figures which are grabs of displays and graphs from the SSS programs. To make such an exercise, you will need a frame grabber for your platform which can save files in GIF format. For the ADDITIONAL EXERCISES in the SSS package (e.g., for "ising") these GIF files are saved in the `sss/lib/ising/figures` directory.

7. If you wish to link to an answer to the question, as is done for the ADDITIONAL EXERCISES, the procedure is similar. The link is created in the `my_exercise.htm` file and the answer given in the newly created `answer` file. The answers for the ADDITIONAL EXERCISES in the SSS package (e.g., for "ising") are stored in the `sss/lib/ising/answers` directory.

The ADDITIONAL EXERCISES files in the SSS package provide a useful model. You may wish to add questions to these pages rather than creating new pages. If you do, be wary of putting in too many questions with links to figures. Loading time and reloading time, on returning from checking an answer, can become annoyingly long. The exercises are easily converted to quizzes by deleting the answer files or restricting their read permissions.

Bibliography

General texts

First we list some standard solid state text books which can be profitably used in conjunction with the SSS package. The group of letters for each reference indicates whether the depth of coverage is adequate for use with the related simulations. A minus superscript suggests a marginal level. These letters are coded according to: X: (2,3) **X**-rays and crystal structure; P: (4,5) **P**honons, lattice vibrations, and heat capacity; E: (6–10) **E**lectrons, bands, and transport; S: (11, 12) **S**emiconductors; M: (13, 14) **M**agnetism; and D: (15) **D**islocations. (The numbers in parentheses refer to chapter numbers.)

[1] N. W. Ashcroft and N. D. Mermin, *Solid state physics*, Saunders College, Philadelphia (1976).

(X P E S M$^-$ D) One of two "standard" texts for graduate level (and upper level undergraduate) courses. Meticulously written and liked by those students who find Kittel [6] too superficial.

[2] J. S. Blakemore, *Solid state physics*, 2nd ed., Cambridge University Press, Cambridge, UK (1985).

(X$^-$ P E$^-$ S$^-$) A text for students with limited mathematical and quantum mechanics background.

[3] G. Burns, *Solid state physics*, Academic Press, New York (1985).

(X$^-$ PESM$^-$) One of the less well-known general texts, this gives a more thorough treatment than most of crystal symmetries and their implications, and of optical properties.

[4] A. Guinier and R. Jullien, *The solid state: from superconductors to superalloys*, Oxford University Press, Oxford (1989).

(P E S M D) This is a lower level undergraduate text, more qualitative than the others in this list. It gives a fine overview of a number of solid state topics and would be a good adjunct to a casual browsing of the simulations of the SSS project, or text for a qualitative introduction to solid state physics.

[5] R. E. Hummel, *Electronic properties of materials*, 2nd ed., Springer-Verlag, Berlin (1993).

(P$^-$ ESM$^-$) A text directed at undergraduates in the engineering disciplines. With a mathematical level lower than in Kittel [6], considerable emphasis is placed on device applications.

[6] C. Kittel, *Introduction to solid state physics*, 7th ed., John Wiley & Sons, Inc., New York (1996).

(X P E S⁻ M D) One of the standard texts for upper level undergraduate (and graduate level) courses in solid state physics, now in its seventh edition. A popular book for students who find Ashcroft and Mermin [1] tough sledding.

[7] H. P. Myers, *Introductory solid state physics*, Taylor and Francis, London (1990). (X P E M D⁻) This holds back on the mathematical formalism, and emphasizes physical concepts. Myers includes brief discussions of many current topics in solid state physics.

[8] J. M. Ziman, *Principles of the theory of solids*, Cambridge University Press, Cambridge, UK (1964). (XPEM) A clean, concise presentation of the essential ideas of solid state theory, at the beginning graduate level. Ziman's is more mathematical than many texts, but he sticks to the bare essentials.

Specific references

The following references include both surveys covering in a general way the indicated sections of the guidebook, and material specific to a particular issue or exploration project. The more general references are listed before the specific, and the more elementary before the advanced.

X-rays and crystal structure

[9] A. Guinier and D. L. Dexter, *X-ray studies of materials*, Interscience Publishers, New York (1963). A delightful small book reviewing all aspects of x-ray diffraction with a minimum of equations. A fine review, or introduction, to a wide variety of ideas.

[10] B. D. Cullity, *Elements of x-ray diffraction*, Addison-Wesley, Reading, Mass. (1978). An introduction to x-ray techniques with an emphasis on experimental aspects. Though targeted at metallurgists it will be useful for anyone interested in learning about techniques. Arguments are based on the Bragg law rather than on an appeal to the concepts of reciprocal space. (The Ewald construction is in an appendix.)

[11] B. E. Warren, *X-Ray diffraction*, Dover Publications, New York (1990). A graduate text with extensive discussion of x-ray diffraction theory. It addresses issues of thermal disorder and imperfections as well as perfect crystal diffraction, with free use of ideas of reciprocal space. (The Ewald construction is on p. 19.)

[12] A. Guinier, *X-ray diffraction*, W. H. Freeman & Co., San Francisco (1963). An advanced level text describing many sophisticated techniques for the x-ray characterization of solids.

[13] J. Drenth, *Principles of protein x-ray crystallography*, Springer-Verlag, Berlin (1994). An opportunity to find out how it's really done!

[14] C. Janot, *Quasicrystals: a primer*, Oxford University Press, Oxford (1994). A fine introduction to the fun ideas of quasicrystals without having to go too deeply.

[15] R. Merlin *et al.*, Quasiperiodic GaAs–AlAs heterostructures, *Phys. Rev. Lett.* **55**, 1768 (1985); and P. Mikulik, *et al.*, X-ray-diffraction on Fibonacci superlattices, *Acta Crystall. A* **51**, 825 (1995).

The first is an early experimental paper, and the second, a recent theoretical one with many references to experimental work, on artificially constructed "one-dimensional" quasicrystals.

[16] E. Sweetland *et al.*, X-ray scattering measurements of the transient structure of a driven charge-density-wave, *Phys. Rev. B* **50**, 8157 (1994); and D. A. DiCarlo *et al.*, Charge-density-wave structure in $NbSe_3$, *Phys. Rev. B* **50**, 8288 (1994).

Here is a beautiful example of the power of modern x-ray techniques, in which both static and time resolved dynamic properties of the charge density wave (CDW) are measured via satellite peaks produced by the CDW's periodic distortion of the host lattice.

Phonons, lattice vibrations, and heat capacity

[17] I. G. Main, *Vibrations and waves in physics*, Cambridge University Press, Cambridge, UK (1993).

An undergraduate text which will be useful for those for whom the early parts of Chapter 4 are new material, not review.

[18] H. Georgi, *The physics of waves*, Prentice-Hall, Englewood Cliffs, NJ (1993).

A second undergraduate text introducing concepts of wave and wave packet propagation in mechanical and electrical systems.

[19] R. M. Eisberg and R. Resnick, *Quantum physics of atoms, molecules, solids, nuclei and particles*, 2nd Ed., John Wiley & Sons, Inc., New York (1985).

Undergraduate text on modern physics with discussion of black body radiation and Wien's law.

[20] M. Born and K. Huang, *Dynamical theory of crystal lattices*, Clarendon Press, Oxford (1966).

The classic treatise on the dynamics of lattice vibrations.

[21] L. Brillouin, *Wave properties and group velocity*, Academic Press, New York (1960); and L. Brillouin, *Wave propagation in periodic structures; electric filters and crystal lattices*, Dover Publications, New York (1972).

Two small volumes about wave propagation, with elegant mathematical argumentation. For the reader looking for deeper insight into the nature of wave motion.

[22] A. A. Maradudin, E. W. Montroll, G. H. Weiss, and L. P. Ipatova, *Theory of lattice dynamics in the harmonic approximation*, Solid State Physics, Suppl. **3**, 2nd ed., eds. F. Seitz and D. Turnbull, Academic Press, New York (1971).

An extensive and advanced treatment of the lattice vibration problem, including the rich collection of complications associated with impurities, but remaining within the harmonic approximation.

[23] L. van Hove, The occurrence of singularities in the elastic frequency distribution of a crystal, *Phys. Rev.* **89**, 1189 (1953).

This is the original paper in which the singularities in the densities of states of phonons were characterized. The treatment is formal and elegant.

[24] J. P. Wolfe, Acoustic wavefronts in crystalline solids, *Physics Today*, **34** (Sept. 1995).

A brief description of current research on phonon propagation in solids with fantastic pictures that have to convince you of the anisotropy of the group velocity in crystalline solids.

[25] G. E. Bacon, *Neutron diffraction*, Oxford University Press, Oxford (1975).

A thorough treatment of the application of neutron diffraction to a variety of problems in solids including magnetic structure determination and phonon spectroscopy.

[26] A. S. Barker, Jr and A. J. Sievers, Optical studies of the vibrational properties of disordered solids, *Rev. Mod. Phys.* **47**, Suppl. 2 (1975).

An extensive review article, treating both theoretical and experimental issues, on the lattice dynamics of imperfect crystals. There is special emphasis on the application of infra-red spectroscopy to the study of disordered materials.

[27] E. R. Cowley, A Born–von Karman model for lead, *Solid State Comm.* **14**, 587 (1974).

Here you see a sophisticated version of the first step of "debye", the calculation of the dispersion relation for lead and a comparison with neutron data.

[28] S.R. Kiselev, S.R. Bickham, and A.J. Sievers, Properties of intrinsic localized modes in one-dimensional lattices, *Comments Cond. Mat. Phys.* **17**, 135 (1995).

A summary of modeling calculations, on a sophisticated level, of anharmonic localized modes.

[29] E. L. Wolf, *Principles of electron tunneling spectroscopy*, Oxford University Press, Oxford (1985).

A thorough and scholarly discussion of a wide variety of applications of electron tunneling to problems of solid state physics. You can learn here, in more detail, of the use of superconducting tunneling to determine phonon densities of states.

[30] Landolt–Börnstein, *Numerical data and functional relationships in science and technology*, Vol. 13, Springer-Verlag, Berlin (1981).

One of a many volume set with extensive tables of a wide variety of properties of materials.

[31] Thermophysical properties research center, Purdue University, *Thermophysical Properties of Matter*, Vol. 4, Plenum Publishing Corp., New York (1970).

Extensive tabulation of the thermal properties, e.g., heat capacity, thermal conductivity, etc., of everything.

Electrons, bands, and transport

[32] D. N. Langenberg, Resource letter OEPM-1 on the ordinary electronic properties of metals, *Am. Jour. of Phys.* **36**, 777 (1968).

An extensive list of references, with comments, to the pioneering papers which served as the basis for our current understanding of electron transport in solids. An excellent launch pad for a more detailed look at electronic transport properties of metals.

[33] J. M. Ziman, *Electrons and phonons*, Oxford University Press, Oxford (1962).

Here is the place to begin a serious study of the electronic transport properties of solids. The phonons are included primarily because of their critical role in the scattering of electrons.

[34] D. Bohm, *Quantum theory*, Prentice-Hall, Englewood Cliffs, NJ (1951).

An introductory quantum text, included for its commentary on the Ramsauer effect.

[35] A. P. Cracknell and K. C. Wang, *The Fermi surface; its concept, determination and use in the physics of metals*, Clarendon Press, Oxford (1973).

Describes the variety of techniques used for experimental Fermi surface determinations and gives illustrations, for a variety of metals, of the shapes of the Fermi surfaces.

[36] T. DeKorsy *et al.*, Bloch oscillations at room temperature, *Phys. Rev. B*, **51**, 17,275 (1995).

Report of the observation of pulse excitation of Bloch oscillations, not in a perfect crystal, but in a man-made periodic structure with a repeat period of 84 angstroms.

[37] A. B. Pippard, A proposal for determining the Fermi surface by magneto-acoustic resonance, *Phil. Mag.* Ser. 8 **2**, 1147 (1957).

The proposal for the first "calipers" experiment using the acoustic wavelength as a gauge.

[38] G. N. Kamm and H. V. Bohm, Magnetoacoustic measurements of the Fermi surface of aluminum, *Phys. Rev.* **131**, 111 (1963).

Report of the first use of the magnetoacoustic "calipers" to measure a diameter of a Fermi surface by matching the cyclotron orbit radius to the acoustic wavelength.

[39] S. Kagoshima, H. Nagasawa, and T. Sambogi, *One-dimensional conductors*, Springer-Verlag, Berlin (1988).

A good introduction, at the undergraduate/graduate level, to the properties of organic crystals and other materials showing highly anisotropic electrical conduction. It is fun to see the new phenomena intrinsic to the one-dimensional system.

[40] F. N. H. Robinson, *Noise in electrical circuits*, Oxford University Press, Oxford (1962).

A thorough review of noise in electrical circuits. Though written at the end of the vacuum tube era, most of it is still relevant to noise in semiconductor circuits.

[41] M. L. Roukes *et al.*, Hot electrons and energy transport in metals, *Phys. Rev. Lett.* **55**, 422 1985.

A first experiment to demonstrate clearly the effects of driving the electrons in a metal out of thermal equilibrium with the phonons, with a resulting dramatic departure from the Ohm's law linearity of current with electric field.

[42] C. Kurdak *et al.*, Electron-temperature in low-dimensional wires using thermal noise measurements, *Appl. Phys. Lett.* **66**, 3203 1995.

Electron heating experiments in wires (0.45 μm wide, 45 μm long) fabricated from a two-dimensional electron gas. How narrow must the wire be to be one-dimensional?

Semiconductors

[43] R. F. Pierret, *I. Semiconductor fundamentals*; G. W. Neudek, *II. The pn-junction diode*; and R. F. Pierret, *IV. Field effect devices*, in *Modular series on solid state devices*, eds. R. F. Pierret and G. W. Neudek, Addison-Wesley, Reading, Mass. (1983).

The three volumes relevant to "fermi" and "poisson" of a series of nine small volumes designed to introduce undergraduate engineering students to semiconductor devices.

[44] R. S. Muller and T. I. Kamins, *Device electronics for integrated circuits*, John Wiley & Sons, Inc., New York (1986).

Undergraduate text on integrated circuit devices and the underlying physics.

[45] S. M. Sze, *Semiconductor devices: physics and technology*, John Wiley & Sons, Inc., New York (1985).

Here is a reduced version of reference [46], designed specifically as an undergraduate text, including problems for each chapter.

[46] S. M. Sze, *Physics of semiconductor devices*, John Wiley & Sons, Inc., New York (1981).

This standard reference tells you how devices work and the physics behind it all. Useful both as an advanced text for a course in device physics and as a reference in the laboratory.

[47] E. H. Nicollian, *MOS (metal-oxide-semiconductor) physics and technology*, John Wiley & Sons, Inc., New York (1982).

Everything you might want to know about the characterization of MOS structures, and probably more. A book likely to be on the shelf beside Sze [46] in any semiconductor electronics laboratory.

[48] T. Ando, A. B. Fowler, and F. Stern, Electronic properties of two-dimensional systems, *Rev. Mod. Phys.* **54**, 437 (1982).

A thorough and impressive review of the experiments and theory of the physics of the two-dimensional electron gas formed in the inversion layer of MOSFET devices.

[49] F. G. Allen and G. W. Gobeli, Work function, photoelectric threshold, and surface states of atomically clean silicon, *Phys. Rev.* **127**,150 (1962).

This work, one of the earliest studies of the influence of surface states on band bending at a semiconductor surface, used the Kelvin method as one technique for determining the work function of a surface cleaved within the high vacuum apparatus.

[50] J. Holzl and F. K. Schulte *Work function of metals*, The Springer tracts in modern physics, Vol. 85, Springer-Verlag, Berlin (1979).

An extensive review of the use of work functions as a characterization tool in the study of metal surfaces.

[51] T. H. Borst, S. Strobel, and O. Weis, High temperature diamond pn-junction: B-doped homoepitaxial layer on N-doped substrate, *Appl. Phys. Lett.* **67**, 2651 (1995), and M. W. Shin *et al.*, Temperature dependence of current-voltage characteristics of polycrystalline diamond FETs, *J. Materials Research – Materials in Electronics* **6**, 111 (1995).

Two articles to show that one can indeed make semiconducting devices using diamond. The *pn*-junction of the first reference having been tested in the temperature range 600–900° C. (Commercial viability is a different issue!)

Magnetism

[52] D. Jiles, *Introduction to magnetism and magnetic materials*, Chapman & Hall, New York (1991).

An up-to-date and quite complete text on magnetism, at the senior/graduate student level.

[53] J. M. Yeomans, *Statistical mechanics of phase transitions*, Clarendon Press, Oxford (1992).

A short text aimed at the introductory graduate level and focused on the problems of microscopic magnetism. An excellent starting point for pursuing further the behavior of the "ising" program. Yeomans addresses many of the issues raised in Chapter 13, and includes discussion of mean field theory, critical exponents, and the Monte Carlo technique.

[54] F. Reif, *Fundamentals of statistical and thermal physics*, McGraw-Hill, New York (1965).

Senior level text in statistical physics, included here primarily for its introduction to the Fourier representation of random functions and the brief treatment of noise, pp. 582–594.

[55] J. F. Dillon, Jr, *Domains and domain walls*, in *Magnetism*, Vol. 3, eds. G. T. Rado and H. Suhl, Academic Press, New York (1963).

The focus here is on experimental techniques for the observation of domain structures. The many pictures will convince you that domains exist; and that "néel" is very naive!

[56] C. Kittel, Physical theory of ferromagnetic domains, *Rev. Mod. Phys.* **21**, 541 (1949); or C. Kittel and J. K. Galt, Ferromagnetic domain theory, in *Solid State Physics*, Vol. 3, eds. F. Seitz and D. Turnbull, Academic Press, New York (1956).

Detailed reviews of domain theory in ferromagnets containing many of the physical arguments that were in an early edition of Kittel's book [6], but which disappeared from the later ones.

[57] D. Turnbull, Phase changes, in *Solid State Physics*, Vol. 3, eds. F. Seitz and D. Turnbull, Academic Press, New York (1956).

A general review of the thermodynamics of phase changes with an extensive section on nucleation.

[58] J. E. Avron et al., Roughening transition, surface tension and equilibrium droplet shapes in a two-dimensional Ising system, *J. Phys. Soc.: Math. Gen.* **15**, L81 (1982).

[59] A. H. Eschenfelder, *Magnetic bubble technology*, Springer-Verlag, Berlin (1981).

Domain structures in garnet films are fascinating, as are the devices that can be made by pushing around bubbles of reverse magnetization. Here is a thorough review of both the physics and the devices.

[60] B. J. Hiley and G. S. Joyce, The Ising model with long range interactions, *Proc. Phys. Soc* **85**, 493 (1965).

Analysis of the transition temperature for the Ising model for a variety of dimensionalities, crystal structures, and power-law interaction potentials.

Dislocations

[61] J. R. Weertman and J. Weertman, *Elementary dislocation theory*, Oxford University Press, Oxford (1992).

Undergraduate introductory level text focused on dislocations, theory and experiment.

[62] S. Amelinkx, Direct observation of dislocations, *Solid state physics*, Suppl. 6, eds. F. Seitz and D. Turnbull, Academic Press, New York (1964).

An extensive presentation of techniques for the observation of dislocations, with many intriguing illustrations.

[63] D. Hull and D. J. Bacon, *Introduction to dislocations*, 3rd ed., Pergamon Press, Oxford (1984).

A solid and readable introduction to dislocations, with the energetics presented in useful detail and plenty of dramatic experimental pictures of dislocations in real systems.

[64] J. P. Hirth, *Theory of dislocations*, 2nd ed., John Wiley & Sons, Inc., New York (1982).

An advanced level source of detailed information on all aspects of dislocation theory. If you can't find it in other references, look here.

Numerical methods

[65] R. L. Burden and J. D. Faires, *Numerical analysis*, Prindle, Weber & Schmidt, Boston (1986).

A great resource for numerical algorithms with no need for concerns over licensing agreements.

[66] J. K. Ousterhout, *Tcl and the Tk Toolkit*, Addison-Wesley, New York (1994).

Gives comprehensive introduction to Tcl/Tk. Unfortunately, the book corresponds to Tcl 7.3 and Tk 3.6 which differs in some important aspects from the (newest) versions 7.6 and 4.2 that we have used. More information on Tcl/Tk as well as the software itself can be found on their web site `http://www.smli.com/research/tcl/`.

Index